DIY Science

Illustrated Guide to Home Biology Experiments

All Lab, No Lecture

First Edition

Robert Bruce Thompson and Barbara Fritchman Thompson

O'REILLY

BEIJING · CAMBRIDGE · FARNHAM · KÖLN · SEBASTOPOL · TOKYO

Illustrated Guide to
Home Biology Experiments

All Lab, No Lecture

by Robert Bruce Thompson and Barbara Fritchman Thompson

Published by Make:Books, an imprint of Maker Media, a division of O'Reilly Media, Inc.
1005 Gravenstein Highway North, Sebastopol, CA 95472.

O'Reilly books may be purchased for educational, business, or sales promotional use.
For more information, contact our corporate/institutional sales department:
800-998-9938 or *corporate@oreilly.com*.

Print History	**Publisher:** Dale Dougherty
April 2012	**Associate Publisher:** Dan Woods
First Edition	**Executive Editor:** Laurie Petrycki
	Editor: Brian Jepson
	Cover Designer: Mark Paglietti
	Interior Designer: Alison Kendall and Ron Bilodeau
	Production Editor: Melanie Yarbrough
	Proofreader: Linley Dolby
	Indexer: Bob Pfahler
	Cover Photograph: Robert Bruce Thompson

ISBN: 978-1-449-39659-6
[LSI]

[1/15]

To Charles Darwin (1809 – 1882), the towering genius whose theory of evolution is the foundation of modern biology.

Contents

Preface

Welcome to *Illustrated Guide to Home Biology Experiments: All Lab, No Lecture*.

We wrote this book in response to the flood of queries we've received from readers of *Illustrated Guide to Home Chemistry Experiments*. (That book was published in 2008 and has become an enduring bestseller, thanks both to homeschoolers and DIY hobbyists.) Most of the queries were on much the same theme: When are you going to do the same thing for biology?

This book is our answer to that question. It took us 18 months to write, and with so much to cover, we had to do some juggling to make sure the most important stuff all made it into the book. Our goals for this book were:

- To write a biology lab manual suitable for a rigorous first-year high school biology course, or for a budding DIY bio enthusiast who wants to learn the fundamentals of biology lab work.

- To cover as broad a range of topics as possible while maintaining useful depth.

- To include only lab sessions that are practical to do at home and do not require the expensive equipment and resources of a formal school laboratory, other than a decent microscope.

- To keep lab costs as low as possible by designing an affordable companion lab kit as we wrote the book.

- To make the lab sessions easy to coordinate with popular homeschool biology texts, such as CK-12, Miller-Levine Biology, A Beka, Apologia, BJUP, and many others.

We think we accomplished those goals. We've never been entirely satisfied with any book we've written. Just ask our editor, who always has to drag the manuscript away from us as we kick and scream and beg for just a little more time. But in this case we think we did what we set out to do.

WHO THIS BOOK IS FOR

This book is for anyone, from responsible teenagers to adults, who wants to learn about biology by doing real, hands-on laboratory work. DIY hobbyists can use this book to learn and master the essential practical skills and fundamental knowledge needed to pursue biology as a lifelong hobby. Home school parents and public school teachers can use this book as the basis of a first-year lab-based biology course.

HOW THIS BOOK IS ORGANIZED

The first part of this book is made up of narrative chapters that cover how to use the book, how to equip your biology lab, and how to work safely. The bulk of the book is made up of more than 30 lab sessions incorporating about 80 separate procedures in the following 11 groups.

Mastering Microscope Skills

In this group, you'll learn the essential skills of using a microscope, mounting specimens (including simple wet mounts, smear mounts, hanging-drop mounts, and sectional mounts), and using simple and Gram staining to reveal the microscopic details of lifeforms.

Building and Observing Microcosms

In this group, you'll build microcosms—miniature worlds contained in soda bottles and populated with diverse microscopic life—and observe those tiny worlds over a period of weeks to months. You'll observe succession—the cycle of life as the microcosms grow and mature—and test the effect of pollution on those microcosms.

Chemistry of Life

In this group, you'll explore the chemistry of life: acids, bases, and buffers, carbohydrates and lipids, proteins, enzymes, and vitamins. You'll also create and observe coacervates, nonliving cell-like structures that may have played a role in abiogenesis—the development of living organisms from nonliving precursors. Finally, you'll extract and visualize actual DNA, build a gel electrophoresis apparatus, and use that apparatus to do simulated DNA analysis. (Or real DNA analysis, if the necessary materials are within your budget.)

Life Processes

In this group, you'll explore some fundamental life processes. You'll observe carbon dioxide uptake in plants, determine the effect of light intensity on photosynthesis rate, and use chromatography to analyze chlorophyll and other plant pigments. You will investigate osmosis, the process by which nutrients and wastes are transported across cell membranes. Finally, you'll observe mitosis, a process by which cells reproduce themselves.

Ecology

You'll begin this group by sampling plant populations in a community, learning how ecologists track changes in the environment by observing changes in plant populations. You'll then learn how different organisms sometimes cooperate to the benefit of both by observing the effect of nitrogen-fixing bacteria on lima bean growth. Next, you'll do air pollution testing for particulates by building, exposing, and observing particle traps. Finally, you'll test soil and water specimens quantitatively for the presence of boron, a common environmental pollutant.

Genetics and Inheritance

In this group, you'll explore Mendelian genetics by testing a sample population for the ability to taste phenylthiocarbamide (PTC). You'll then analyze your data by charting that ability against family relationships to determine whether PTC tasting ability is heritable in strict Mendelian terms.

Cells and Unicellular Organisms

You'll begin this group by observing specialized eukaryotic cells, comparing and contrasting their features and structures. You'll then prepare culturing media, culture mixed bacteria species, and prepare pure broth cultures of each of those species. You'll then test each species for sensitivity to a range of antibiotics, culture an antibiotic-resistant strain, and retest the sensitivity of that resistant strain to the same range of antibiotics.

Protista

In this survey group, you'll observe four members of kingdom *Protista*—*Spirogyra* (algae), *Euglena*, *Amoeba*, and *Paramecium*—which are simple organisms that have some of the characteristics of plants and fungi and some of the characteristics of animals. You'll use both live specimens—

which allow you to observe the organisms going about their business and interacting with their environments—and prepared slides, which allow you to observe more of the fine details of their structure.

Fungi

In this survey group, you'll observe the structures and features of members of kingdom *Fungi* in the phyla *Zygomycota*, *Ascomycota*, and *Basidiomycota* using both live specimens and prepared slides.

Plantae

In this survey group, you'll observe members of kingdom *Plantae* (plants), beginning with the structures and features of the simplest plants, mosses and ferns. You'll observe germination of an angiosperm (seed plant) and compare and contrast the similarities and differences in the root, stem, leaf, and reproductive structures of seed plants.

Animalia

In this final group of survey lab sessions, you'll observe the structures of members of kingdom *Animalia* (animals), beginning with the simple invertebrates—*Porifera* (sponges) and *Cnidaria*—and then through *Platyhelminthes*, *Nematoda*, and *Annelida* (three classes of worms) to *Arthropods* (bugs and related creatures). In the final lab session, you'll investigate the four types of vertebrate tissues—epithelial, connective, muscle, and nerve—by observing vertebrate organs and structures made up of these tissue types.

ACKNOWLEDGMENTS

Although only our names appear on the cover, this book is very much a collaborative effort. It could not have been written without the help and advice of our editor, Brian Jepson, who contributed numerous helpful suggestions. As always, the O'Reilly design and production staff, who are listed individually in the front matter, worked miracles in converting our draft manuscript into an attractive finished book.

Special thanks are due to our technical reviewer, Dr. Richard H. Kessin, Professor of Pathology and Cell Biology at Columbia University. Rich's comments and suggestions made this a better book than it might otherwise have been.

Finally, we want to acknowledge the Wikicommons illustrators, many of whose graphics we've used in this book. We understand the science, but are hopeless as illustrators. They both understand the science and are skilled illustrators. To all of them, but especially to Lady of Hats (*http://commons. wikimedia.org/wiki/User:LadyofHats*), a big thank you. And, in particular, thanks for releasing your work into the public domain, where it is freely usable by anyone who wants to learn (or teach) about the science.

HOW TO CONTACT US

We have verified the information in this book to the best of our ability, but you may find things that have changed (or even that we made mistakes!). As a reader of this book, you can help us to improve future editions by sending us your feedback. Please let us know about any errors, inaccuracies, misleading or confusing statements, and typos that you find anywhere in this book.

Please also let us know what we can do to make this book more useful to you. We take your comments seriously and will try to incorporate reasonable suggestions into future editions. You can write to us at:

Maker Media, Inc.
1005 Gravenstein Hwy N.
Sebastopol, CA 95472

(800) 998-9938 (in the U.S. or Canada)
(707) 829-0515 (international/local)
(707) 829-0104 (fax)

Maker Media is a division of O'Reilly Media devoted entirely to the growing community of resourceful people who believe that if you can imagine it, you can make it. Consisting of Make Magazine, Craft Magazine, Maker Faire, and the Hacks series of books, Maker Media encourages the Do-It-Yourself mentality by providing creative inspiration and instruction.

For more information about Maker Media, visit us online:

MAKE: *www.makezine.com*
CRAFT: *www.craftzine.com*

Maker Faire: *www.makerfaire.com*
Hacks: *www.hackszine.com*

To comment on the book, send email to:

bookquestions@oreilly.com

The O'Reilly website for *Illustrated Guide to Home Biology Experiments* lists examples, errata, and plans for future editions. You can find this page at:

http://shop.oreilly.com/product/0636920017691.do

For more information about this book and others, see the O'Reilly website:

http://www.oreilly.com

To contact the authors directly, send mail to:

robert@thehomescientist.com
barbara@thehomescientist.com

We read all mail we receive from readers, but we cannot respond individually. If we did, we'd have no time to do anything else. But we do like to hear from readers.

We also maintain a dedicated landing page on our main website to support *Illustrated Guide to Home Biology Experiments*. This page contains links to equipment kits customized for this book, corrections and errata, supplemental material that didn't make it into the book, and so on. Visit this page before you buy any equipment or chemicals and before you do any of the experiments. Revisit it periodically as you use the book.

www.thehomescientist.com/biology.html

THANK YOU

Thank you for buying *Illustrated Guide to Home Biology Experiments*. We hope you enjoy reading and using it as much as we enjoyed writing it.

AUTHOR BIOS

Robert Bruce Thompson is the author of numerous articles, training courses, and books about computers, science, and technology, including many co-authored with his wife, Barbara. He built his first home lab as a teenager, and went on to major in chemistry in college and graduate school. Robert maintains a home laboratory equipped for doing real chemistry, forensics, biology, earth science, and physics.

Barbara Fritchman Thompson is, with her husband Robert, the co-author of numerous books about computers, science, and technology. With her Masters in Library Science and 20 years' experience as a public librarian, Barbara is the research half of the writing team.

Introduction

1

We really mean the "No Lecture" part of the title. This book is not intended to teach you all you need to know about basic biology. It's focused on providing an affordable, intense, reasonably comprehensive introductory biology lab experience. The background material we do provide is intended solely as a reminder of some important points that you've studied previously. For example, in one of the early lab sessions on biologically important molecules, we spend a couple pages covering carbohydrates and lipids and then jump into the actual lab work. If you start that lab work knowing only what we've told you about carbohydrates and lipids, you'll learn less than you should from doing the labs.

To get the most benefit from this book, use it in conjunction with a standard biology text, such as CK-12 *Biology* (*www.ck12.org/flexbook/book/2537*), which is freely downloadable, or Miller & Levine *Biology* (*www.millerandlevine.com*), both of which we highly recommend. Otherwise, you'll miss a lot of *really* important stuff. And you'll never know what you missed.

We sometimes refer in passing to things that you may be completely unfamiliar with. For example, in one early lab session we talk about aldehyde and ketone functional groups. If you've not yet taken first-year chemistry, let alone organic chemistry, you probably don't know much about aldehydes and ketones other than what little information we provide in the text. That's fine. For our purposes, all you really need to know is that aldehydes and ketones are different types of organic chemical compounds.

If you want to learn more about such topics, check Wikipedia (*www.wikepedia.org*). Although Wikipedia often presents a biased viewpoint in articles on controversial subjects, most science articles are well-written and reasonably comprehensive.

USING THIS BOOK WITH YOUR CURRICULUM

There are many ways to organize a first-year biology book. Some books take a generally top-down approach, starting with organisms and working their way down through organs and tissues to cells and cell structures and eventually to molecular biology. Other books take a generally bottom-up approach, starting with the molecular basis of life and working their way up to larger structures.

This book takes the latter approach, because we believe that to understand the whole it's best to first understand the parts that make up that whole. But regardless of which approach your textbook takes, you should have no difficulty in correlating the lab sessions in this book with topics in your text. With few exceptions, which are noted in the lab sessions themselves, you can do the lab sessions in this book in any order.

Feel free to pick and choose among the lab sessions and procedures to fit your curriculum, interests, and available time. For example, the third group of lab sessions covers the chemistry of life. This group has seven lab sessions, which incorporate 17 individual procedures. Ideally, you should do all of the sessions and procedures, but if time is limited, you can eliminate procedures or even entire lab sessions as necessary.

We designed this book with the intention of covering as much useful material as possible in an intense year-long laboratory course. That's not to say that you need to complete this material in two semesters or any other arbitrary period. One of the huge advantages of homeschooling is that you can take things at your own speed, fitting the course material to the student rather than shoehorning students into predefined standard-length courses as public schools must do. If it takes you 18 months or two years to cover all of the material, so be it.

We do strongly recommend doing the first and second group of lab sessions before anything else. The first group of lab sessions covers basic microscope skills, which you'll need in many later lab sessions. We assume those skills in later lab sessions. For example, in a later lab session we may tell you to make a smear mount and stain it without providing details about how to do so, because we assume you've already learned those skills in the first group of lab sessions. The second group of lab sessions involves building microcosms, which are self-contained soda bottle worlds of pond life. You should do those sessions early, both because you'll observe these microcosms over a period of weeks to months and because you'll use some of the organisms growing in these microcosms in later lab sessions.

Conversely, do not be afraid to incorporate other laboratory sessions from your biology text or elsewhere, either as a supplement to the lab sessions in this book or as replacements for some of them. In our experience, students who learn—really learn—laboratory sciences do so by spending lots of time doing actual lab work rather than just reading about it. The more lab work, the better.

> The lab sessions in this book correlate well with most homeschool biology curricula. For a correlation guide, visit *www.thehomescientist.com/biology.html*.

PLANNING AND SCHEDULING

Nature runs on its own schedule and, other than building greenhouses and other artificial environments, there's not much we can do about it. For example, if you plan to study leaf structures in January, unless you live in a warm area, there probably won't be many leaves on the trees. Of course, you can work around this problem, at least to some extent, by using leaves from house plants, the florist, a garden center, and so on, but that's less than an ideal solution. If you plan ahead, you can gather many different leaf specimens during the spring and summer and preserve them by pressing them between sheets of absorbent paper.

Similarly, biology lab work doesn't lend itself to nice, self-contained lab sessions that you can begin and finish in one lab period. Living things take time to grow, mature, and senesce. If you want to observe the life cycle of a particular organism, you may find yourself making observations over a period of weeks to months, or longer. If you decide to complete all of the lab sessions in this book, expect that at times you may have several sessions in progress. Some of those may require observation or other attention daily, others perhaps only weekly or monthly, but it's important to keep track of what needs to be done when. (We use the calendar/to-do list on our computer to track action items and prompt us when it's time to do something.)

We recommend that before you begin any lab work you first skim through this book to decide which lab sessions you intend to do over the course of the year, in what order, and (if possible), approximately when you intend to do them. That way, you can

be sure that you'll have everything you need, when you need it. Also decide what you intend to buy from science supply vendors versus making yourself or obtaining locally. For example, rather than collecting live protozoa from a pond, you might decide to order a mixed live protozoa culture from Carolina Biological Supply (CBS). That involves some delay for order processing and shipping, which you'll need to take into account when scheduling the lab session(s) that require that culture. Furthermore, that culture may have a lifetime measured in days, so you want to make sure to order it to arrive just before you start the lab session.

Also, if you intend to make rather than buy, allow time to collect specimens, prepare and stain slides, and so on. Of course, these activities can themselves be a part of the lab experience, but it's important to factor in the time needed and have the specimens available on lab day.

WORK AREAS

Give careful consideration to your work areas. You will, at various times, be working with chemicals that are toxic, corrosive, flammable, or otherwise hazardous. Biological stains do exactly what their name suggests—staining anything they come into contact with, sometimes indelibly. You may even decide to risk working with potentially pathogenic microorganisms.

Although the risks to your person are small and manageable, the same cannot always be said for the risks to your furniture, countertops, and floors. If you get a Sudan III stain on your hands, for example, it will eventually wear off. But if you spill Sudan III on your antique dining room table or hardwood floor or composite countertop, the stain may never come out, short of sanding down the surface and refinishing it.

If you have a well-lit, well-ventilated basement workshop or similar utility area, great. That's an ideal location for doing the messier work involved in a biology lab course. For many people,

though, it's the kitchen table or nothing. That's workable if it's the only realistic option, but you'll probably want to take some precautions:

- It's a bad idea to keep science equipment, chemicals, cultures, and other related materials in an area where food is prepared and consumed, so have these items in the kitchen only while you're actually doing the lab sessions. Store them elsewhere, secure from children and pets.

(As a young teenager, Robert learned this lesson the hard way when his mother screamed after spotting a 10 cm centipede crawling across the kitchen table.)

- When you finish a lab session, immediately wash and dry the equipment separately and put it away. Do not, for example, put a used beaker in the dishwasher with ordinary dishes, even if the beaker contained nothing hazardous and you rinsed it thoroughly. Doing so is a bad habit to get into. In fact, don't run science equipment through the dishwasher at all, even separately. Wash it by hand and keep it segregated.

- Protect the kitchen table or other work surfaces against spills. A cheap plastic tablecloth is good insurance. Cover the tablecloth with newspaper, old towels, or something else that will absorb spills, and always have a roll of paper towels handy.

- Photography darkrooms are always organized with a wet bench and a dry bench. Use the same principle in your lab work. For example, you might do all wet work in the kitchen, but keep your microscope in another room, safe from spills.

MAINTAINING A LABORATORY NOTEBOOK

A *laboratory notebook* is a contemporaneous, permanent *primary record* of the owner's laboratory work. In university and corporate labs, the lab notebook is often a critically important document, for both scientific and legal reasons. The outcome of zillion-dollar patent lawsuits often hinges on the quality, completeness, and credibility of a lab notebook. Many organizations have detailed procedures that must be followed in maintaining and archiving lab notebooks, and some go so far as to have the individual pages of researchers' lab notebooks notarized and imaged on a daily or weekly basis.

If you're just starting to learn about lab work, keeping a detailed lab notebook may seem to be overkill, but it's not. Developing the habit of keeping comprehensive records of all lab work is a critical skill for any STEM student and certainly for any working scientist, and such habits are best developed early. If you're using this book to prepare for college biology, and particularly if you plan to take the Advanced Placement (AP) Biology exam, you should keep a lab notebook. Even if you score a 5 on the AP Biology exam, many college and university biology departments will not offer you advanced placement unless you can show them a lab notebook that meets their standards.

Always keep at least one spare lab notebook on hand. If you complete all of the procedures in this book and document all of your work properly, you can expect to fill several such notebooks. We order the Mead composition books by the case from Costco.

If you're looking for a better quality lab notebook, look no further than the Maker's Notebook, also published by MAKE. Although the Maker's Notebook costs significantly more than cheap composition books—comparable to "official" lab notebooks—it's also better bound, uses better paper, and incorporates other nice features.

LABORATORY NOTEBOOK GUIDELINES

Use the following guidelines to maintain your laboratory notebook:

- The notebook must be permanently bound. Looseleaf pages are unacceptable. Never tear a page out of the notebook. We use inexpensive Mead hardbound 100-sheet composition books, available at drugstores, Walmart, Costco, and so on.

- Use permanent ink. Pencil or erasable ink is unacceptable. Erasures are anathema.

The one departure from this rule that we consider acceptable is using colored pencils for making sketches of your observations. Colored marking pens are simply too crude a tool for recording fine details and subtle gradations of color.

- Before you use it, print your name and other contact information on the front of the notebook, as well as the volume number (if applicable) and the date you started using the notebook.

- Number every page, odd and even, at the top outer corner, *before* you begin using the notebook.

- Reserve the first few pages for a table of contents.

- Begin a new page for each experiment or observing session.

- Use only the right-hand pages for recording information. The left-hand pages can be used for scratch paper. (If you are left-handed, you may use the left-hand pages for recording information, but maintain consistency throughout.)

- Record all observations as you make them. Do not trust to memory, even for a minute.

- Do not record information that you don't actually have. For example, if you observe a protist that you are certain is an *Amoeba* and believe to be *Amoeba proteus*, do not record your observation with more specificity than justified. Rather than recording the protist as "*Amoeba proteus*," record it as *Amoeba sp.* to indicate that the genus is known but the species is uncertain. You can record uncertain information in the form "*Amoeba sp.* (poss. *proteus*)" or "*Amoeba sp.* (prob. *proteus*)" to indicate your belief as to species and the level of your uncertainty.

- Print all information legibly, preferably in block letters. Do not write longhand.

- If you make a mistake, draw one line through the erroneous information, leaving it readable. If it is not otherwise obvious, include a short note explaining the reason for the strikethrough. Date and initial the strikethrough.

- Do not leave gaps or whitespace in the notebook. Cross out whitespace if leaving an open place in the notebook is unavoidable. That way, no one can go back in and fill in something that didn't happen. When you complete an experiment, cross out the white space that remains at the bottom of the final page.

- Incorporate computer-generated graphs, charts, printouts, images, and similar items by taping or pasting them into the notebook. Date and initial all add-ins on the add-in itself.

- Include only procedures that you personally perform and data that you personally observe. If you are working with a lab partner and taking shared responsibility for performing procedures and observing data, note that fact as well as describing who did what and when.

- Remember that the ultimate goal of a laboratory notebook is to provide a permanent record of all the information necessary for someone else to reproduce your experiment and replicate your results. Leave nothing out. Even the smallest, apparently trivial, detail may make the difference.

LABORATORY NOTEBOOK FORMAT

Use the following general format for recording an experiment in your lab notebook:

Introduction

The following information should be entered before you begin the laboratory session:

Date

Enter the date at the top of the page. Use an unambiguous date format, for example 10 June 2012 or June 10, 2012 rather than 10/6/12 or 6/10/12. If the experiment runs more than one day, enter the starting date here and the new date in the procedure/data section at the time you actually begin work on that date.

Experiment title

If the experiment is from this or another laboratory manual, use the name from that manual and credit the manual appropriately. For example, "Investigating Bacterial Antibiotic Sensitivity (Illustrated Guide to Home Biology Experiments, Lab VII-4)." If the experiment is your own, give it a descriptive title.

Purpose

One or two sentences that describe the goal of the experiment. For example, "To investigate the sensitivity of the bacteria *Bacillus subtilis*, *Micrococcus luteus*, and *Rhodospirillum rubrum* to amoxicillin, chlortetracycline, sulfadimethoxine, and neomycin."

Introduction (optional)

Any preliminary notes, comments, or other information may be entered in a paragraph or two here. For example, if you decided to do this experiment to learn more about something you discovered in another experiment, note that fact here.

Chemical notes

For investigations in which chemical reactions play a prominent role, include balanced equations for all of the reactions involved in the experiment, including, if applicable, changes in oxidation state. Record important information about all chemicals used in the experiment, including, if appropriate, physical properties (melting/boiling points, density, etc.), a list of relevant hazards and safety measures (from the Material Safety Data Sheet), and any special disposal methods required. Include approximate quantities, both in grams and in moles, to give an idea of the scale of the experiment.

> Always read the MSDS (Material Safety Data Sheet) for each chemical before you begin work. The MSDS is a concise listing of the hazards involved in using that chemical, steps to take to minimize risk, exposure limits, and other important information. If an MSDS was not supplied with the chemical, search Google for "<chemical name>" + MSDS.

Organism notes

For investigations that focus on a particular organism or organisms, record the particulars about the organism(s), including type, binomial name, the reason (if any) that particular organism was chosen, and so on.

Planned procedure

A paragraph or two to describe the procedures you expect to follow.

Main body

The following information should be entered as you actually do the experiment:

Procedure

Record the procedure you use, step by step, as you actually perform the procedures. Note any departures from your planned procedure and the reasons for them.

Data

Record all data and observations as you gather them, in-line with your running procedural narrative. Do not attempt to organize or tabulate the data here; simply record it in-line with your running narrative.

Sketches and/or images

Make sketches or, if you have the necessary equipment, shoot images of what you observe. Label significant features. For microscopic observations, indicate the magnification used, any special staining protocols, and so on. Always include a dimensional scale to indicate the approximate size of individual features in the image.

Calculations

Include any calculations you make. If you run the same calculation repeatedly on different data sets, one example calculation suffices.

Table(s)

If appropriate, construct a table or tables to organize your data. Copy data from your original in-line record to the table or tables.

Graph(s)

If appropriate, construct a graph or graphs to present your data and show relationships between variables. Label the axes appropriately, include error bars if you know the error limits, and make sure that all of the data plotted in the graph is also available to the reader in tabular form. Hand-drawn graphs are preferable. If you use computer-generated graphs, make sure they are labeled properly and tape or paste them into this section.

Conclusion

The following information should be entered after you complete the experiment:

Results

Write a one or two paragraph summary of the results of the experiment.

Discussion

Discuss, if possible quantitatively, the results you observed. Do your results confirm or refute the hypothesis? Record any thoughts you have that bear upon this experiment or possible related experiments you might perform to learn more. Suggest possible improvement to the experimental procedures or design.

Answer questions

If you've just completed a lab exercise from this or another book, answer all of the post-lab questions posed in the exercise. You can incorporate the questions by reference rather than writing them out again yourself.

Equipping a Home Biology Laboratory

2

Other than a microscope and accessories, it doesn't take much special equipment to learn about biology. You'll need some general lab equipment, chemicals, and so on, but much of what you need can be improvised or substituted for by items inexpensively available from the drugstore or hardware store. If you keep a close eye on your budget, you can complete most or all of the lab sessions in this book for surprisingly little money.

You'll have decisions to make that balance cost versus time versus quality. For example, many lab sessions call for prepared microscope slides. If low cost is your top priority, you can prepare many of those slides yourself for a few cents each in materials, but at the expense of significant time and effort and possibly lower quality. Conversely, if you want top quality and cost is a low priority relative to your time, you can purchase very high quality prepared slides, although those may cost $5 to $20 or more apiece. Or you can compromise by purchasing inexpensive prepared slides for a buck or two apiece. Their quality won't be as good as that of the expensive prepared slides, but they'll probably be good enough for your purposes, and buying them will certainly save you a lot of time and effort.

If you're pursuing biology as a hobby, your budget may range from next to nothing to essentially unlimited. Many golfers get along just fine with a $250 set of clubs, but there's no shortage of golf enthusiasts with $1,500 drivers and $5,000 iron sets in their bags. And you'll often find that a kid with a $250 set of clubs outplays a guy lugging around $10,000 worth of clubs. (As a teenager, Barbara was a scratch golfer, and regularly embarrassed middle-aged rich guys.)

DIY biology enthusiasts are no different. If you attend a DIY Bio meet-up, you'll find kids who've accomplished amazing things on next to no budget gathered heads-down with doctors and lawyers and executives who've spent $100,000 or more to turn their garages into serious biotechnology labs. It's not about how much equipment you have; it's what you do with the equipment you do have.

That said, lack of equipment *can* limit what you can accomplish. Or, more precisely, lack of functionality. If you need to spin down a plasmid mini-prep, for example, there's no alternative: you need an ultracentrifuge. If you have $15,000 burning a hole in your pocket, great. Go buy a commercial ultracentrifuge. We couldn't afford that, so we built our own functional equivalent for less than $150. And for what we need to do, it's actually just as good as that $15,000 commercial unit.

> $150? Seriously? Yep. See the discussion of our "Dremelfuge" later in this chapter. That's what it cost us to buy a Dremel MotoTool (which we actually already had), a miniature drill-press mount for it, and the custom-made centrifuge head that fits on the Dremel in place of a standard bit.
>
> That centrifuge accepts standard 1.5 mL "Eppie" polypropylene micro centrifuge tubes, which it can spin fast enough to disintegrate the tubes. Even on a middle speed setting, it produces more than enough centripetal acceleration for pelleting organelles and other fine particulates and for gradient separations.
>
> You won't need one of these babies unless you get seriously into DIY biology. If you do, it's nice to know that the functionality is quite affordable.

In this chapter and throughout the book, we've tried to focus on getting a lot done on as small a budget as possible. That doesn't mean you should never use commercial products when there's a cheaper alternative. For example, gel electrophoresis is used to separate and purify DNA, proteins, and other biologically important molecules. If you're a home schooler, you'll need a gel electrophoresis apparatus to complete one or two lab sessions and then just move on to the next lab sessions. You probably want your gel electrophoresis apparatus to be as inexpensive as possible, so we'll show you how to build a usable apparatus for $10 (about $9 of which is for 9V batteries). But if you're a DIY Bio enthusiast, you'll probably be using gel electrophoresis frequently. You'd soon tire of replacing expensive 9V batteries every few runs, so it makes sense to spend $300 or so on a commercial gel electrophoresis tank and power supply that minimizes the running costs.

If you're on a tight budget, you may need to skip some of the lab sessions or at least some parts of some lab sessions, but try to make that a last resort. We'll try to point out as we go along where you can improvise and substitute to get the job done. We also recognize that it can be very expensive to buy many different items piecemeal. For example, you may need only one gram of a particular chemical for one of the lab sessions in this book, but the minimum you can buy from a science supplies vendor is, say, a 30 gram bottle for $5. That wouldn't be too bad if you needed only that one chemical, but since you need many different chemicals the cost adds up fast.

Accordingly, we've put together a customized kit that includes many items that are difficult to find, hard to substitute for, or expensive to purchase piecemeal. In order to avoid retail markups and keep the cost to you as low as possible, the kits are available only direct from our own company, The Home Scientist, LLC (*www.thehomescientist.com/biology.html*). We can ship the kits to all 50 states, but shipping regulations make it impossible for us to ship them to other countries. Sorry.

MICROSCOPES AND ACCESSORIES

The one piece of equipment most closely associated with biology is, of course, the microscope, and rightly so. Biology as a modern science would not exist without the microscope, and good microscopes are essential day-to-day tools for most biologists.

Choosing a suitable microscope is a nontrivial task, so we devote a significant amount of space in this section to explaining the things you need to know to choose the right microscope for you. Before we get into that, though, we'll offer some advice about how to go about acquiring the microscope you decide best fits your needs.

For most people, buying a suitable microscope is a major purchase. You don't want to pay more than you need to, but neither do you want to paint yourself into a corner by buying too little microscope.

Microscopes range in price from $25 toys to professional models from German and Japanese manufacturers that cost from $3,000 or $4,000 to $25,000 or more. If you can afford a top-tier microscope, great. Buy a suitable model from Leitz, Zeiss, Fujinon, or one of the other German or Japanese microscope makers. Your credit card will be smoking, but you'll have one of the finest optical instruments on the planet, and one that will last a lifetime.

Most of us aren't that lucky, but fortunately there are affordable alternatives. The best Chinese microscopes offer 90% of the optical and mechanical quality of the top-tier models at 20% of the price. We'll make specific recommendations by brand and model later in this chapter, but for now be aware that although the best of the Chinese microscopes are very, very good, most Chinese microscopes are of very poor quality. It's impossible to tell the difference just by looking at the microscopes or comparing prices, so the key to getting a good one is to buy from a reputable vendor.

The first thing to decide is whether you want to keep the microscope indefinitely or use it only for a short period, such as a school year. Once you decide that, you can decide whether to buy a new microscope, buy a used microscope, or rent the microscope.

Buy a new microscope

If you intend to keep the microscope, buying new is usually the best option. You'll pay more than you would for the same model used, but you'll get exactly the microscope you want with exactly the options you want. You'll also get a warranty, which for most better models is a limited lifetime warranty. (Don't overvalue the warranty; if treated properly, good microscopes almost never need to be repaired, other than trivial failures like bulbs and fuses.)

Buy a used microscope

A good microscope that has been well cared for is as good now as the day it was made. Unfortunately, the converse is also true: a bad microscope will never be anything but a bad microscope, so you have to be very careful buying used.

Pricing for used microscopes is all over the map. Inexpensive no-name microscopes have essentially no resale value. House-brand models from Home Science Tools and similar vendors may on average sell for 33% to 50% of the current selling price for the same model new, but we've seen prices listed for such scopes that range from 10% to 100% or more. The best Chinese scopes, such as the midrange and better National Optical models, may sell for 70% to 80% of their current new selling price. Current top-tier models may sell for 90% or more of

their current new selling price. In fact, some models are so popular that you may have to join a waiting list to get one, and these may actually sell for more used than their current list prices. Discontinued older top-tier models may sell for 80% or more of the current selling price of the equivalent replacement model.

The advantage of buying a used microscope locally is that you can actually see and touch it before buying. The disadvantages are that the selection is likely to be limited, and you'll have to negotiate the price with the seller. If you want to buy locally, check Craigslist and your local homeschool group. The advantages of buying a used microscope online from a reputable vendor are that you'll get what you pay for (although never more than you paid for, as can happen buying locally), that the selection will be much better, and that the vendor will do at least some minimum screening and usually provide at least a short warranty.

Rent a microscope

If you need the microscope for only a short period, renting is another option. The advantages to renting are that it requires the least cash outlay and you can select among many models. The disadvantage is that you may have to pay as much as 50% of the current selling price of the scope to have the use of it for only a year or less. The rental vendor will charge your credit card for the price of the scope initially—often at list price—and then refund your money less the rental fee once you return the microscope. Also, some rental vendors are very picky about the condition of the returned scope, so you may end up paying a higher rental fee than you expected to cover "damage" such as minor scratches.

On balance, for those who intend to keep the microscope indefinitely, we recommend buying a new or used model from an online vendor, depending on your comfort level with buying used. If you need the microscope for a limited time, such as a school year (or several school years), buy a new or used model from an online vendor. Make sure it's a respected brand name— National Optical, Swift, Motic, and Leica are the best brand names in mid-priced scopes—and then resell it to a local home schooler when you no longer need it. (Never sell the microscope back to an online vendor. You'll get only a small fraction of what you would by selling it locally.)

All of that said, let's take a detailed look at what you need to know to make an informed purchase.

MICROSCOPE TYPES

Broadly speaking, three types of microscopes are useful for studying biology. (Well, there's also the electron microscope, but few home scientists can afford one of those.)

Compound microscope

A *compound microscope* is what most people think of as a microscope. It's used to view small specimens, usually by transmitted light, at three or four medium to high magnifications, typically 40X, 100X, 400X, and sometimes 1,000X.

Stereo microscope

A *stereo microscope*, also called a *dissecting microscope*, operates at low magnification, typically 10X to 40X. Some models offer only one magnification, others two, and zoom models offer continuously variable magnifications. Stereo microscopes are used to view medium to large specimens, usually by reflected light.

Portable microscope

A *portable microscope*, also called a *field microscope*, is small enough to carry in your pocket on field trips. Some models offer fixed magnification, often 30X, while others provide zoom magnifications to 100X or more. Most models include a battery-powered LED illuminator and use reflected light only. Some models make provision for using standard slides to examine specimens by transmitted light as well.

In the following sections, we'll take a closer look at each type.

COMPOUND MICROSCOPE

A *compound microscope*, shown in Figure 2-1, is what most people think of as a microscope. You use it to view small specimens, usually by transmitted light, at three or four medium to high magnifications, typically 40X, 100X, 400X, and sometimes 1,000X.

Figure 2-1: *A typical compound microscope (image courtesy National Optical & Scientific Instruments, Inc.)*

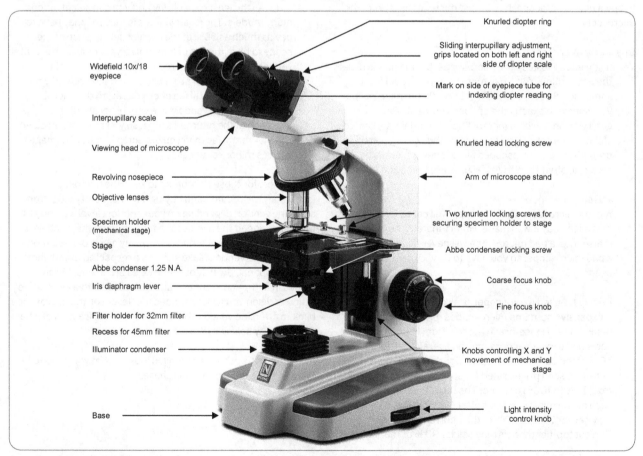

The following sections describe some factors to consider when you choose a compound microscope:

HEAD STYLE

Compound microscopes are available in the four head styles shown in Figure 2-2.

Monocular

A *monocular head*, shown on the left in Figure 2-2, provides only one eyepiece. This is the least expensive of the four head styles, and is suitable for general use.

Dual-head

A *dual head*, shown left-center in Figure 2-2, provides two eyepieces, one vertical and one angled. The second eyepiece allows two people to view a specimen simultaneously, for example a teacher and a student. A dual head is also very convenient if you want to mount a still or video camera to image specimens. Dual head models typically cost $50 to $100 more than comparable monocular models.

Binocular

A *binocular head*, shown right-center in Figure 2-2, provides two eyepieces to allow viewing specimens with both eyes. One eyepiece is individually focusable to allow the instrument to be set up for one person's vision. The advantage of a binocular head is that it's less tiring to use over long periods and may allow seeing more detail in specimens. The disadvantage is that the focusable eyepiece must be adjusted each time a different person wants to use the scope. Binocular models typically cost $150 to $250 more than comparable monocular models.

Trinocular

A *trinocular head*, shown on the right in Figure 2-2, provides two eyepieces for binocular viewing and a separate single eyepiece for viewing by a second person or for mounting a camera. Trinocular models typically cost $300 to $400 more than comparable monocular models.

At any particular price point, a monocular-head model offers the maximum bang for the buck. You'll get better optical and mechanical quality with the monocular head than with any of the multiple-head models.

Regardless of head style, most better models allow the head to be rotated through 360° to whatever viewing position you prefer. The left image in Figure 2-2 shows the traditional viewing position, with the support arm between the user and the stage. The other three images show the reversed viewing position, with the stage between the user and the support arm. Most people prefer the latter position, which makes it easier to manipulate slides, change objectives, and so on.

Figure 2-2: *Monocular, dual-head, binocular, and trinocular head styles, left to right (images courtesy National Optical & Scientific Instruments, Inc.)*

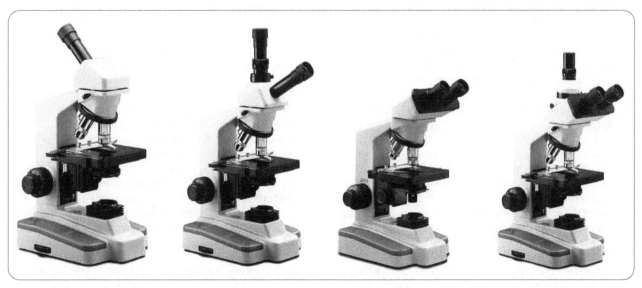

ILLUMINATION TYPE AND POWER SOURCE

The next issue to consider is the illumination type and power source.

Early microscopes and some inexpensive current models have no built-in illuminator. Instead, they use a mirror to direct daylight or artificial light up through the stage and into the objective lens. Because any mirror small enough to fit under the microscope stage gathers insufficient light to provide bright images at high magnifications, such scopes are limited to use at low and medium magnifications unless they are equipped with an accessory illuminator. Most microscopes include built-in illuminators of one of the following types, roughly in order of increasing desirability:

Tungsten illumination

Tungsten illumination is the least expensive type, and the most common on low-end microscopes. Tungsten illuminators use standard incandescent light bulbs. They are relatively bright, but they produce a yellowish light and considerable heat. In particular, as the light is dimmed, it shifts further toward orange. This warm color balance can obscure the true colors of specimens. The heat produced by the incandescent bulb may kill live specimens and quickly dries out temporary wet mounts made with water. Lamp life is relatively short.

Fluorescent illumination

Fluorescent illumination costs a bit more than tungsten, and was quite popular before the advent of LED illuminators. Fluorescent illuminators provide bright light that appears white to the human eye, but is actually made up of several different discrete colors that are mixed to appear white. Accordingly, color rendition can differ significantly from the true color rendition provided by daylight. Fluorescent bulbs emit much less heat than incandescent bulbs, and so are well suited to observing live specimens. Some fluorescent illuminators are battery-powered, but most use AC power. Lamp life is relatively long.

LED illumination

LED illumination costs about the same as fluorescent illumination. LED illuminators have become very popular, largely replacing fluorescent illuminators. LED illuminators have the same color-rendition problems as fluorescent illuminators, but are otherwise ideal for many purposes. LED illuminators draw very little power and emit essentially no heat. Their low power draw means they're the best choice for a battery-powered microscope, and are ideally suited for portable microscopes that can be used in the field. Lamp life is essentially unlimited.

Quartz-halogen illumination

Quartz-halogen illumination is the most expensive type, and the one preferred by most experienced microscopists. Quartz-halogen provides the brilliant white light needed for work at high magnification that reveals the true colors of specimens. Unfortunately, quartz-halogen lamps also produce more heat than any other type of illuminator. Their high power draw means they are AC-only. Lamp life is relatively short.

Choose quartz-halogen if it is available for the microscope model you purchase. Otherwise, choose LED. Fluorescent illumination is obsolete, and tungsten is appropriate only for an entry-level scope.

ILLUMINATION METHODS

The next thing to consider is which illumination method or methods the microscope supports.

Brightfield illumination

Brightfield illumination is supported by all compound microscopes, and is the only method available with most. With brightfield illumination, you view specimens by transmitted light that passes directly through the specimen. Scattered light reduces contrast and detail, and is minimized by restricting the diameter of the light cone passing through the specimen to match the field of view of the objective lens being used to view it.

The advantage of brightfield illumination is that it provides very bright, clear images. The drawback is that many types of specimens have so little inherent contrast that it's very difficult to discriminate detail in the specimens. Staining specimens provides color contrast that reveals these otherwise-hidden details, but staining can be time-consuming.

Darkfield illumination

Darkfield illumination is the exact opposite of brightfield illumination. Rather than viewing the specimen by transmitted light and blocking as much scattered light as possible, you block transmitted light and view the specimen only by scattered light. This is accomplished by using an opaque circular mask below the specimen that prevents light from passing through the specimen. By adjusting the condenser diaphragm, you produce a bright ring surrounding the specimen that illuminates the specimen from just outside the field of view.

At first glance, the image provided by darkfield illumination appears to be a simple negative of the image provided by brightfield illumination, but that is not the case. Some things are visible with darkfield illumination that are invisible with brightfield. For example, live microorganisms in a drop of water may be invisible with brightfield because their transparency and refractive index is so close to that of water. With darkfield, those microorganisms are revealed brightly lit against a dark background.

Unfortunately, darkfield illumination also has several drawbacks. First, as you might expect, the image can be quite dim, particularly at high magnification, and images of even colorful specimens present in essentially monochrome. Second, darkfield demands extraordinary cleanliness and care in sample preparation. Every extraneous spec of dust shows as a blazing spot of light in the field. Specimens must be extremely thin and even, much more so than for brightfield. For best results, you must use top-quality slides and coverslips, and the best slides for use with darkfield are thinner (about 0.7 to 1.0 mm) and more fragile than those typically used with brightfield illumination.

Some microscopes include darkfield stops as standard equipment, or offer a darkfield kit as an inexpensive option. Even if your microscope does not, you can still use darkfield as long as your microscope has a filter holder beneath the substage condenser lens. One way is to use darkfield stops or a darkfield kit supplied by a different vendor. (The size of the stop should be matched to the objective lens, ranging from 20 mm for a 4X objective down to 8 mm with a 100X objective.) Another is simply to use an opaque circle (such as a peel-and-stick dot) pasted to the center of a colorless or colored glass filter placed between the lamp and the bottom of the substage condenser.

Phase-contrast illumination

Phase-contrast illumination is a complex illumination method in which small phase shifts in the light passing through a transparent specimen are converted to visible amplitude (contrast) differences. Phase-contrast illumination reveals the low-contrast details that would otherwise be invisible without staining. This has two huge advantages. First, it eliminates the time spent preparing and staining specimens, which can be significant. Second, it allows live microorganisms to be observed in their natural environment, without the possible damage caused by fixing and staining them. (There are some stains, called *vital stains*, that can be used on living cells, but the choices are quite limited.)

Many professional-grade microscopes include phase-contrast illumination as a standard feature, and it is available as an option on some other models.

Unfortunately, it's an expensive option. A phase-contrast kit for a standard microscope typically boosts the price of the microscope by $800 to $3,000.

Köhler illumination, devised by August Köhler in 1893, is a type of brightfield illumination that is extremely even and provides the highest possible contrast. Unfortunately, using Köhler illumination requires physical features that are not present on many microscopes, including a positionable lamp and a focusable lamp condenser.

Fortunately, the alternative, called *critical illumination* or *standard brightfield illumination*, is perfectly usable for most visual work. In fact, many experienced microscopists prefer critical illumination to Köhler illumination for visual work at high magnification. The extreme evenness of Köhler illumination is important for professional quality results if you are shooting images through the microscope, but otherwise critical illumination works fine.

NOSEPIECE AND OBJECTIVE LENSES

The next features to think about are the nosepiece and objective lenses, which are one factor in determining the magnifications that are available with a particular microscope.

The *nosepiece*, also called the *turret*, is a rotating assembly that holds three, four, or five *objective lenses*. By rotating the nosepiece, you can bring a different objective lens (usually just called an *objective*) into position and change the magnification you use to view the specimen. Inexpensive microscopes use friction-bearing nosepieces; better models use ball-bearing nosepieces with positive click-stop detents. Figure 2-3 shows a typical nosepiece with three objectives visible.

The nosepiece may be mounted in the forward position (tilted away from the support arm) or reverse position. If you use the scope in the forward viewing position (with the support arm between you and the stage), having the nosepiece mounted in the forward position makes it a bit easier to change objectives. If you use the reverse viewing position, it's easier to use a nosepiece mounted in the reverse position.

Objective lenses are usually color-coded to make it obvious which one is currently being used. The standard color codes are red (4X), yellow (10X), green (20X), light blue (40X or 60X), and white (100X). Not all manufacturers follow this standard.

Figure 2-3: *A typical microscope nosepiece with objective lenses*

Most inexpensive microscopes—those that sell for less than $300 or so—provide only three objective lenses, 4X, 10X, and 40X. Used with a 10X eyepiece, such microscopes provide 40X, 100X, and 400X magnification. Although such microscopes may be suitable for most first-year biology lab work, we think it's a bad idea to buy one unless you are absolutely certain it will suffice for the work you intend to do, now and in the future. The problem is that these 400X microscopes do not provide the higher magnifications needed to observe bacteria and other tiny specimens, which is required for second-year biology and other more advanced work.

Microscopes of the quality needed for advanced biology studies and beyond include a fourth, 100X, objective lens, which provides 1,000X magnification with a 10X eyepiece. Some models have a five-position turret. The fifth position may be empty, or it may include a fifth objective, often 60X.

The quality of the objective lenses is as important as their number and magnification. Microscope objective lenses differ in two major respects, color correction and flatness of field.

Color correction

The level of *color correction* is specified as either *achromatic* or *apochromatic*. Achromatic lenses are corrected for chromatic aberration at two specific wavelengths of light, usually red and green. An achromat brings those two wavelengths to the same focus, with other wavelengths very slightly out of focus. An apochromat is corrected for three specific wavelengths of light—usually red, green, and blue—and brings those three wavelengths to the same focus, providing slightly sharper images than an achromat. Apochromatic objectives are extremely

expensive, some costing more than $10,000, and are found only on professional-grade microscopes. Any microscope affordable for a home lab uses achromatic objectives.

WARNING

Used microscopes, even very old models, can be excellent bargains. As a teenager in the '60s, Robert used a WWI-era Zeiss microscope that was superb, both optically and mechanically. But you have to be careful and know what you're doing.

Some old microscopes have apochromatic objectives, but those objectives were optimized for photomicrography with black-and-white film rather than visual use. Their correction is superb in the blue, violet, and ultraviolet range, but when used for visual observing, their image quality is actually inferior to a modern achromat.

Flatness of field

Standard objectives have limited correction for spherical aberration, which means that only the central 60% to 70% of the field of view is in acceptably sharp focus. Semi-plan objectives have additional correction that extends the sharp focus area to the central 75% to 90% of the field of view. Plan objectives extend the area of sharp focus to 90% or more of the field. This additional correction for flatness of field is completely independent of color correction. You can, for example, buy semi-plan apochromatic objectives and plan achromat objectives.

Finally, some vendors offer optional upgrades to superior lens coatings, often under such names as Super High Contrast or something similar. These superior coatings don't improve color correction or flatness of field, but they increase image contrast noticeably and are worth having.

For most home lab use, ordinary achromatic objectives provide perfectly acceptable images and are the least expensive choice. Our own microscope, a Model 161 dual-head unit shown left center in Figure 2-2, has the upgraded ASC objectives, which we purchased for their higher contrast and superior visual image quality.

All but toy microscopes are *parfocal* and *parcentered*. Parfocal means that all objectives have the same focus. When you focus a specimen at 40X, for example, and then change to 100X, the specimen remains focused. (You may have to touch up the focus with the fine-focus knob, but the focus should be very close to start with.) Parcentered means that if you have an object centered in the field of view with one objective and you change to a different objective, the object remains centered in the field of view. Professional-grade microscopes provide adjustments for both parfocality and parcentrality, but student- and hobbyist-grade microscopes are set at the factory and cannot be adjusted by the user. That means it's important to check these settings as soon as you open the box of your new microscope.

To check parfocality, place a flat specimen (a thin-section or smear slide is good, if you have one; otherwise, any flat specimen) on the stage and focus critically on it at the lowest magnification. Then change to your next highest magnification and check the focus. It should be in focus or nearly so, requiring at most a partial turn of the fine-focus knob to bring it into critical focus. Change to your next higher magnification and again check the focus. Again, it should require at most a small tweak with the fine-focus knob to bring the specimen into sharp focus.

To check parcentrality, center an object in the field of view at the lowest magnification and then switch objectives to the next-higher magnification. The object should remain centered, or nearly so. Repeat until you are viewing the object at your highest magnification. Because it's easier to judge whether an object is centered at high magnification, center the object at your highest magnification and then work your way down to lower magnifications. If the object remains centered (or nearly so), your parcentrality is acceptable. If the position of the object in the field of view shifts dramatically when you change objectives, the parcentrality is off. The only solution is to return the microscope for a replacement.

EYEPIECES (OCULARS)

The *eyepiece* (or *ocular*) is the second factor that determines the overall magnifications available with a particular microscope. The eyepiece magnifies and focuses the image provided by the objective lens and presents it to your eye. Multiplying the objective magnification by the ocular magnification yields the overall magnification. For example, using the 100X objective with a 10X eyepiece provides 1,000X overall magnification.

Magnifications are specified linearly rather than areally. For example, if you use 1,000X magnification, a square object that is actually 1 μm (micrometer, one millionth of a meter or one thousandth of a millimeter) on a side appears to be 1,000 μm (or 1 mm) on a side. The image of that object at 1,000X magnification therefore covers 1,000,000 times as much area as the actual object.

Standard microscope ocular barrels are either 23.2 mm (usually abbreviated to 23 mm) or 30 mm in diameter, which means it's easy to exchange oculars if you need a different magnification range. The standard ocular magnification factor is 10X, but 15X oculars are readily available to increase the range of magnifications available to you. Avoid zoom oculars, all but the most expensive of which produce inferior images.

Most standard oculars are unobstructed, but some have a standard or optional pointer or reticle (grid or graduated scale). A pointer is primarily useful in a teaching or collaborative environment, where one person can place the pointer on an object of interest so that the other person can identify it unambiguously. A graduated reticle is useful for measuring the size of objects in the field of view, and a grid reticle is useful for counting large numbers of small objects in the field of view.

You might think you can achieve any magnification you want simply by combining different eyepieces and objective lenses. For example, you could combine a 25X eyepiece with a 100X objective to give you 2,500X magnification. Why, with a 30X eyepiece, you can get to 3,000X magnification, and with a 50X eyepiece to 5,000X!

Well, you can, but you won't actually see any more detail than you would using that objective with a 10X or 12.5X eyepiece. The image scale will be larger at 2,500X to 5,000X, of course, but the image will also be much dimmer and fuzzier. You will see no more detail than you can see at 1,000X or 1,250X, and probably less. That's why such high magnifications are called *empty magnification*.

Unfortunately, the laws of optics mean there is an upper limit to the useful magnification for any particular objective, even if it is of the best possible optical quality. Straying above that limit is possible, but the image quality will be degraded.

As a rule of thumb, determine *maximum useful magnification* by multiplying the numeric aperture (NA) of the objective by 1,000. For example, a 100X objective with an NA of 1.25 has a maximum useful magnification of (1.25 X 1,000) = 1,250X.

FOCUSER

The next consideration is the focuser, which may sound trivial to those who've never used a microscope. It's anything but. A microscope with a smooth focuser is a joy to use, and makes it easy to achieve the best possible focus to reveal the maximum amount of detail. A microscope with a poor focuser is almost unusable.

Microscopes use one of two methods for focusing. Most older models and some current models keep the stage in fixed position and move the head up and down to achieve focus. Most current models and some older models reverse this, keeping the head in a fixed position and moving the stage up and down to achieve focus. Either method works fine for general use, but if you plan to mount a camera on a dual-head or trinocular model the fixed-stage method may be problematic because the weight of the camera makes it more difficult to focus and may cause focus drift.

Inexpensive microscopes have one focus knob that changes focus at intermediate rate, which makes it difficult or impossible to achieve critical focus. Midrange models have separate coarse-focus and fine-focus knobs. More expensive models usually have a coaxial focus knob, often one on each side of the microscope, with the coarse focus on the inner knob and fine focus on the outer knob, as shown in Figure 2-4. Coarse focus tension should be adjustable.

You use the coarse-focus knob to bring the specimen into reasonably close focus, and then use the fine-focus knob to tweak the focus slightly to achieve the sharpest possible focus. If you are viewing a three-dimensional object, particularly at higher magnifications, you'll find that you can't bring the entire depth of the object into focus at the same time. You use the fine-focus knob to adjust focus slightly as you're viewing the object to view different "slices" of it in depth.

Many coaxial focus knobs, including the one in Figure 2-4, provide a graduated scale. One obvious use for this scale is in a collaborative situation. One person can focus critically, note the scale setting, and then turn over the microscope to the second person, who refocuses as necessary. When the first person returns to the eyepiece, merely resetting the fine-focus knob to the original setting puts the specimen back into critical focus. A less obvious use of the graduated scale is to determine relative depths of parts of a specimen. By setting a baseline focus on one level of the specimen and then noting how much change in scale units is needed to refocus on parts of the specimen at different depths, you can get a relative idea of the differences in depth of different parts of the specimen.

Figure 2-4: *Coaxial focusing knob, with coarse focus (inner knob) and fine focus*

MECHANICAL STAGE

Inexpensive microscopes use a pair of clips to secure the microscope slide to the stage. Although (barely) usable at low magnifications, this method becomes increasingly difficult as you increase magnification. The problem is that a very small movement of the microscope slide translates into a huge movement in the field of view. At low magnification, the smallest movement you can make manually shifts the object significantly within the field of view. At high magnification, the smallest movement you can make manually may move the object completely out of the field of view. If you're viewing a living, moving object (such as a paramecium), it can be almost impossible to keep the object in the field of view.

The solution to this problem is a mechanical stage, shown in Figure 2-5. With a mechanical stage, you clamp the slide into an assembly that provides rack-and-pinion geared movements. Turning the control knobs, shown in Figure 2-6, moves the slide continuously along the X-axis (left or right) and the Y-axis (toward or away from you) in extremely small increments.

Figure 2-5: *A typical mechanical stage (note the verniers on the X and Y axes and the top lens of the Abbe condenser below the stage)*

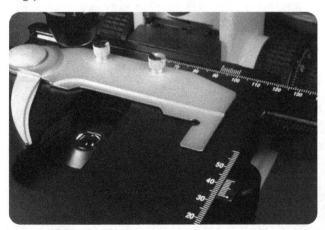

With a mechanical stage, centering an object in the field of view becomes trivially easy, as does keeping a moving object in the field of view. Because the mechanical stage provides X-axis and Y-axis verniers, it's easy to return to a specific location on the slide even after you've moved it completely outside the field of view. We wouldn't even consider using a microscope without a mechanical stage. Life is too short.

Figure 2-6: *Knobs control the X- and Y-axis movements of the mechanical stage*

SUBSTAGE CONDENSER, DIAPHRAGM, AND FILTER HOLDER

Despite the fact that they're located below the stage (and therefore below the specimen), the *substage condenser* (the top lens of which is visible in Figure 2-5) and the *diaphragm* and filter holder, shown in Figure 2-7, have a significant effect on image quality.

Figure 2-7: *The diaphragm control lever (note the swing-out filter holder to the lower left)*

The condenser sits between the diaphragm and the stage, focusing light from the illuminator onto the specimen to provide a bright, sharp image. Toy microscopes have no condenser. Somewhat better microscopes use a simple fixed-focus condenser, usually rated at 0.65 NA (where the NA of the condenser must be at least as high as the NA of the objective lenses it is to be used with. A 0.65 NA condenser can be used with at most a 40X objective. Oil-immersion 100X objectives with a 1.25 NA rating require a 1.25 NA condenser.) Midrange microscopes use a focusable Abbe condenser, usually of 0.65 NA and usually with a spiral focusing arrangement. Better models provide a rack-and-pinion focusable Abbe condenser with a 1.25 NA for use with any objective up to a 100X oil-immersion objective.

The diaphragm is used to control the diameter of the light cone where it intersects the specimen being viewed. Ideally, you want the diameter of the light cone to be the same size as the field of view of the objective lens you're using. At low magnification, where the field of view is relatively large, you want a larger light cone; at higher magnification, where the field of view becomes correspondingly smaller, you want a smaller light cone. If the light cone is smaller than the field of view, the field is not completely illuminated. If the light cone is larger than the field of view, "waste" light from outside the field of view reduces contrast and image quality.

Toy microscopes have no diaphragm. Basic models have a disc diaphragm, which is simply a metal disc with several (usually five or six) holes of different diameters that can be rotated into position. Disc diaphragms provide only compromise settings, but are generally quite usable. Better microscopes have iris diaphragms, which can be set continuously to provide any size of aperture, from pinhole to wide open.

Many sources, including some that should know better, state that the diaphragm should be used to control image brightness. Wrong. Adjusting the diaphragm does dim or brighten the image, of course, but that is not its primary purpose. To control image brightness, use the illuminator switch to dim or brighten the lamp.

The diaphragm controls the diameter of the light cone from the condenser, which affects the contrast of the image and how evenly the specimen is illuminated. There is no set rule as to what diaphragm setting to use at a particular magnification. The optimum diaphragm setting depends on the nature and transparency of the specimen, the magnification you are using, and your own preferences.

As a starting point for general work (using the entire field of view), match the light cone approximately to the field of view of the objective lens you're using. If the outer edge of the field is noticeably dimmer than the center, the diaphragm aperture you're using is too small. There are times when it's worth using a very small aperture and accepting darkening at the edge of the field to get a better view of the center of the field.

As a rule of thumb, adjust the diaphragm for optimum contrast and the illumination dimmer switch for optimum brightness.

Most microscopes provide a swing-out filter holder immediately beneath the diaphragm, and often include a few round glass filters—often blue, yellow, and clear. Filters serve two purposes. First, they can be used to enhance minor color contrasts in a specimen. For example, if your specimen is mostly colorless but with some very pale yellow structural detail, using a blue filter darkens the pale yellow, making detail more visible. Second, using filtered light can help you wring out that last bit of detail if you're using achromatic objectives. Remember that achromats don't bring all colors of light to the same point of focus, so when you view a specimen with full-spectrum light, there is some blurring. Using a color filter effectively makes the illumination monochromatic, which means that even an achromatic objective can bring it to sharp focus.

Polarizing microscopes are specialized instruments. They're used mostly by geologists but have many applications in biology. Few home biologists can justify spending several hundred dollars or more for a dedicated polarizing microscope. The good news is, you don't have to.

For a few dollars, you can buy two pieces of polarizing film from Edmund Scientific or another lab supplies vendor. (We use a set of filters mounted in 35 mm slide holders.) Place one of the polarizing films flat on the filter holder and hold the other between the eyepiece and your eye. Rotate that film to observe the specimen by polarized light. No, it doesn't do everything that a real polarizing microscope does, but it does a lot and you can't beat the price.

THE FINAL DECISION

So, with all of that said, which compound microscope should you get? Obviously, that depends on both your needs and your budget, but we can offer some advice to help you make a good decision.

Entry-level 400X microscope (~$240 street price)
  If you need an entry-level 400X scope, we recommend the National Optical Model 131-CLED with the optional mechanical stage. This scope can serve a student from elementary school through high school, excepting advanced biology. The optics and mechanicals are good

to very good. The only major missing feature is the 100X oil-immersion objective, which is needed for cell biology studies in high school advanced biology courses.

Basic 1,000X microscope (~$350 street price)

If you need a basic 1,000X scope, we recommend the National Optical Model 134-CLED. This scope is excellent for hobby use, and is the only scope a student will need from middle school or junior high school through high school advanced biology. This scope is essentially a Model 131 upgraded to include a 100X oil immersion objective, a focusable 1.25 NA Abbe condenser, an iris diaphragm, and a standard mechanical stage.

Mainstream microscope (~$460 to $1,600+ street price, depending on model and options)

If you need a mainstream microscope, we recommend one of the National Optical 160-series models, the Model 160 (monocular), Model 161 (dual-head), Model 162 (binocular), or Model 163 (trinocular). The only major feature missing from these 160-series scopes is support for Köhler illumination. The model 165 (binocular) and 166 (binocular plus a camera port) add support for Köhler illumination, as well as a five-position turret and a Siedentopf head on the binocular models.

Achromatic objectives are standard on the 160, 161, 162, and 163, with upgrades to ASC (Achromatic Super Contrast) or plan objectives. ASC objectives are standard on the 165 and 166 models, with plan objectives optional. Phase-contrast objectives are available as a factory option on the 162, 163, 165, and 166 models.

Any of these 160-series microscopes is a superb choice for hobby use, and is the only scope a student will need from middle school or junior high school all the way through university and graduate school, particularly if you opt for phase-contrast support. The optics and mechanicals are excellent, and the feature list is impressive. Even people who use professional-grade microscopes every day are invariably surprised by the level of mechanical and optical quality the 160-series microscopes provide at this price point.

Does National Optical pay us to say this stuff? Nope. Nor do we own stock in them, nor have any other financial interest. They've never sent us so much as a free mouse pad. We use National Optical scopes, but we paid for them with our own money. Our only connection with National Optical is as satisfied customers.

We recommend National Optical models because they provide the quality control that is lacking in most Chinese microscopes. Chinese factories are famous for huge variations in product quality. The same factory can turn out both truly excellent microscopes and visually identical models that are fit only as boat anchors, and all on the same day. National Optical rides close herd on its Chinese factories, making sure they meet National Optical quality standards, and hand-inspecting every scope before it's shipped to the customer.

National Optical microscopes are widely distributed, as you'll find if you do a quick Google search for the model or models you're interested in. National Optical also imports microscopes under the Motic and Swift Optical brand names, both of which also offer excellent bang for the buck.

Finally, educational microscopes sold under the Leica brand name are available from many vendors. These microscopes are actually made in China (in the same factory that makes National Optical and Swift models, we're told) but are backed by the Leica name. Leica models are somewhat more expensive than NO or Swift models with comparable features, but the Leica name undoubtedly adds to their resale value.

STEREO MICROSCOPE

A *stereo microscope*, shown in Figure 2-8, is also called a *dissecting microscope* or an *inspection microscope*. It uses two eyepieces, each with its own objective lens, to provide a 3D image of the specimen.

These microscopes are used to examine relatively large solid objects at low magnification, usually by reflected rather than transmitted light. Most stereo microscopes provide a top illuminator that directs light downward onto the specimen. Better models often also provide a bottom illuminator that allows specimens to be viewed by transmitted light.

Stereo microscopes operate at low magnification, usually in the 10X to 40X range. Some models have fixed magnification, usually 10X, 15X, 20X, or 30X. Other models offer a choice of two magnifications, often 10X, 15X, or 20X and 30X or 40X. Zoom models offer continuously variable magnification.

Figure 2-8: *A typical stereo microscope (image courtesy National Optical & Scientific Instruments, Inc.)*

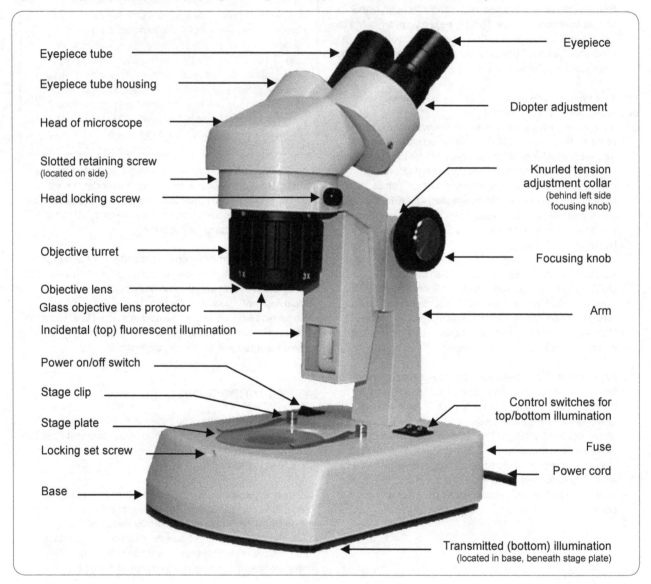

Eyepiece tube

Eyepiece tube housing

Head of microscope

Slotted retaining screw
(located on side)

Head locking screw

Objective turret

Objective lens

Glass objective lens protector

Incidental (top) fluorescent illumination

Power on/off switch

Stage clip

Stage plate

Locking set screw

Base

Eyepiece

Diopter adjustment

Knurled tension
adjustment collar
(behind left side
focusing knob)

Focusing knob

Arm

Control switches for
top/bottom illumination

Fuse

Power cord

Transmitted (bottom) illumination
(located in base, beneath stage plate)

A stereo microscope is very helpful for dissections because it leaves both hands free for using the dissection instruments. It's also useful for examining relatively large specimens such as leaves. As helpful as it can be, we don't consider a stereo microscope to be essential for most biology lab work. Buy one if you can afford it, but don't skimp on the compound microscope. It's better to buy a good compound microscope and no stereo microscope than to buy inexpensive models of each. If you don't have a stereo microscope, you can often substitute a magnifier or pocket microscope, or in some cases simply use your compound microscope at its lowest magnification.

The best (and most expensive) stereo microscopes are made by German and Japanese companies such as Leitz, Zeiss, Fujinon, and Nikon. Affordable high-quality models are available from National Optical, Motic, and Swift. For an entry-level stereo microscope, we recommend the NO 400TBL-10-2, which is available for $175 or so in 20X, 30X, and 40X models. For a dual-magnification model, we recommend the NO 409/410/411, which is available for $270 or so in 10X/20X, 10X/30X, and 20X/40X models. (Note that all of these magnifications assume you are using the standard 10X eyepieces. You can buy extra eyepiece pairs in 5X, 15X, or 20X to expand your available range of magnifications.)

PORTABLE MICROSCOPES AND MAGNIFIERS

As useful as standard microscopes are when you're working in your lab, you'll probably also want a portable microscope like the one shown in Figure 2-9 to take along on field trips. You can pay anything from $10 or $15 up to $500 or more for a portable scope. Having lost more than one of these to accidents and dunkings, we prefer the cheap ones.

It's easy to think of these $15 microscopes as children's toys—and in fact they are excellent gifts for children—but they're also serious scientific instruments. A typical $15 model might provide 30X fixed magnification, battery-powered illumination, and little more. Somewhat more expensive models might offer higher magnification, sometimes 100X or more, white LED illumination, and provision for using standard microscope slides. Some models offer zoom magnification, and may even be convertible to a tiny telescope. The best models—those

that cost $100 to $500 or more—have optical and mechanical quality similar to that of a desktop microscope that sells for about the same price.

Figure 2-9: *A typical portable microscope (image courtesy United Scientific)*

Particularly if you don't have a stereo microscope, you'll probably want at least one magnifier. These range from standard magnifying glasses that are sold for a few bucks at Walmart to the high-quality magnifiers intended for scientific and industrial use that are sold by Edmund Optics and similar vendors. Our favorite midrange magnifier is the 5X Viewcraft Lupe, which sells for $20 or so. It has excellent optics for the price, and its clear acrylic base means you don't ordinarily need a separate illuminator. It also has built-in millimeter scale, which is quite useful for measuring specimens. We use it often in the lab and in the field. In fact, we used it with the Live View feature of our DSLR to focus many of the images we shot through our microscope for this book.

MICROSCOPE ACCESSORIES

Even the best microscope is useless without some essential accessories. It's easy to go overboard with microscope accessories, so we'll try to limit our advice to the must-have items. Okay, maybe we'll talk about some nice-to-have items as well.

CLEANING EQUIPMENT

It's impossible to do good work with a dirty microscope. Microscopes are dust magnets, and of course under magnification, a grain of dust looks like a boulder.

Your first line of defense is a dust cover, which should always be in place unless you are actually using the microscope. Most microscopes come with a fitted vinyl dust cover. If yours did not, either buy a suitable one or substitute a plastic wastebasket liner. A soft bristle brush and canned air are excellent for getting rid of any dust that still manages to accumulate.

You'll also need a supply of lens cleaning fluid and lens cleaning tissues. Don't substitute household items for either of those. Window cleaner or eyeglass cleaner may contain chemicals that may damage or smear your lenses, and facial tissue or toilet paper leaves fibers on the lenses and may scratch them as well.

Follow the microscope manufacturer's instructions for cleaning. Don't over-clean. In ordinary use, your objective lenses—other than your oil-immersion objective—may not need to be cleaned from one year to the next, and unnecessary cleaning risks wear or damage to lens coatings. The eyepiece lens may accumulate skin oils transferred by your eyelashes. Clean it as often as necessary. Clean the oil-immersion objective each time you use it.

IMMERSION OIL

Most objectives of 40X or lower magnification are designed to be used with only air separating the bottom of the objective lens from the top of the coverslip. *Oil-immersion objectives*, including the 100X objectives common on high school and university grade microscopes, are designed to be used with a drop of oil bridging the gap between the objective lens and the coverslip.

Avoid using an oil-immersion objective without oil, because image quality is noticeably poorer if only air separates the objective lens from the coverslip. There are two exceptions to this dictum. First, at times you will need to change magnifications back and forth repeatedly, for example, to locate objects at low magnification and then zoom in on them at high magnification. Second, oil-immersion works best with permanently mounted coverslips. With temporary wet mounts, the coverslip is free to slide around on a thin layer of water or other mountant. With a drop of oil linking the loose coverslip to the objective lens, the coverslip tends to stick to the oil and objective lens, so only the slide itself moves. In either of these situations, it may be necessary to use the oil-immersion objective without oil, accepting lower image quality.

A typical microscope includes 4X, 10X, and 40X objectives designed to be used dry and a 100X oil-immersion objective. In casual writing or conversation, the first two objectives are often called "low" and "medium." To avoid ambiguity between the 40X and 100X objectives, the former is usually called "*high and dry*" or "*high-dry.*"

Don't use just any oil. An oil-immersion objective is optimized for oil with a refractive index of about 1.54, which is the RI of commercial oil-immersion oil. Commercial oil-immersion oil is also formulated to avoid damaging the objective (specifically, the adhesives used to glue the lens elements) and to be nongumming. For all these reasons, you should use only commercial oil-immersion oil with your microscope, ideally *Cargille Type A immersion oil*. It's usually sold in small containers, often 5 to 10 mL, because you use only a drop at a time. Even a small bottle lasts a long time.

> Okay, we admit it. We've sometimes found ourselves fresh out of Cargille Type A oil and substituted another oil. In a pinch, we've raided the kitchen for ordinary olive oil (RI ~1.47) and found it an acceptable substitute.
>
> That said, if you use a substitute oil and your objective lens falls off, don't blame us.

WARNING

NEVER allow immersion oil to remain on the lens after you finish a session. Even the best oil will eventually gum up your objective lens, making it difficult or impossible to clean thoroughly. Worse still, that oil can eventually penetrate the lens mounting and get between the individual lenses in the objective. If that happens, the only realistic alternative is to replace the objective.

After you finish your session, clean the oil-immersion objective with an approved lens-cleaning solution and dry it with lens-cleaning tissues. Follow the precise cleaning instructions provided in your microscope manual. If you were using a prepared slide, don't forget to clean the slide as well.

PHOTOMICROGRAPHY EQUIPMENT

Photomicrography is the process of recording images through a microscope. (Microphotography, conversely, is the process of making very tiny photographs, such as the microdots formerly used by spies.) Using a camera to record the specimens you observe with your microscope is a very useful adjunct to narrative descriptions and sketches of your observations.

The ideal setup for photomicrography is a microscope with a built-in digital camera, many of which can record high-resolution video in addition to still images. Such a setup vastly simplifies getting good images, particularly with respect to focusing and exposure. With many models, you can view the live image on a computer display as you record it. The downside is cost. Even the least expensive built-in cameras, despite their relatively low image quality, add hundreds of dollars to the cost of the microscope. Better models add thousands. And, of course, because the camera is intimately tied to the microscope, it's not easily upgradeable, and failures can be very expensive and time-consuming to have repaired.

The second option is an add-on camera specifically designed to be used with a microscope. These cameras are installed in place of the vertical eyepiece on a dual-head or trinocular microscope, although some can be used with binocular models by installing the camera in place of one of the eyepieces. The least expensive models cost less than $100, but typically offer only 640X480 resolution. Better models cost several hundred to several thousand dollars.

If you have a digital SLR, you already have the core of an excellent photomicrography setup. A typical DSLR, even an entry-level model, has better resolution and more capabilities than even most high-end dedicated microscope cameras. For example, few dedicated microscope cameras have resolutions greater than 8 megapixels, support the superior RAW image format, or have autobracket or extended dynamic range (XDR) features.

To use a point-and-shoot camera with your microscope, check the manufacturer's website for a microscope adapter. For a DSLR, either purchase the dedicated microscope adapter made by the camera manufacturer or mount the camera to the microscope with a T-ring and a generic adapter. We used the Edmund Optics Microscope Adapter (#41100) and a Pentax K-r DSLR with a K-mount T-ring adapter to shoot many of the photomicrographs in this book.

Even if you don't have a dual-head microscope or a camera adapter, it's possible with patience and trial-and-error to shoot usable photomicrographs merely by setting a point-and-shoot digital camera to macro mode and holding it up to the microscope eyepiece. If you attempt this method, you may get better results if you use a short length of cardboard or plastic tube between the eyepiece and the camera lens to block extraneous light and make alignment easier.

SLIDE PREPARATION EQUIPMENT AND SUPPLIES

Even if you intend to use mostly purchased prepared slides in your studies, you'll still need the basic equipment and supplies required to make your own slides. If you're a student, you need to learn how to make slides as a basic lab skill. If you're involved in DIY bio or doing other research of your own, you may be making a lot of slides, so you might as well start from day one with all of the equipment and supplies you need to make your life easier.

The amount of time and effort needed to make a slide varies widely. You can make a simple stained smear mount of bacteria or other microorganisms in a few minutes. Histology tissue specimens take much longer. If you need to dehydrate a specimen, embed it in paraffin, section it, and stain it, you may be looking at an hour or more of actual work, possibly spread over several days. But no matter what type of slides you're making, having the right equipment minimizes the amount of time and work required.

SLIDES

Microscope slides are rectangular pieces of thin, flat, transparent glass or plastic that are used to mount and view small specimens with a compound microscope. The standard and by far most common size is 1x3" (25x75 mm), although slides for specialized purposes can also be purchased in 2x4", 4x4", and other sizes. Standard glass slides are nominally 1.1 mm thick, with acceptable variation of ± 0.1 mm. (The older standard was 3/64" thickness, which translates to about 1.19 mm, so 1.2 mm slides are quite common.)

> Plastic microscope slides and coverslips are readily available, but that doesn't mean you should use them. They eliminate the danger of getting glass slivers in your skin, but that's about the best that can be said about them. Plastic is perfect for elementary school students but completely unsuitable for any serious work. Even the best plastic slides and coverslips are optically inferior to glass.

You might expect something as simple as a microscope slide to be a fungible commodity. That's far from true, however. A good slide must meet numerous requirements. It must be flat and of even thickness, without any bubbles, streaks, unevenness, or other flaws in the glass. The glass itself must be colorless and homogeneous. Sharp edges must be polished smooth, and the slide must be precleaned of all dust or other contaminants. The cheapest slides fail on some or all of these counts. Fortunately, even high-quality slides—most of which are also made in China or India—cost only a few dollars per box of 72.

We prefer to use slides that have one end frosted for labeling with a pencil. With plain glass slides, the alternative is to use sticky labels or a glass marking pen.

In addition to a good supply of flat slides, you'll probably want at least a few *well slides*. These are considerably more expensive than flat slides, but they're useful for tasks like observing live microorganisms in a drop of pond water. They're also useful for making permanent whole mounts of specimens such as small insects.

There are two types of well slides:

Shallow-well slides

Shallow-well slides, often called *depression slides*, are of standard thickness (~1.2 mm) and have one or more shallow circular depressions. To use them, you simply place a drop or two of your pond water or other specimen in the well and cover it with a coverslip. Sooner or later— sooner if you're using quartz-halogen illumination—the water drop will evaporate. You can delay that by using a dab of glycerin or petroleum jelly to make a seal between the slide and the coverslip.

Deep-well slides

Deep-well slides, often called *cavity slides*, are about three times thicker than standard slides and have one or more deep cylindrical cavities. These slides are ideal for confining small live insects (such as ants) and similar tasks. The wells in these slides are deep enough to allow using the *hanging-drop method* for observing the motility of protozoa, as described in Lab I-2. If you're going to have only one type of well slide, go for the deep-well version. You can substitute it for a shallow-well slide, but not the converse.

If you don't have a well slide handy when you need one, make your own. Use super glue (or even petroleum jelly) to affix a small rubber or plastic washer to a flat slide. Use a standard coverslip.

> **WARNING**
>
> Never buy flat microscope slides by the dozen. First, most slides that are sold by the dozen are very poor quality. Second, you'll pay a significant premium versus buying them in standard 72-slide (half gross) or 144-slide (gross) boxes. Third, you can easily use up a dozen slides in one session.

Slides have one other important use that's completely unrelated to microscopy. They're perfect for making homemade *thin-layer chromatography (TLC) plates*, which have numerous uses in a biology lab.

Commercial TLC plates are readily available, but expensive. TLC plates the size of a microscope slide sell for a buck or two each, and you usually have to buy them in boxes of 50 or 100. With just some microscope slides, a few inexpensive chemicals, and some basic labware, you can make up your own TLC plates for about a dime each. And most of that is in the cost of the slide, which can be recycled over and over to make new TLC plates.

Making the plates is easy, but a bit messy. Robert has a video on his YouTube channel that illustrates the process:

http://www.youtube.com/watch?v=pNDQkM3jasA

With a supply of homemade TLC plates and a Coplin jar (described later in this section), you're prepared to do TLC. With a bit of care, your results should be about as good as if you used expensive commercial TLC plates.

COVERSLIPS

A microscope *coverslip*, also called a *coverglass*, is a square, rectangular, or circular piece of extremely thin, flat, transparent glass or plastic that is used to cover a specimen that has been mounted to a slide. In addition to protecting the specimen (and the objective lens from contacting the specimen), coverslips actually function as part of the optical train—microscope objectives are designed on the assumption that a coverslip will be positioned between them and the specimen—so you should never observe a specimen without a coverslip in place.

> The truth is that we frequently observe slides without coverslips, particularly for quick looks at bacteria smears. Yes, we give up some image quality by not using a coverslip, but sometimes the time saved is worth the trade-off.

Coverslips are available in four thicknesses, denominated #0 (0.083 to 0.13 mm), #1 (0.13 to 0.16 mm), #1.5 (0.16 to 0.19 mm), and #2 (0.19 to 0.25 mm). #0 coverslips are specialty items that are not stocked by most vendors. #1 and #2 coverslips are widely available, but #1.5 coverslips are harder to find. Fortunately, #1 coverslips are acceptable for nearly any purpose.

Many microscope objectives are labeled with the recommended coverslip thickness or range of thicknesses. High-end objectives may be adjustable to optimize them for a range of thicknesses, typically something like 0.11 to 0.23 mm. Less expensive objectives are not adjustable but are designed to use coverslips of one particular thickness, typically 0.17 mm, or about the middle of the acceptable variation for #1.5 coverslips.

Some texts recommend using only #1 coverslips, and there's actually a good argument in favor of that. If your objectives are designed for use with #1.5 (or even #2) coverslips, #1 coverslips cause very little image degradation.

The advantage of using #1 coverslips may become apparent the first time you use your oil-immersion objective. At that high magnification, depth of focus is nearly nil. If you're using a #2 or even a #1.5 coverslip, you may find that you cannot focus sharply on the specimen—particularly a specimen that has depth—without ramming the objective into the coverslip. For that reason, we use #1 coverslips by default, particularly for permanent mounts, even though our objective lenses are designed for #1.5 slips.

Most objective lenses are labeled with the coverslip thickness they're designed to use. For example, Figure 2-3 shows the objective lenses of a typical microscope. That "0.17" label visible on the objective lenses indicates that these objectives are designed to use a 0.17 mm coverslip.

Using a coverslip that is thinner or thicker than the objective is designed for introduces chromatic aberration and reduces the quality of the image. For objectives with an NA of 0.4 or less, coverslip thickness is not a major issue. With 60X to 100X oil-immersion objectives, coverslip thickness becomes more important.

Coverslips are available in a variety of sizes and shapes. The most common and least expensive for any given quality level are square, and are readily available in sizes from 18 mm to 24 mm. (Larger sizes allow larger specimens, but conversely they also give you more area to search, particularly if you are using high magnification.) Rectangular coverslips, such as 22x40 mm, are less common but still readily available. These are useful for mounting very large or multiple specimens on a single slide.

> Although it appears odd to most people at first glance, there's nothing wrong with using two or three coverslips to mount multiple specimens on one slide. Multiple mounts on one slide can make some tasks a lot faster, because you can quickly flip back and forth between specimens without changing slides.

Round coverslips are much less common now than they were a few decades ago. Before about 1970, many slides were prepared using Canada balsam (essentially, highly refined pine tree sap) as a mounting fluid. Even exposed to air, Canada balsam remains tacky for some time, so the final step in slide preparation was to place the slide on a turntable and rotate the slide while trimming off any excess Canada balsam that had squooshed out from under the coverslip. That's easy to do if the coverslip is circular, but impossible if it's rectangular or square.

Although Canada balsam is still available (and still preferred by some microscopists), it has been largely replaced by Permount and similar synthetic mounting fluids that dry without tackiness. Consequently, trimming excess mountant is no longer necessary (although it still makes the final slide look neater), and round coverslips are going the way of the dodo. It doesn't help that round coverslips typically sell for several times the price of square ones of similar size and quality.

Slides and coverslips are inexpensive enough that we often use them once and discard them, but many people do recycle them. If you want to do that, swirl them in sudsy tap water, rinse them with distilled water, and then use a final rinse of pure acetone. Place slides in a slide-storage box to drain and dry. Place coverslips on a clean sheet of lint-free paper or cloth to dry.

Some microscopists store slides and coverslips in jars of acetone or absolute alcohol, which keeps them clean and sterile. When you remove one from the jar, the liquid evaporates rapidly, leaving you with a clean, dry slide or coverslip.

BUTANE LIGHTER OR ALCOHOL LAMP

A disposable *butane lighter*, *alcohol lamp*, or other flame source is needed to *heat-fix* bacterial smear slides. Heat-fixing kills any live bacteria present and causes them to stick to the microscope slide. (Otherwise, they'd be washed away during staining or mounting.) To heat-fix a slide, allow the smear to dry and then pass the slide—bacteria side up—several times through the naked flame. If you don't heat it long enough, the bacteria won't adhere. If you heat it too long, the bacteria explode or char. You'll soon get a feel for it. A butane lighter is convenient for heat-fixing one slide at a time. If you're making many smear slides, an alcohol lamp is better.

FORCEPS AND SLIDE TONGS

Forceps are what nonscientists call tweezers. They're available in many sizes and styles in plastic or metal. For manipulating coverslips, we prefer the plastic variety. *Slide tongs* are similar to test tube tongs, but designed to grip a standard size microscope slide. They're used to hold a slide during heat fixing or to manipulate the slide during staining procedures. They can also be used to manipulate coverslips.

MOUNTANTS

A *mountant* or *mounting fluid* is a liquid that is introduced between the slide and the coverslip. An ideal mountant is clear, colorless, and (for best image quality) has a refractive index very close to that of the crown glass used in coverslips (RI = ~1.54). If it is to be used to make permanent mounts, it should dry without shrinking, cracking, or bubbling. If it is to be used to make temporary mounts, it should not evaporate quickly. In any case, it must not rupture, fade, or otherwise damage specimens.

The most common mountant by far among amateur biologists is ordinary distilled water. Although its refractive index, about 1.33, is lower than ideal, it's much better than having an air gap (RI = 1.00) between the slide and coverslip.

If distilled water is unsuitable, you can substitute *glycerin*, also called *glycerol*, in straight or diluted form. Glycerin evaporates much more slowly than water, days to weeks versus minutes to hours.

If you're observing live protozoa (such as paramecia), you'll need some way to slow them down. In water, a paramecium can travel 10 or more body lengths per second, so just keeping it in the field of view can be challenging. The best solution to this problem is to mount the slide with an aqueous solution of *methyl cellulose*, which is much more viscous than water.

Methyl cellulose is a *physical immobilizer*, which simply means that protists have a hard time swimming through this viscous fluid. (Glycerin is another immobilizer in this class.) A *chemical immobilizer* slows down protists by fatiguing them. Two common chemical immobilizers are a 3% solution of copper(II) acetate and a 1% solution of copper(II) sulfate.

If you have the kit for this book and you want to try a chemical immobilizer, mix a few drops of Barfoed's reagent—which is 6% copper(II) acetate—with the same amount of distilled water. Or simply use the Barfoed's reagent without dilution, but make sure there's a least a couple drops of water in your wet mount and use only a small drop of Barfoed's.

Finally, if you want to make permanent slides, you'll need a permanent mountant such as *Permount* or *Canada balsam*. (We think Permount is better in every way than Canada balsam, but some traditionalists continue to prefer the latter.) These mountants dry clear and transparent, with little or no color, essentially cementing the coverslip to the slide. You can substitute colorless nail polish, such as Sally Hansen's Hard As Nails. We have slides five years old or older that were mounted with Sally Hansen's Hard As Nails and they are as good now as when they were made.

BIOLOGICAL STAINS

Biological stains are natural or synthetic dyes that selectively bind to different parts of cell structures. For example, one stain may bind strongly to the cell wall, coloring it intensely, while leaving the nucleus untouched. Another stain may do the converse.

Most people don't realize how important stains are to biologists. The first problem microscopists faced was the fact that most of the things they wanted to look at were opaque. They solved that problem by cutting very thin sections with microtomes. But that left another serious problem. Most of those thin sections were essentially transparent, with almost no contrast between different parts of the cell structure. Back in 1858, someone had the cunning idea of using *carmine*, a dye obtained from female

cochineal beetles, to stain sections. Sure enough, carmine was differentially absorbed by different parts of the cell structure, increasing contrast immensely.

Ironically, despite the fact that biologists have tried literally tens of thousands of different stains in the intervening 150 years, the most widely used stains today are the first two stains that came into common use after carmine. They're *hematoxylin*—which for some reason many people (including some biologists) pronounce he-muh-TOX-uh-linn rather than the correct he-MAT-oh-ZYE-linn—a dye derived from logwood, and *eosin*, one of the first synthetic dyes. They were first used in about 1865 in the so-called *H&E staining protocol*, and are still used that way today. In fact, H&E is probably the most commonly used staining protocol even today, with the possible exception of the *Gram staining protocol*.

> For school work or if you're just experimenting with staining, it's cheaper and easier to buy small amounts of prepared stains individually or in kits rather than making them up yourself. If you use stains in larger volumes, as some DIY biology hobbyists do, making up the stains yourself from raw chemicals is both cheaper and more flexible. Instructions for making up various stains are readily available online.
>
> Although there is considerable overlap, biological stains are broadly grouped into *bacteriology stains*, used to stain bacteria, and *histology stains*, used to stain tissue sections from plants and animals. Most bacteria are so tiny that the goal of staining them is sometimes simply to make them visible against the background clutter. Histology stains, conversely, are used to selectively stain different cell components to reveal internal cell structures. The Gram staining protocol is one example of a bacteriology staining protocol; the H&E staining protocol is an example of a histology staining protocol.

There are literally scores of biological stains and staining protocols in common use, and hundreds in occasional use, although most home biologists can get by with just a few stains. Even with our tendency to accumulate neat stuff, we have "only" 31 different stains at our microscopy station. We consider two stains to be essential for any microscopist:

Methylene blue

Methylene blue is a general primary stain used for nuclear material and other acidic cell components. It's one of the most widely used primary stains, and is effective on animal, plant, bacteria, and blood specimens. Various types of methylene blue stains are available commercially, from simple aqueous solutions of the dye to versions that incorporate sodium hydroxide or potassium hydroxide to versions that use various alcohols and other organic solvents instead of water. For routine use, any of them suffice. You can substitute methylene blue in most staining protocols that call for carmine or Janus green B.

Eosin Y (0.1% - 1% aqueous)

Eosin Y, usually supplied as an aqueous solution, is a general primary stain for cytoplasm material, which it stains pink or red. It's also one of the two stains used in the popular H&E (hematoxylin & eosin) staining protocol. Eosin Y may be substituted in most staining protocols that call for congo red or neutral red.

For some staining protocols you can substitute erythrosin B, also known as FD&C Red #3. Results are generally inferior to those obtained with eosin Y, but are still quite usable. The red food coloring dye sold in grocery stores is usually a mixture of erythrosin B and Allura Red AC (FD&C Red #40). We've used this mixture in diluted form with reasonably good results.

Your next priority should be the following three stains, which are needed for the bacteriology Gram staining protocol.

Hucker's crystal violet

Hucker's crystal violet is the primary stain in the Gram staining protocol. Gram-positive bacteria retain this stain after decolorizing, while Gram-negative bacteria do not. A 1% crystal violet solution is sold in drugstores, usually under the name gentian violet. This solution is usable for Gram staining, but inferior to Hucker's crystal violet.

Gram's iodine

Gram's iodine functions as a mordant in the Gram staining protocol, binding the crystal violet stain to Gram-positive bacteria, but it is also useful as a primary stain in its own right. Gram's iodine stains the starch in plant cells dark blue or black and the glycogen in animal cells red. You can substitute drugstore 2% iodine tincture diluted one part tincture to five parts distilled water.

Safranin O

Safranin O is a red counterstain usually used for Gram staining, and is also used as a primary stain for cartilage, mast-cell granules, and mucin. You can substitute eosin for most purposes, including Gram counterstaining, although safranin O is preferable for the Gram protocol.

A *primary stain* is used to differentiate the structural element that is of primary interest. A *secondary stain* (or *counterstain*) is one of contrasting color to the primary stain that is used to provide additional color contrast by staining other structural elements. For example, in the H&E protocol, hematoxylin is the primary stain, which stains nuclei (and a few other structures) blue. Eosin is the counterstain, and stains other structures red, pink, or orange. A stain may be used as a primary stain in one staining protocol and as a counterstain in another.

With just these stains, you're equipped to use the most important staining protocols. If you have room on the shelf for a few more, we'd choose *toluidine blue* (for staining nuclear material during mitosis and also as a fast alternative to H&E staining), *Sudan III* (for staining lipids), and *Wright's blood stain* (for staining, uh, blood). If you intend to explore histology in any depth, you'll need *hematoxylin* to use the H&E staining protocol. (You can substitute methylene blue for hematoxylin in the H&E protocol, although the differentiation is usually inferior to that obtained with hematoxylin.)

> If you want to try staining without purchasing special biological stains, you can try various stains that are readily available locally. You can view an excellent article at *www.crscientific.com/microscope-stain.html* about using standard food coloring dyes as biological stains.

In addition to the stains, you'll also need various other chemicals that are used for *destaining*, also called *decolorizing* or *clearing*, which is the procedure used to remove excess stain from a specimen. Common destaining agents include distilled water, dilute hydrochloric acid, ethanol, isopropanol, and acetone. Although all of these are available in very pure form, called *histology grade*, we've used ordinary lab grade, USP grade (from the drugstore), and even technical grade (from the hardware store) destaining chemicals without any problems.

> *Vital stains*, also called *supra stains* or *supravital stains*, are those that can be used to stain living organisms without harming them. Many common stains can be used at high dilutions as vital stains, including methylene blue and eosin Y. When used for staining living organisms, the stains are highly diluted, usually in the range of 1:5,000 to 1:500,000 (versus typically 1:100 to 1:1,000 for nonvital staining).

The Gram staining protocol was devised by Danish physician Hans Christian Gram in 1884. It involves first staining a specimen with a solution of crystal violet (also called gentian violet), which stains all of the bacteria purple. An iodine-iodide solution is applied next to fix the violet stain. Rinsing with alcohol removes the stain from some but not all bacteria. The specimen is then counterstained with safranin O, dried, and viewed. Those bacteria that retain the crystal violet are stained blue or violet, and are called Gram-positive bacteria; those that did not retain the crystal violet are stained pink or red by the safranin O counterstain, and are called Gram-negative bacteria.

Gram staining was a revolutionary technique for bacteriologists and epidemiologists worldwide. Its value was that it allowed scientists to quickly discriminate between some pathogenic (disease-causing) bacteria and other harmless bacteria that appeared visually similar to the pathogens. Even today, Gram staining is one of the two most widely used staining protocols and is routinely used for quick screening.

After Gram introduced his eponymous staining protocol, other scientists immediately began tweaking it, trying slightly different formulations and concentrations. Over the next 45 years, dozens of variants were announced, but it turned out that none of them had any great advantage over the original protocol. One thing all of these variants had in common was using aniline, then called analin oil, as a component of the violet stain. Eventually, a more-or-less standard Gram staining protocol that incorporated the best of the variations came into common use.

That changed in 1929, when Yale undergraduate bacteriology student Thomas Hucker introduced a variant that immediately became the standard and has been used universally ever since. It all started when Hucker's professor asked him to prepare a paper to present at a biology conference. Hucker asked his professor for topic suggestions, and the professor recommended that Hucker contact hospitals and university labs to ask them about their specific methods for using Gram staining. Hucker intended to compare and contrast these protocol variants in his paper.

Replies soon began to arrive in the mail, but Hucker was disappointed to find that every organization he contacted was using essentially identical protocols. With one exception. Dartmouth College reported that instead of anilin oil they were using ammonium oxalate in the crystal violet stain. Hucker tried this ammonium oxalate variant, and found that it indeed yielded much better results than the methods using anilin oil. He wrote up his paper, sent a copy to the Dartmouth bacteriology department, and thanked them for their advice.

A few days later, Dartmouth called Hucker and told him that he must be mistaken. They were using the same protocol, with analin oil, that everyone else had reported using. They'd never heard of using ammonium oxalate. Hucker was puzzled, but he knew what he'd seen, so he presented the paper at the conference, to great acclaim.

After returning from the conference, Hucker decided to find out what was going on. He boarded a train and visited Dartmouth. He met the head of the bacteriology department, who once again denied having any idea what Hucker was talking about. He'd been on vacation when Hucker's original letter arrived, and denied having even seen the letter, let alone responded to it. Deciding to get to the bottom of this puzzle, the bacteriology chairman started walking around the building, collaring people and asking if they had any idea what had happened.

He finally got his answer in the chemistry department, which was situated next door to his own department. Apparently, one of the chemists had gotten Hucker's original query letter and had taken it upon himself to respond. He wandered next door to the bacteriology department and noticed that a bottle of the stain had "A.O." listed among the contents on the label. Of course, that was shorthand for "anilin oil," but the chemist for some reason decided that "A.O." must be an abbreviation for ammonium oxalate. He so informed Hucker by return mail, and the rest is history.

As pretty as this story is, we wonder if it's really true. A biologist might conceivably abbreviate analin oil as "A.O." but no chemist would assume that "A.O." was an abbreviation for ammonium oxalate. "NH4 Ox" perhaps, or even "Amm Ox." But not "A.O." On the other hand, if it didn't happen that way, how did it happen? Whether the story is true or not doesn't really matter. What matters is that, by pure accident, Hucker discovered a much superior Gram staining protocol, which has been in universal use since 1929.

COPLIN STAINING JAR

Although it's not essential for some staining tasks, a *Coplin staining jar*, shown in Figure 2-10, is a handy item to have around. These jars are available in glass or polypropylene and have built-in grooves to support microscope slides vertically, usually 10 slides in five back-to-back pairs. To stain a slide, you simply fill the jar a bit more than halfway with the staining solution and slide the slide into one of the grooves.

A Coplin jar is particularly useful in two situations. First, if you have many slides to be stained with the same stain, it's much more convenient to batch-stain them than to stain each slide individually. Second, although some stains operate very quickly—a few seconds to a minute or two—some other stains must be left in contact with the specimen much longer, sometimes as long as several days. We keep half a dozen Coplin jars on hand for just these reasons.

A Coplin staining jar is also a perfect chromatography jar for developing the homemade TLC plates described earlier in this section.

Figure 2-10: *A Coplin staining jar (image courtesy United Scientific)*

SLIDE STORAGE

Wet mounts are temporary. You make the mount, observe the specimen, and then discard or recycle the slide and coverslip. But if you intend to make permanent mounts, you'll need an organized way to store the slides.

The most popular solution is a slide storage box. These are available in wood or plastic from most lab supplies vendors, often in many colors. Boxes typically store 10, 12, 25, 50, or 100 slides, each in its own numbered slot. A label area provides a numbered line for each slide.

Slide storage books or wallets are a bit harder to find, but some microscopists (including us) prefer them. Each page stores (typically) a half-dozen to a dozen slides, in one or two columns, with room for detailed labels rather than just one-line descriptions. Although these books are more expensive and occupy more space than slide boxes, the slide labels are much more visible and it's much easier to remove and replace the slides.

WARNING

Whichever method you use, if possible, store slides flat rather than vertically. Back in the Bad Olde Days when we made permanent mounts with Canada balsam, slides stored vertically for months to years were sometimes ruined when the coverslip detached or simply slid down the slide. Almost as bad is when your nicely centered specimen floats down to the bottom edge of the coverslip under the force of gravity. That's not supposed to happen with modern mountants like Permount, but we'd rather not find out the hard way.

If you have imaging equipment, you may not need to store slides at all. One slide may require a dozen or more images to show all of the features of interest on that slide, but you can store hundreds to tens of thousands of high-resolution images on something as small and durable as a USB flash drive. Just make sure to back it up.

CULTURING EQUIPMENT AND SUPPLIES

Culturing is the procedure used to grow microorganisms in a controlled environment. Many microorganisms reproduce very quickly. Under ideal conditions, some can double their numbers every 20 minutes or less. In other words, if a single microorganism is present to begin with, after 20 minutes there are two, after 40 minutes there are four, after 60 minutes there are eight, and so on. That may not sound impressive until you realize that after 10 hours (30 doubling periods) have elapsed, that single microorganism has become more than a billion microorganisms.

All of that assumes that the microorganisms have space to grow and food to feed them. That's the purpose of culturing. Culturing provides surface area (or volume) and the nutrients necessary for growth. We'll use culturing to produce large numbers of microorganisms for various lab sessions in this book. But culturing isn't limited to lab procedures. Wine and beer making uses culturing on a large scale, as do cheese making, biofuel production, and many other endeavors. In fact, you've probably unintentionally cultured bacteria yourself. If you've ever opened a bottle of milk and found it'd soured, you were holding a bottle of a bacteria culture.

AUTOCLAVE

If you're working with microorganisms, it's important to be able to kill them reliably. For example, if you've cultured an unknown bacterium, it's irresponsible to toss it out with the household garbage unless you're certain no live bacteria remain.

There are many ways to kill microorganisms, including chemical action, flame, dry or wet heat, ultraviolet light, and ionizing radiation such as X-rays or gamma rays. In a home biology lab, the first three methods are the most practical.

> To *sterilize* something means to kill all microorganisms present. To *disinfect* or to *sanitize* something means to kill almost all of them. (That's why we call that goopy stuff hand sanitizer rather than hand sterilizer.)

Chemical disinfectants, such as chlorine laundry bleach or Lysol, may kill 99.999% or more of the live microorganisms present, but they are not 100% effective. For many purposes, that's good enough. For example, if you need to dispose of a live bacteria culture growing in a Petri dish, the traditional method is to soak the dish overnight in a 1:4 solution of chlorine bleach and water. That kills very nearly 100% of the live bacteria present—not to mention viruses, protists, molds, fungi, and any other

living things—at which point there is no longer a biohazard and the culture can be discarded with the household trash or flushed down the drain.

The problem is, some bacteria form spores, which are resistant to chemical disinfectants, including bleach. If you're disposing of the material, spores aren't usually a problem. There are zillions of spores surrounding you, on every surface in your lab and floating on dust motes in the air. Spores become a problem only when you're making up culture plates or culture tubes, which must be sterile. If any spores are present in the culture media or on the plates or tubes, they'll germinate and contaminate the colonies you're attempting to grow.

Flame kills spores quickly, and is commonly used to sterilize inoculating loops and the mouths of culture tubes when you're doing a transfer. But flame can't be used to sterilize the culture media itself. That's where the third method comes in: heat.

Most spores are very resistant to moderately high temperatures. Boiling water or steam at 100 °C kills most spores, eventually, but it may take literally hours to days to do so. Fortunately, boosting the temperature slightly and using steam—121 °C steam is the standard used for sterilization in biology labs—reduces the time needed to a few minutes.

An *autoclave* sterilizes contaminated materials by exposing them to high temperatures, usually in the presence of pressurized steam. Autoclaving can be done dry, but that requires higher temperatures and longer treatment, and can be used only for glassware and other heat-resistant materials. Steam autoclaving can be used to sterilize autoclavable plastic items, which would melt at the temperature needed for dry autoclaving.

Commercial autoclaves cost hundreds to thousands of dollars. Fortunately, you don't need a commercial autoclave. A pressure cooker from Walmart or Target works just as well. If your pressure cooker doesn't come with a rack, also purchase an oven rack that fits inside the pressure cooker. You'll need that to keep containers from directly contacting the hot bottom of the pressure cooker.

INCUBATOR

An *incubator* is a device used to maintain a specific environment for culturing. Many environmental factors can affect the efficiency and effectiveness of culture growth, including humidity, level and type of light present, carbon dioxide concentration, and so on. But for most culturing the single most important factor is temperature.

Many microorganisms have evolved to have strong preferences as to their environment. For example, human pathogens (disease-causing microorganisms) evolved in human hosts, so they grow best at normal human body temperature, 37 °C. (That's why you develop a fever when you have an infection; your body is doing its best to make the infecting bacteria unhappy by boosting the temperature of their environment.)

Some microorganisms flourish across a wide range of temperatures—often growing faster as the temperature increases, but only to some maximum value—but many grow best only within a narrow temperature range. An incubator allows you to maintain the temperature best suited to growing whatever you're culturing.

You can purchase commercial incubators from lab supply vendors or on eBay, but that's overkill for most home lab tasks. For a few dollars, you can make your own incubator from a disposable foam cooler, an inexpensive thermometer, an incandescent light bulb, a plastic picnic plate or dish, and an oven rack or kitchen cupboard rack. If you want to get fancy, you can add a small fan to circulate the air and a dimmer switch to control the brightness (and heat output) of the light bulb.

Depending on the size of the foam cooler and the ambient temperature, you might need anything from an incandescent Christmas-tree bulb to perhaps a 25-watt bulb to maintain an internal temperature of 37 °C (the most common temperature for incubation of bacteria). To maintain humidity, place a dish of water beneath the rack.

CULTURING CONTAINERS

You can grow cultures in any imaginable container. Microorganisms, after all, have no problems growing in the wild. However, for best results, you'll probably want to keep at least a few special culturing containers on hand. There are three common types of culturing containers, each used for slightly different purposes: Petri dishes, culture tubes, and flasks.

PETRI DISHES

A *Petri dish* is a flat dish with a matching cover of slightly larger diameter. Petri dishes are available in glass and plastic, in a variety of diameters and depths. Some Petri dishes are divided into two, three, or four isolated compartments, which allows you to grow more than one culture in a single Petri dish.

Petri dishes are used with gelling culturing media, such as agar. Typically, you would prepare a warm liquid agar solution, fill the Petri dish to a depth of a few millimeters, and place it in the autoclave to sterilize it. Once the dish, with its sterile contents, cools down, the agar solidifies into a gel that resembles Jell-o. The dish is then carefully inoculated with the microorganism by streaking or flooding, and placed in the incubator.

Petri dishes, shown in Figure 2-11, are useful for growing pure colonies—separate groups of specific microorganisms that will subsequently be cultured in broth to produce large numbers of a single microorganism—and for such tasks as testing antibiotic sensitivity with treated disks. Petri dishes are fragile and bulky relative to the alternatives. We keep a few on hand for when they're really needed, but otherwise we use culture tubes or flasks.

Many people prefer plastic Petri dishes, which are sold for a few dollars in sleeves of 10 to 25. Each dish is individually wrapped and sterile. Unfortunately, these dishes are sterilized at the factory using radiation but cannot be resterilized at home. (The plastic melts at the temperature needed to sterilize them.) After use, the dish is put in a bleach bath to kill any microorganisms present and then discarded. If you culture infrequently, plastic Petri dishes are a good choice. They're convenient, require no sterilization, and are reasonably economical.

If you culture frequently, you'll want at least a small supply of glass Petri dishes. These must be sterilized before each use, but can be reused repeatedly. Brand-name Petri dishes, such as Pyrex and Kimax, are very expensive. Depending on the size and type, one Petri dish may cost $10 or more. Similar no-name models made in China or India sell for a fifth to a tenth the price and are perfectly usable. Incidentally, there's no need for borosilicate glass in Petri dishes, which are not subjected to thermal shock. Soda-lime glass is fine, and cheaper.

Figure 2-11: *Petri dishes (image courtesy United Scientific)*

CULTURE TUBES

A *culture tube* is essentially just a test tube with some means of capping it. Bespoke culture tubes are available in a huge range of sizes, shapes, materials, graduations, screw-top and plug closures, and so on. But simple test tubes with cotton-ball plugs work fine for a home lab, and are very inexpensive.

Culture tubes may be used with solid (gel) culturing media or with liquid (broth) culturing media. In either case, the tubes are filled with nonsterile culturing media, plugged with cotton balls, racked, and then placed in the autoclave for sterilization. For broth media, the culture tubes are allowed to cool and then used immediately or placed in the refrigerator for short term storage. For gel media, the culture tube is removed from the autoclave while still warm and placed at an angle to cool. The agar hardens in place, increasing the available surface area, and producing a *slant tube*. We use culture tubes for about 95% of our culturing.

You'll also need a *wet/dry tube rack* that fits your culture tubes. These racks are made of polypropylene plastic, which can be autoclaved at 121 °C without melting. They're also dense enough that they won't float in a water bath. (If you don't have a rack, you can simply stand the tubes in a glass or polypropylene beaker while they're in the pressure cooker.)

CULTURE FLASKS

A *culture flask* is essentially a large-volume culture tube that is used only with broth media. Traditionally, standard Erlenmeyer flasks are used for this purpose. They are filled with culturing broth, stoppered with cotton balls, and sterilized. Like culture tubes, they are best used as soon as possible, but they can be refrigerated for several days without harm.

Culture flasks are used to produce large populations of one microorganism. Unlike gel media, where the microorganisms grow almost exclusively on the surface of the gel, broth media allows growth to occur throughout the media, greatly increasing the number of microorganisms produced and reducing the time needed to do it.

You might use broth culturing, for example, to produce a large sample of bacteria for DNA analysis rather than using PCR to replicate DNA from a smaller sample. Alternatively, you may be less interested in the bacteria themselves than in something they produce. The first example of large-scale broth culturing for this purpose was the early production of penicillin, which took place in industrial-scale culturing vats. Many antibiotics are produced the same way today.

For home lab work, the most convenient culturing flasks are 125 mL, 150 mL, or 250 mL Erlenmeyer flasks, which are readily available inexpensively. If you have the choice, buy the narrow-mouth versions, which are more easily stoppered, rather than wide-mouth versions. You can often get a significant discount by buying a case of 12 or 24.

CULTURING MEDIA

Some microorganisms aren't picky about what they eat. They'll grow on almost anything that is moist and has some food value to it. Other microorganisms are very selective about the media they'll grow on.

Culturing broth is simply a liquid that contains nutrients. Microorganisms require carbohydrates (sugars or starch), nitrogen, and trace nutrients. For culturing many bacteria species, diluted chicken or beef broth (strained or dipped to remove oils and grease) with some added table sugar works well.

Culturing gel is made up from agar, a material derived from algae. Like gelatin, agar is mixed with hot water and then cooled to form a gel. (No, you can't use gelatin for culturing; bacteria eat it and turn the gel into a runny mess.) Agar itself is merely an inert substrate. With the exception of a few marine bacteria, few microorganisms actually feed on agar. To provide food for the microorganisms, you add various nutrients to the agar mixture as you're making it up.

Food-grade agar is sold in supermarkets (often in the vegetarian section) for a few dollars an ounce. *Culturing-grade agar* is sold by lab supply vendors, and is more expensive. *Agarose* is refined from agar by removing agaropectin to leave pure agarose. It is used for gel electrophoresis and may cost a dollar a gram in small quantities. Is there really a difference, or can you use food-grade agar for culturing and electrophoresis? Well, it depends...

All agar is derived from algae, but most food-grade agar is produced from the algae *Gracilaria lichenoides*, while culturing-grade agar is produced from a different algae, *Gelidium sesquipedale*. Culturing-grade agar gels at about 35 °C versus about 42 °C for food-grade agar, but the important difference is in matrix strength. Gels formed with culturing-grade agar are about twice as rigid as those formed with food-grade agar, which means culturing-grade agar can be used in concentrations as low as 0.5% (0.5 grams/100 mL) and still form a usable gel. Food-grade agar must be used at twice or more the concentration to form a gel with similar qualities. That said, we have used food-grade agar successfully in the past and will continue to do so.

Agarose is recommended for gel electrophoresis because the agaropectin present in food- and culturing-grade agar restrains the movement of fragments through the gel, reducing the sharpness of the separation. If you're doing serious gel electrophoresis, no question, use agarose whatever the price. But if you're simply doing gel electrophoresis as a learning experience, food-grade agar is perfectly acceptable.

It's less expensive to make up your own culturing broths and gels, but many biologists prefer to use commercial products. These are available from Carolina Biological Supply and other vendors as dehydrated powders that contain a mixture of agar (if it's intended to produce a gel rather than a broth) and nutrients. You make them up just as you would plain broth or agar, but without adding any additional nutrients.

If you get seriously involved with culturing bacteria and other microorganisms, you'll probably need many different types of agar gels and nutrient broths. For example, at one time or another we've used standard nutrient agar, plate-count agar, potato-dextrose agar, beef-extract agar, beer agar, Luria agar, tryptic-soy agar, McConkey agar, tri-sugar iron (TSI) agar, brain-heart infusion (BHI) agar, Sabouraud dextrose agar, malt-extract agar, and probably half a dozen or more others we've forgotten.

Rather than purchasing these as agar powder products or even powdered nutrients—many of which have limited shelf lives—it's cheaper and more efficient just to stock the necessary individual nutrients such as peptone, dextrose, and so on, and make up the various types of agar or nutrient broths as needed.

Premade agar gel of various types is available in sterile bottles, which may require refrigeration before use. To use the premade gel, you place the bottle in a hot water bath to remelt the gel and then pour it into sterile containers. Premade agar gels are convenient but expensive.

CULTURING ACCESSORIES

In addition to the major items already described, you'll need a few accessories. Most of those—a beaker and hotplate or kitchen stove burner (or a microwave oven) for making up agar, graduated cylinder for measuring liquids, and so on—are items found in most homes or are general labware, but there are a couple of special items.

An *inoculating loop* (or *inoculating needle*) is a short piece of nichrome wire in a metal, wood, or plastic handle. It's flame-sterilized before each use and is used to streak culture plates or tubes or to transfer material to culturing broth. You can make your own inoculating loop by embedding a sewing needle in a wooden dowel. Some people prefer disposable plastic inoculating loops/needles, which are sold inexpensively in individually wrapped sterile packages.

To minimize contamination by airborne microorganisms while you are inoculating culture containers, keep one plastic *spray bottle* filled with water and a second filled with Lysol or a similar disinfectant handy. Before you inoculate your Petri dishes or culture tubes, dampen the work surface slightly with Lysol spray and spray the air in the work area liberally with water. The water mist causes dust in the air to clump and settle out, carrying bacteria and other airborne microorganisms with it.

Finally, keep a *disposal container* on hand. We use a plastic pail with a tight-fitting lid. Fill it with a solution of one part 5.25% chlorine laundry bleach to four parts water. When you finish using a Petri plate, culture tube, disposable inoculating needle or whatever, immerse it in the bleach solution, making sure that no air bubbles prevent the solution from reaching all surfaces of the item. Allow the items to remain in the bleach bath overnight or longer, and then remove them for disposal or for washing and reuse. The bleach bath doesn't need to be replaced often, but it will eventually accumulate agar scum and other dead but unsightly contaminants.

HISTOLOGY EQUIPMENT AND SUPPLIES

Histology is the study of the microscopic structure of the cells and tissues of plants and animals, which are prepared and examined in the form of thin slices, called *sections*. Histologists typically use the following steps to prepare a specimen:

Fixing

Living specimens, once dead, begin to decay. *Fixing* is the process of killing the specimen, if it is not already dead, and preserving it either by chemical means or by freezing. The goal of fixation is to prevent decay and to preserve insofar as is possible the original structure of the cells and tissues. Fixing is sometimes combined with the following step by using the first, dilute, ethanol bath as the fixative.

Dehydration

During the *dehydration* phase, water is removed from the specimen because water interferes with later processing steps. This is usually done by immersing the specimen in a series of ethanol baths of increasing concentration, typically 30%, 50%, 70%, 95%, and 100% (anhydrous). The ethanol draws water out of the specimen, gradually in the early baths to prevent damage. The final bath of anhydrous ethanol removes nearly all of the remaining water.

Clearing

During the *clearing* phase, the specimen is immersed in an organic solvent, usually xylene or toluene, which removes the alcohol from the specimen.

Infiltration

During the *infiltration* phase, the dried and cleared specimen is soaked in a bath of the liquid embedding material—usually molten paraffin—which drives off any remaining clearing agent and replaces it with the embedding material.

Embedding

The *embedding* phase consists of transferring the infiltrated specimen to a mold, which is then filled with liquid embedding material. Once that embedding material solidifies, the block containing the embedded specimen is removed from the mold and trimmed to remove excess embedding material.

Sectioning

During *sectioning*, the embedded specimen is placed in a microtome, which is used to cut very thin sections of the specimen. Depending on how the specimen is oriented in the embedding material and the microtome, these sections can be longitudinal (lengthwise) sections, cross-sections, or tangential sections.

Slide preparation

The thin section is transferred to a microscope slide, which is warmed to melt the embedding material and cause the section to adhere to the slide. The specimen is then stained and mounted in the usual manner.

As you may imagine, preparing histology specimens can be complicated and time-consuming, and not only in terms of physical preparation. Obtaining suitable specimens can itself be time-consuming, challenging, and expensive.

Accordingly, some people decide histology isn't worth the effort and instead use prepared slides exclusively. At anything from $2 to $50 apiece, good prepared slides aren't inexpensive—the cheapest ones aren't worth having—but a well-chosen set of high quality prepared slides can often be resold for most of what it cost after you have finished using them. For learning purposes, we think the best compromise is often to make your own bacteriology slides and some of your histology slides, but purchase prepared histology slides for specimens that would be too difficult, time-consuming, or expensive to prepare yourself.

If you do decide to prepare your own histology slides, you'll need some specialized equipment and materials.

MICROTOME

A *microtome* is a device for cutting very thin sections of material. Professional microtomes cost hundreds to thousands of dollars, but manual versions like the one shown in Figure 2-12 are suitable for a home lab and are available for $50 or less. Although there are workarounds—we have used a paraffin-filled drinking straw and a razor blade to cut cross-sections—there's really no getting around it: if you want to make histology (tissue section) slides, you really need a microtome. If your focus is on microorganisms, you don't.

Figure 2-12: *A manual microtome (image courtesy United Scientific)*

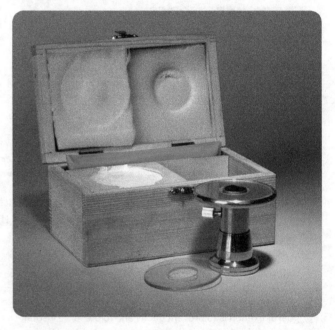

BUILD A $1 MICROTOME

Buy a real microtome if you plan to do much sectioning, but if you'd just like to play around with sectioning, you can build a usable microtome for less than a buck with just a few minutes' work. You'll need the following items:

- A large, empty wooden or hard plastic thread spool. These are becoming hard to find, as thread is sold mostly on plastic foam spools nowadays. You can find wooden spools at flea markets or ask your parents or grandparents if they have a wooden thread spool they're willing to contribute to the cause.

- Several flat metal washers, ideally with a diameter close to that of the thread spool and the diameter of the bolt, with a central hole slightly larger than the bore of the thread spool.

- A bolt, with fine (UNF/NF) or extra-fine (UNEF) threads if possible, that's about the same diameter as the thread spool bore. It should slide easily in and out of the thread spool bore with as little slack as possible, and should have a threaded length a bit longer than the thread spool bore. Take your spool to a hardware store and find a bolt that fits.

- A flat nut to fit the bolt.

- Some epoxy cement or super glue.

- A scalpel or single-edge razor blade.

- A candle.

To make the microtome, proceed as follows:

1. Glue a stack of washers together concentrically. The assembled stack should be 0.5 cm to 1 cm thick.

2. Apply glue to one side of the washer stack.

3. Slide the bolt into the thread spool bore until it protrudes from the other side of the bore.

4. Slide the washer stack down over the bolt and press the glued side of the washer stack against the top surface of the thread spool until it adheres, with the hole in the washer centered over the thread spool bore. Remove the bolt from the thread spool bore.

5. Carefully apply glue to one surface of the nut, making sure to keep it out of the threads.

6. Center the hole in the nut over the thread spool bore and press the nut against the bottom surface of the thread spool until it adheres.

7. Allow the glue to set, fixing the nut to the bottom of the thread spool and the washer stack to the top.

8. Thread the bolt into the nut until the tip of the bolt is just above the bottom surface of the washer stack, leaving a well deep enough to contain the specimen.

9. Drip hot paraffin wax onto the tip of the bolt and, using forceps, immediately press the prepared specimen into the wax before the wax solidifies completely. Keep the specimen centered in the well.

10. Drip additional wax into the well until the well is full and the wax is slightly above the surface of the washer stack. Allow the wax to cool and set.

11. Turn the bolt to force wax up through the washer until the top of the specimen is slightly above the surface of the washer stack.

12. Position the scalpel blade or single-edge razor blade against the surface of the washer stack at a very slight angle (as nearly parallel to the washer surface as possible).

13. Press the edge of the blade through the wax to trim it off flush to the surface of the washer. Use straight pressure; do not attempt to saw through the wax. Discard this wax plug. The wax should now be flush with the top surface of the washer.

The procedure for using the $1 microtome is a bit different than the one for using a commercial microtome, because the $1 microtome has no specimen clamp. To use the $1 microtome, take the following steps:

1. Turn the bolt a fraction of a turn to force more wax (and embedded specimen) above the surface of the washer stack. (Your goal is to produce slices that are so thin they are transparent. You'll soon learn by trial and error how much to turn the bolt to produce sections of the desired thickness.)

2. Use the scalpel to slice the thin section and transfer it to a slide. Examine the specimen at 40X to determine if it is suitable. If not, continue cutting sections until you have one that shows the detail you want. Once you have a suitable specimen, pass the slide (specimen side up) through a flame to melt the wax slightly and adhere the specimen to the slide. You can then observe the specimen normally.

DEHYDRATION BATHS

If you're processing one or a few small specimens, you can use a stoppered test tube as a *dehydration bath*. Partially fill it in sequence with the different concentrations of ethanol you plan to use. For multiple or larger specimens, substitute a series of jars of the appropriate size, each containing one of the ethanol concentrations. Allow each specimen to soak for the indicated time in each concentration, drain it, and transfer it to the jar with the next higher concentration.

To avoid handling damage to specimens, we use "teabags" made from lengths of string and a piece of muslin or cheesecloth. Place the specimen in the center of the cloth, lift the corners, and tie off the bag to secure the specimen. Suspend the bag in each jar, using the lid to keep the string out of the liquid. To change jars, just lift the teabag, allow it to drain, and transfer it to the next jar.

EMBEDDING BLOCKS

An *embedding block*, shown in Figure 2-13, is simply a mold that is used to hold the infiltrated specimen while it is covered with embedding material in liquid form. Some embedding blocks are made of soft material and are designed to be sectioned right along with the specimen. Others are made of harder material, and can be reused repeatedly. With the latter type, after the embedding material solidifies, the portion containing the specimen is popped free of the block or pried out with a needle and then sectioned.

Figure 2-13: *Embedding blocks (image courtesy United Scientific)*

GENERAL LABORATORY EQUIPMENT

In addition to equipment used solely or primarily for biology, you'll need some general laboratory equipment.

Balance

A *balance* is not essential for most tasks, but is useful for making up solutions accurately, weighing specimens, and so on. An inexpensive battery-powered electronic balance with 100 gram capacity and 0.01 gram (centigram) resolution is sufficient, and can be used both in the lab and in the field. Better still is a dual-range balance that offers centigram resolution to 100 grams or so and decigram (0.1 gram) resolution to 500 grams or so.

Beakers

A *beaker* is a cylindrical container used for making up solutions, holding hot or cold water or ice baths, and so on. Beakers are available in glass and polypropylene, either of which can be used for most tasks, although the polypropylene models cannot be heated with direct flame. You'll want a range of sizes from at least 50 mL to 250 mL. If you plan to heat solutions in a microwave oven, polypropylene is fine. If you'll use a stove burner or hotplate, have at least one borosilicate glass beaker on hand.

Centrifuge tubes

Although they're intended for use in a centrifuge, standard *centrifuge tubes*, shown in Figure 2-14, are also useful for many other tasks around the lab, including storing solutions and specimens. Buy the polypropylene versions with screw caps, which are autoclavable. We keep a supply of 15 mL and 50 mL centrifuge tubes on hand. The conical end versions are usable for most purposes. Self-standing versions, which have a skirt around the conical end, stand upright on a flat surface. The 50 mL self-standing versions are also excellent containers for developing chromatography paper strips.

Figure 2-14: *50 mL standard centrifuge tubes (image courtesy United Scientific)*

Smaller *microcentrifuge tubes*, shown in Figure 2-15, usually called *Eppendorf tubes* or *Eppie tubes*, are available in sizes from 0.25 mL to 1.5 mL or more. They're intended for use in micro-ultracentrifuges like the Dremelfuge described later in this chapter, and are also useful for storing small solid or liquid specimens. Buy the polypropylene versions with snap caps, which are autoclavable.

Figure 2-15: *1.5 mL microcentrifuge tubes (image courtesy United Scientific)*

Flasks

An *Erlenmeyer flask*, also called a *conical flask*, is essentially a conical beaker with a narrow neck. Erlenmeyer flasks are useful as culture vessels, gas generating bottles, and for temporary storage of solutions and liquid specimens. For general use, keep at least one 125 mL or 250 mL Erlenmeyer flask with solid, one-hole, and two-hole stoppers to fit it. If you're doing large scale broth culturing, buy as many in whatever size or sizes as you need.

A *volumetric flask* is used to make up solutions to accurately known volumes. It has a long, narrow neck with an index line to indicate the nominal volume. In use, the solution is made up to slightly less than the final volume, transferred to the flask, and the flask is then filled to the index line. The most useful sizes in a home biology lab are 25 mL, 100 mL, and 500 mL. For most purposes you can substitute a graduated cylinder without much loss of accuracy.

Graduated cylinders

A *graduated cylinder* is a tall, narrow container with graduation lines to indicate the volume it contains. Polypropylene graduated cylinders are suitable for most purposes other than measuring strong organic solvents, and are considerably less expensive than glass models. You'll want at least one 10 mL and one 100 mL graduated cylinder.

Polyethylene pipettes

Polyethylene pipettes, also called *beral pipettes* or *disposable pipettes*, are essentially one-piece plastic eyedroppers. They're available in various capacities from 1 mL to 5 mL or more. Buy models with graduated stems. We prefer the 3 mL models with stems graduated to 1 mL by 0.25 mL, which can be interpolated to 0.05 mL or less. These pipettes are usually sold in 10-packs for a dollar or so, or in bags of 100 for $5 to $8. Although they're referred to as "disposable," in fact they can be used over and over simply by rinsing them with distilled water after each use.

Graduated pipettes

Graduated pipettes, also called *Mohr pipettes* or *serological pipettes*, are used to measure small volumes of liquids with very high accuracy. The most useful sizes in a home biology lab are 1.0 mL x 0.01 mL and 10.0 mL x 0.1 mL. You should never pipette by mouth, so you'll also need a *pipette bulb* or a *pipette pump*. For most purposes, you can substitute a polyethylene pipette, either by using its graduations or by calibrating your polyethylene pipettes with your balance and then counting drops.

Hotplate and wire gauze

A *hotplate* (or kitchen stover burner) is useful for heating solutions, making up agar culturing medium, and so on. Use a *ceramic wire gauze* to protect the bottom of the beaker or flask from the hotplate or burner surface. For most purposes, you can substitute a microwave oven.

pH meter and pH test paper

A *pH meter* is used to measure pH with high accuracy. Professional-grade pH meters are accurate to 0.001 pH units or better but cost hundreds of dollars. Models accurate to 0.1 pH unit are available for $25 from Cynmar and other retailers. Models that sell for $50 to $100 may be accurate to 0.01 pH units and may include a digital thermometer function that is accurate to 0.1 °C or better. For times when you just need a quick, rough indication of pH, use *pH test paper*.

Reaction plate

A *reaction plate* is a small rectangular plastic plate with an array of cylindrical wells, usually 2x3, 3x4, 4x6, or 8x12. Reaction plates are often used for testing one sample against many reagents or testing many samples against one reagent, while using only small volumes of both the samples and reagents. They're also useful for making up small amounts of various concentrations of chemicals, organizing small specimens, and so on. Sterile plates can be used for culturing many samples simultaneously in small volumes. The 24-well and 96-well versions are the most useful in a home biology lab.

Thermometer

You'll need at least one *thermometer*, unless your pH meter also measures temperature. For general lab use, the least expensive alternative is one or more glass-tube thermometers. Ideally, you should have one thermometer for general use with a range of -20 °C to 110 °C and resolution of 1 or 2 °C, and a second thermometer with a narrower range, at least 0 °C to 50 °C, but better resolution.

Test tubes, brush, clamp, rack, and stoppers

In addition to serving as culturing tubes, *test tubes* are used in many routine lab procedures. We buy them by the case, usually 72 tubes, but they're not much more expensive by the dozen. Keep at least half a dozen on hand. We find the 16x100 mm size most useful. You'll also need a *test tube brush* to fit the tubes, a *test tube clamp* for heating them, and a *test tube rack* to hold them. Buy a wet/dry rack, which you can also use for sterilizing them when you're using them as culturing tubes. Purchase at least a few rubber *stoppers* to fit the tubes, in solid and one-hole versions.

Wire gauze

A *wire gauze* is used to isolate a glass beaker or flask from exposure to direct flame or high temperature.

MAJOR EQUIPMENT

There are several pieces of major (read "expensive") equipment that we'll mention for completeness. Few homeschoolers will have the need (or budget) for these items, but all of them are important for DIY biology enthusiasts. In some cases, there are usable workarounds or inexpensive hacks that avoid the need for these items, often at the expense of producing less accurate results or requiring more time and effort to use. We use many of these hacks ourselves, but if biology is your hobby and you can afford the real items, by all means buy them.

SPECTROPHOTOMETER

A *spectrophotometer* measures the *transmittance* (or its inverse, *absorbance*) of light at various wavelengths, which may range from the ultraviolet through the visible spectrum and into the infrared. The simplest models, called *colorimeters*, measure transmission at only a few discrete wavelengths, usually in the visible spectrum. More expensive models measure transmission at very narrow intervals (often 2 nm or less) across a wide range of wavelengths. Alas, the least expensive spectrophotometers cost between $500 and $1,000 new, and even an inexpensive commercial colorimeter costs $200 or more. Used models are available on eBay for less, but there's often a good reason why they're being sold on eBay. (Hint: they're usually worn out or damaged.)

Fortunately, the human eye is very good at discriminating differences in the color and intensity of light between two samples, so with a bit of extra time and effort, it's often possible to get good data with no special equipment simply by visually comparing an unknown solution against a series of solutions of known concentrations. That's the procedure we'll use in this book, but if you own or can beg access to a real spectrophotometer, you'll obtain more accurate data with less time and effort.

CELLPHONE SPECTROPHOTOMETER

If you're willing to do a bit of hands-on hacking, check out the spectrophotometer project by Alexander Scheeline and Kathleen Kelley of the Department of Chemistry at the University of Illinois at Urbana-Champaign. Using only a few dollars' worth of readily available parts and a cellphone camera or a point-and-shoot digital camera, they've put together an actual spectrophotometer, including the (free) software needed to calibrate it and capture data from the digital image files.

www.asdlib.org/onlineArticles/elabware/Scheeline_Kelly_Spectrophotometer/index.html

ULTRACENTRIFUGE

A *centrifuge* is a device that uses centrifugal force to separate suspensions and colloids, which remain homogeneous under ordinary gravity. (The Brownian motion of the solvent molecules is sufficient to keep the tiny solid particles suspended rather than precipitating out.) Tubes filled with the suspension to be separated are placed in the centrifuge head, which is then spun at high speed. The resulting centrifugal force causes the denser solid particles to be accelerated away from the center of rotation, where they accumulate in the bottom of the centrifuge tubes.

Hand-cranked manual centrifuges are available for less than $100, and entry-level motorized centrifuges for less than $250. Unfortunately, neither spins fast enough to be useful in a biology lab, where centrifuges are used to separate or purify DNA, proteins, microorganisms, and so on. Because these materials are both extremely tiny and not much denser than water, very high speeds (accelerations) are needed to precipitate them.

The solution is an *ultracentrifuge*. Standard centrifuges operate at a few hundred to a couple thousand RPM. Ultracentrifuges operate at 10,000 RPM or more and can achieve accelerations of 10,000 or more times the force of gravity. Alas, commercial ultracentrifuges cost several thousand dollars and up.

Fortunately, Irish DIY bio hacker extraordinaire Cathal Garvey developed a usable ultracentrifuge for those of us on a budget. His Dremelfuge is simply a micro-ultracentrifuge head. You supply the motor power (with a standard Dremel Moto-Tool or electric drill) and the shielding (Garvey uses a metal cooking pot). You can watch the Dremelfuge in operation at *www.youtube.com/watch?v=86WnXeTZO_Y* and purchase one at *www.shapeways.com/shops/labsfromfabs*. (Note that there are two models available, one for a Dremel Moto-Tool and one for a standard chuck drill.)

WARNING

Garvey sells the DremelFuge as an ornamental item only, and warns that he's not responsible if you use it as an actual centrifuge. And with good reason. Centrifuge accidents are no joke. If the DremelFuge fails catastrophically, fragments will fly in all directions at Warp Factor 6, possibly causing serious injury or death.

If you decide to use the DremelFuge (we do), we strongly recommend you clamp it vertically—we use a Dremel 220-01 Workstation—with the power cord disconnected and the head in position. Cover the entire assembly with one of those five-gallon plastic buckets sold at DIY supercenters. (We also wear safety goggles **and** a face shield.) Stand way back and sweet-talk someone you don't like into plugging in the power cord for you. After you've run it for the required time, disconnect the power, wait for it to spin down, and only then remove the bucket.

Seriously, this is no joke. Running the DremelFuge without taking these precautions is simply foolish.

The DremelFuge accepts standard 1.5 mL plastic Eppendorf tubes, so you'll need a supply of those. Depending on what you're doing, the tubes may or may not be reusable. You may use two, four, or six tubes at once, and if you run the Dremelfuge at high speed you're likely to crack tubes occasionally, so order a reasonable number.

DNA PROCESSING EQUIPMENT AND SUPPLIES

Although you can isolate DNA with equipment and materials found in most kitchens, doing really interesting things with DNA requires some special equipment and reagents. You can make some of the necessary equipment and reagents yourself, but for advanced work some must be purchased.

For serious DNA work, the first requirement is a DNA gel electrophoresis apparatus with power supply. Edvotek (*www.edvotek.com*) sells complete setups starting at $139 for the apparatus (shown in Figure 2-16) and $145 for a power supply, which is the best price we've found. This is an entry-level DNA gel electrophoresis setup, but quite adequate for doing serious work. As supplied, the kit produces 7x7 cm gels. (You can buy an optional gel casting tray for 7x14 cm gels.) The power supply is fixed at 70 VDC, which is a good compromise for most purposes.

If you plan to do a lot of DNA work, you'll want a setup that allows you to run larger gels and possibly multiple gels simultaneously. You'll also want a variable-output power supply, ideally with a range of at least 50 to 200 VDC and sufficient amperage to run as many gels (or tanks) as you plan to run simultaneously. Such setups are available from Edvotek and other vendors.

For learning purposes, you don't need to buy a commercial DNA gel electrophoresis setup. We once bet someone that we could produce usable DNA gels with only items commonly found around a home, and do it all for less than $10. We won that bet, too. We used plain gelatin to make the gels, which we cast in a trimmed-down Gladware container with masking tape to form the casting dams. We made up the loading buffer and running buffer with table sugar, table salt, and baking soda, used food coloring as the marker dye, developed the gel in a flat Tupperware container with a stack of five 9V transistor batteries wired in series using aluminum foil for electrodes, and visualized our developed gel with some methylene blue from the aquarium supplies. It wasn't pretty but, hey, it actually worked.

Figure 2-16: *Edvotek M6+ DNA gel electrophoresis setup (image courtesy Edvotek)*

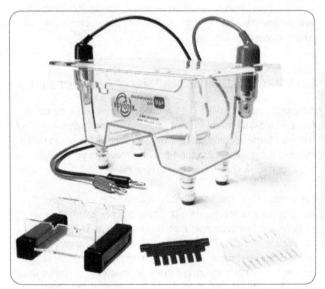

You'll also need some way to measure and transfer small volumes of liquids. For educational use, you can use inexpensive polyethylene pipettes, whose drop sizes range from 50 drops/mL (20 μL/drop) to 25 drops/mL (40 μL/drop). For serious work, you'll need a *micropipetter* like the one shown in Figure 2-17. These are available in fixed-volume and variable-volume

models, with variable-volume models starting at around $150. (Multichannel models can fill eight or more wells in one operation, but cost several hundred dollars and up.)

Figure 2-17: *Single-channel variable volume micropipette (image courtesy United Scientific)*

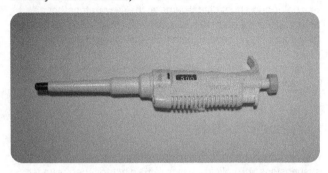

If you need something better than a disposable pipette but don't want to spend $150 or more, consider using a $20 fixed-volume *minipipette* like the ones shown in Figure 2-18. These are typically a bit less accurate than more expensive models, but more than good enough for most work. Whichever type you use, you'll need a supply of disposable tips to fit it.

Figure 2-18: *Single-channel fixed volume minipipettes (image courtesy United Scientific)*

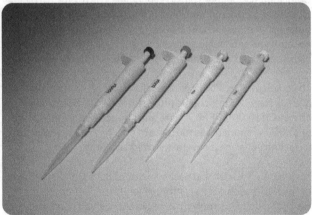

For many tasks, you can simply extract as much DNA as you need to work with. For example, you can obtain sufficient DNA from one strawberry or carrot to run hundreds of gels, or you can run a broth culture to obtain large amounts of bacterial DNA. But sometimes you can obtain DNA only in amounts too tiny to analyze.

The solution is a procedure called *PCR* (*polymerase chain reaction*), which is used to amplify (replicate) tiny amounts of DNA into the larger amounts needed to analyze it by causing the DNA to replicate exact copies of itself. The procedure involves mixing the DNA sample with a solution that contains enzymes, amino acids, and buffers and then exposing the mixture to a series of precisely controlled temperatures for specific durations. That series may require dozens to hundreds of repetitions, which is where a *PCR thermal cycler* comes in.

A thermal cycler provides either a receptacle for a standard 96-well plate or a grid—usually 3x3, 4x4, or 5x5—that accepts small microtubes. You place your DNA sample(s) in the well plate or microtubes, turn on the thermal cycler, and wait for it to complete however many cycles you've specified. When the cycles complete, each well or tube contains much more DNA than was originally present. That DNA can then be separated, purified, and subjected to analysis.

Commercial PCR thermal cyclers are extremely expensive. The least expensive model we found was the Edvotek model, at about $1,800. Fortunately, the DIY bio hacking community has come to the rescue. The OpenPCR initiative (*http://openpcr. org/*) offers a complete kit for $512, with the hard parts pre-assembled. We've so far avoided the need for PCR in our own work, but one of these is definitely on our wish list.

If you're very patient, you can actually do thermal cycling manually. All the equipment you'll need is some foam cups, a way to heat water, a thermometer, and a clock or watch with a second hand. You simply fill the cups with water at the required temperatures (keeping the temperatures at the necessary levels by adding hot water as needed) and move the Eppie tubes from each cup to the next after the required time has elapsed.

We actually did this once, just to be able to say we'd done it. We ran through two dozen iterations before we lost patience. Unfortunately, real-world PCR may require hundreds of iterations, and we're not that patient.

For some work, you'll need a freezer, but not just any freezer. Ordinary home freezers operate at about -20 °C, which isn't cold enough for some work. You can buy a *-80 °C freezer* from a lab supply vendor, but it'll cost a bundle. Fortunately, there's a cheap and practical alternative for a home lab. Obtain a small foam cooler and a supply of dry ice from the supermarket or an ice cream store. Place your specimens and some dry ice in the cooler, with the lid fitted loosely on top to allow gas to escape. Place the cooler in the regular freezer. The dry ice will keep the contents of the cooler at -80 °C for anything up to a week, and the dry ice can be replaced as necessary.

Finally, you'll need various supplies, including agarose for gels, loading and running buffers, enzymes, and so on. The supplies you need depend on what exactly you're doing. Get a catalog from Edvotek and visit the website of Carolina Biological Supply (*www.carolina.com*).

PREPARED SLIDES

Many of the lab sessions call for prepared slides, which you can make yourself or purchase individually or in sets. Cynmar (*www.cynmar.com*) and eNasco (*www.enasco.com*) carry a wide range of sets and individual slides at good prices. Carolina Biological Supply offers many slides that are not available from Cynmar or eNasco, but has generally higher prices. Home Science Tools offers a reasonable selection of slide sets and individual slides at prices that are generally between those of Cynmar/eNasco and CBS.

Before you purchase slides, check with your local home school group or co-op. Some have prepared slides available for members to borrow or rent inexpensively, and other members may have unneeded slide sets from prior years that they're willing to sell. Even if those slide sets don't include all of the slides you need and include some you don't need, you can use those sets as a starting point and add other individual slides as needed. And, of course, when you are finished using the slides you can sell them on to someone else in the group.

To decide which slides you need to buy, whether in sets or individually, we recommend scanning through the book to determine first which lab sessions you intend to do and which slides you want to have available for them. Then decide which slides you can prepare yourself and which would be better to purchase. For example, you probably don't want to purchase a prepared slide of squamous epithelial cells, because you can make your own in about 15 seconds by scraping the inside of your cheek with a toothpick, transferring the cells to a slide, and adding a drop of methylene blue stain. Similarly, it's easy to prepare your own slides of whole-mounted creatures from pond water, plant stems and leaves, and so on. Conversely, if the specimen itself is difficult to obtain (or hazardous) or if it requires complex histological preparation, sectioning, and staining, most people will understandably prefer to buy the prepared slide.

If you do intend to prepare some or all of your own slides, take scheduling into account, both in terms of availability of the specimens (which may vary by season) and in terms of the time and resources needed to prepare the slides. Rather than devote precious lab time to preparing slides, other than simple wet mounts of whole specimens, it's often best to prepare slides the day or the week before you need them for a session. Water-based wet mounts dry out quickly, but if you substitute glycerol as the mounting fluid, the slide should remain usable for at least a week and possibly much longer. You can, of course, make permanent mounts using Permount or a similar commercial mounting fluid, or even colorless nail polish.

As an alternative to spending a lot of money on prepared slides, the customized kit for this book, available from The Home Scientist, LLC (*www.thehomescientist.com/biology.html*), includes a disc that contains high-resolution color images of most of the microscopy images used in this book. You can view these on your computer and see more detail than in the printed images. We'll admit it up front: we're not expert photomicrographers and our work will probably never be published by *Nature*, but these images are a reasonably good substitute for using prepared slides. Not perfect, by any means, because no image can show you everything you can see while viewing an actual slide through an actual microscope, where you're able to move the specimen around, change magnifications to zoom in or zoom out on particular parts of the specimen, and so on. Finally, most specimens have depth, which means it's impossible to focus critically on all parts of the specimen at one time, particularly at higher magnifications.

SPECIMENS

Many of the lab sessions require live or preserved specimens or live microorganism cultures. Many of these specimens, such as a human hair or pond water or tree leaves, are easy to obtain locally at little or no cost. Some, such as bacteria cultures, must be ordered from a science supplies vendor if you intend to complete those lab sessions. Wherever possible, we point out free alternatives to purchased specimens.

Here are the sources we recommend for specimens:

Carolina Biological Supply

CBS offers a huge variety of live and preserved specimens and live cultures, many of which are available to individuals and homeschoolers. (Sales of some potentially hazardous organisms are restricted to high schools or universities.) Their prices and shipping charges tend to be on the high side, but their quality is excellent. If you can't find something anywhere else, chances are good that CBS offers it. 800-334-5551 or *www.carolina.com*.

eNasco

eNasco is generally homeschool-friendly, and offers a wide selection of specimens. Some items are sold only to schools and businesses, although that's easy enough to get around. Although its selection is not as large as CBS's, eNasco generally has somewhat lower prices and their quality is generally excellent. eNasco is the first place we look for specimens. 800-558-9595 or *www.enasco.com*. eNasco also has an excellent FAQ on live materials at *www.enasco.com/science/page/livematerialFAQ*.

DISSECTION SPECIMENS

We don't do any dissections in this book because we think there are better ways to spend precious lab time. If you want to make dissections part of your curriculum, you can purchase kits from Carolina Biological Supply, Cynmar, eNasco, and other vendors that include various specimens, dissecting instruments, and illustrated instructions.

As an alternative to doing actual dissections, you can purchase high-resolution color charts that cover dissecting various types of specimens from BioCam Charts (*www.biocam.com*). Many schools now use these charts exclusively, and not just to save money. Using these charts avoids the "yuck factor" that has killed a lot of students' interest in biology. No surprise, if one's first experience with biology is cutting up something dead, smelly, squishy, and generally disgusting. They don't call it gross anatomy for nothing.

Laboratory Safety

3

This is a short chapter, but a very important one. Many of the procedures described in this book use chemicals, such as strong acids and bases, that are dangerous if handled improperly. Some procedures use open flame or other heat sources, and many use glassware.

Doing biology lab work at home has its dangers, but then so does driving a car. And, just as you must remain constantly alert while driving, you must remain constantly alert while doing lab work at home.

It's important to keep things in perspective. Every year, millions of students do biology labs. There are, of course, occasional small injuries, such as minor cuts and burns. More serious injuries are extraordinarily rare and occur almost exclusively when students fail to take basic precautions, such as wearing gloves and goggles when working with hazardous chemicals. For students who do take these precautions, the likelihood of suffering a serious injury is probably about the same as being struck by lightning.

The primary goal of laboratory safety rules is to prevent injuries. Knowing and following the rules minimizes the likelihood of accidents, and helps ensure that any accidents that do occur will be minor ones.

Here are the laboratory safety rules we recommend:

PREPARE PROPERLY

- All laboratory activities must be supervised by a responsible adult.

  Direct adult supervision is mandatory for all of the activities in this book. This adult must review each activity before it is started, understand the potential dangers of that activity and the steps required to minimize or eliminate those dangers, and be present during the activity from start to finish. Although the adult is ultimately responsible for safety, students must also understand the potential dangers and the procedures that should be used to minimize risk.

- Familiarize yourself with safety procedures and equipment.

  Think about how to respond to accidents before they happen. Have a fire extinguisher and first-aid kit readily available and a telephone nearby in case you need to summon assistance. Know and practice first-aid procedures, particularly those required to deal with burns and cuts. If you have a cell phone, keep it handy while you work.

  One of the most important safety items in a home lab is the cold water faucet. If you burn yourself, immediately (seconds count) flood the burned area with cold tap water for several minutes to minimize the damage done by the burn. If you spill a chemical on yourself, immediately rinse the chemical off with cold tap water, and keep rinsing for several minutes. If you get any chemical in your eyes, immediately turn the cold tap on full and flood your eyes until help arrives.

WARNING

Everyone rightly treats strong acids with great respect, but many students handle strong bases casually. That's a very dangerous practice. Strong bases, such as solutions of sodium hydroxide, can blind you in literally seconds. Treat *every* chemical as potentially hazardous, and *always* wear splash goggles.

Keep a large container of baking soda on hand to deal with acid spills, and a large container of vinegar to deal with base spills.

- Always read the MSDS for every chemical you will use in a laboratory session.

The MSDS (Material Safety Data Sheet) is a concise document that lists the specific characteristics and hazards of a chemical. Always read the MSDS for every chemical that is to be used in a lab session. If an MSDS was not supplied with the chemical, locate one on the Internet. For example, before you use toluidine blue stain in a procedure, do a Google search using the search terms "toluidine blue" and MSDS.

- Organize your work area.

Keep your work area clean and uncluttered—before, during, and after laboratory sessions. Every laboratory session should begin and end with your glassware, chemicals, and laboratory equipment clean and stored properly.

DRESS PROPERLY

- Wear approved eye protection at all times.

Everyone present in the lab must at all times wear splash goggles that comply with the ANSI Z87.1 standard. Standard eyeglasses or shop goggles do not provide adequate protection, because they are not designed to prevent splashed liquids from getting into your eyes. Eyeglasses may be worn under the goggles, but contact lenses are not permitted in the lab. (Corrosive chemicals can be trapped between a contact lens and your eye, making it difficult to flush the corrosive chemical away.)

- Wear protective gloves and clothing.

Never allow laboratory chemicals to contact your bare skin. When you handle chemicals, particularly corrosive or toxic chemicals or those that can be absorbed through the skin, wear gloves of latex, nitrile, vinyl, or another chemical-resistant material. Wear long pants, a long-sleeve shirt, and leather shoes or boots that fully cover your feet (NO sandals). Avoid loose sleeves. To protect yourself and your clothing, wear a lab coat or a lab apron made of vinyl or another resistant material. Wear a disposable respirator mask when you handle chemicals that are toxic by inhalation or potentially pathogenic microorganisms.

AVOID LABORATORY HAZARDS

- Avoid chemical hazards.

Never taste any laboratory chemical or sniff it directly. (Use your hand to waft the odor toward your nose.) When you heat a test tube or flask, make sure the mouth points in a safe direction, in case the liquid suddenly boils and is ejected forcefully from the container. If you heat a liquid in a microwave oven, be extremely cautious. The liquid may become superheated (hotter than the boiling temperature, but not yet boiling) and any disturbance may cause the liquid to suddenly boil violently. Never carry open containers of chemicals or biologicals around the lab. Always dilute strong acids and bases by adding the concentrated solution or solid chemical to water slowly and with stirring. Doing the converse can cause the liquid to boil violently and be ejected from the container. Use the smallest quantities of chemicals that will accomplish your goal.

- Avoid biological hazards.

  Always remember that all lifeforms have defenses, and take steps to protect yourself from those defenses. Always wear at least exam gloves when handling live specimens of any species, and wear heavier protective gloves if you handle a specimen that bites, stings, or is otherwise dangerous to handle. Also wear goggles if the specimen presents any danger, however slight, to your eyes.

  When you culture microorganisms—bacteria, protists, or fungi—always assume that the microorganisms are lethal to humans (even if you "know" they are not) and act accordingly. Learn aseptic procedures, and follow them rigorously. Wear gloves and goggles at all times, and if you have even the slightest doubt whether the microorganism is pathogenic, wear an N100 particulate mask. Do not culture unknown microorganisms unless you have the proper equipment and are absolutely certain that you know how to do so safely. (Even then, doing so is dangerous, as many experienced biologists have learned the hard way.)

  When you finish using a culture, sterilize it either by autoclaving the culture container and all tools that have come into contact with live microorganisms or by immersing them in a chlorine bleach bath overnight. Disinfect your work area before and after using live microorganisms by spraying it thoroughly with a strong solution of Lysol or a similar disinfectant.

- Avoid fire hazards.

  Never handle flammable liquids or gases in an area where an open flame or sparks might ignite them. Extinguish burners as soon as you finish using them. Do not refuel a burner until it has cooled completely. If you have long hair, tie it back or tuck it up under a cap, particularly if you are working near an open flame.

- Avoid glassware hazards.

  Assume all glassware is hot until you are certain otherwise. Examine all glassware before you use it, and particularly before you heat it. Discard any glassware that is cracked, chipped, or otherwise damaged.

DON'T DO STUPID THINGS

- Never eat, drink, or smoke in the laboratory.

  All laboratory chemicals should be considered toxic by ingestion, and the best way to avoid ingesting chemicals is to keep your mouth closed. Eating or drinking (even water) in the lab is very risky behavior. A moment's inattention can have tragic results. Smoking violates two major lab safety rules: putting anything in your mouth is a major no-no, as is carrying an open flame around the lab.

- Never work alone in the laboratory.

  No one, adult or student, should ever work alone in the laboratory. Even if the experimenter is adult, there must at least be another adult within earshot who is able to respond quickly in an emergency.

- No horsing around.

  A lab isn't the place for practical jokes or acting out, nor for that matter for catching up on gossip or talking about last night's football game. When you're in the lab, you should have your mind on lab work, period.

- Never combine chemicals arbitrarily.

  Combining chemicals arbitrarily is among the most frequent causes of serious accidents in home labs. Some people seem compelled to mix chemicals more or less randomly, just to see what happens. Sometimes they get more than they bargained for.

Laboratory safety is mainly a matter of common sense. Think about what you're about to do before you do it. Work carefully. Deal with minor problems before they become major problems. Keep safety constantly in mind, and chances are any problems you have will be very minor ones.

Using a Microscope

EQUIPMENT AND MATERIALS

You'll need the following items to complete this lab session. (The standard kit for this book, available from *www.thehomescientist.com*, includes the items listed in the first group.)

MATERIALS FROM KIT

- Goggles
- Forceps
- Ruler (millimeter scale)

MATERIALS YOU PROVIDE

- Gloves
- Lamp or book light
- Microscope and illuminator
- Scissors
- Slide, prepared (bacteria or diatoms)
- Specimen: notebook or copy paper
- Specimen: snippet from cover of this book

BACKGROUND

Biology as a modern science would not exist without the microscope, just as modern astronomy would not exist without the telescope. Both perform the same function: making the invisible visible. Without microscopes, biologists would be literally blind to the vast majority of the world around them.

When we took our first high-school biology courses more than 40 years ago, they focused on survey segments—such as examining different types of plants for gross similarities and differences—and dissections of frogs and other unfortunate specimens. We were lucky to get in a few minutes a week using a microscope.

That wasn't because surveys and dissections were the best way to learn biology. It was a matter of necessity. Microscopes were very expensive, and most school budgets were too small to provide enough of them. Many biology classrooms had only one microscope. Or none. Nowadays, although excellent inexpensive microscopes are available, most schools still come up far short, requiring each microscope be shared by two, three, or even four students. That's no way to learn biology. Every student needs his or her own microscope.

It's not that a student will be using the microscope every moment of every day, but that the microscope needs to be available on the spur of the moment. Every time you encounter something new to you, your first thought should be, "I wonder what this looks like under the microscope." And your first action should be to get it under the microscope to find out.

If at all possible, you should dedicate a work area to microscopy, where the microscope and its accessories are safe from lab spills and always ready to use at a moment's notice. For example, although our main lab is downstairs, Robert's microscope station, shown in Figure I-1-1, is a large desk next to his (even larger) main office desk. (Yes, this is an actual un-retouched image of a real working microscope bench. We didn't clean up or move anything before we shot this image, other than removing the cover from the microscope.) Switching from using the computer to using the microscope takes literally seconds, which means there's never an excuse not to check something out under the microscope.

Different models of compound microscopes are more similar than different, but some differences do exist from model to model. The two most obvious visible differences you're likely to encounter are the presence or absence of an inclination joint, which allows the microscope to be tilted, and whether the microscope focuses by moving the stage or the body tube.

Whatever the particular configuration of your microscope, it's important that you be able to identify each feature and understand how it is used properly. With practice, operating your microscope should become second nature. For example, you shouldn't have to think about which direction to rotate the focus knob to open or close focus or which direction to turn the mechanical stage control to move a specimen to the right in your field of view.

In addition to learning to use the microscope itself, you also need to learn some basic microscopy techniques, including basic slide making, making whole mounts, smear mounts, and

sectional mounts, determining the dimensions of specimens, simple staining, and differential staining. There's always more to learn, but mastering these fundamentals provides a firm foundation.

The only way to master your microscope and basic microscopy techniques is with practice and more practice. So, let's get started.

Figure I-1-1: *Robert's microscopy workstation*

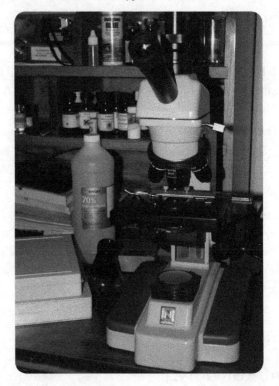

PROCEDURE I-1-1: A MICROSCOPE TOUR

To begin, place your microscope on a flat, stable surface. Remove the dustcover, and plug in the power cord, if applicable. Refer to the diagram in your microscope manual that identifies the key parts of your microscope and then locate and identify those components on your microscope.

OPENING AND CLOSING FOCUS

To avoid confusion, we use the terms *open focus* and *close focus* rather than the more common "focus up" or "focus down" throughout this book. Open focus means to increase the separation between the stage and the objective lens; close focus means to decrease that distance.

The potential for confusion arises because different microscopes use different means of focusing. Some keep the objective lens fixed and raise or lower the stage to focus; others keep the stage fixed and raise or lower the objective lens. With the first type, lowering the stage opens focus; with the second type, lowering the objective lens closes focus.

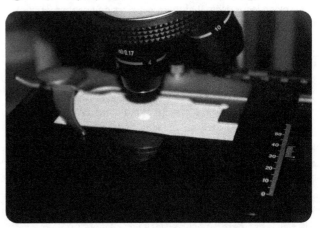

Figure I-1-2: *Paper positioned on the microscope stage*

CONTROLLING AND FOCUSING THE LIGHT

1. Observing from the side of the microscope, use the coarse-focus knob to open focus on your microscope until the stage and objective lenses are as far apart as possible.

2. Turn on the illuminator and adjust it to maximum brightness.

3. Place a piece of thin paper (such as copy paper) on the stage, positioned to cover the hole in the stage. Secure the paper flat against the stage using the stage clips or mechanical stage. A bright illuminated circle should be visible through the paper, as shown in Figure I-1-2.

4. If your substage condenser is focusable, rack the focus up and down and note the effect on the size and sharpness of the illuminated circle on the paper.

5. Locate the diaphragm control. If your microscope has an iris diaphragm, this control is probably a lever or knob. Adjust that control to open the diaphragm to maximum. If your microscope has a disk diaphragm, rotate the disk to place the largest opening in position over the substage condenser lens. Close the diaphragm gradually (or rotate each smaller disk opening into position) and note the effect on the size and brightness of the illuminated circle. When you finish, open the diaphragm again to its widest setting.

6. Use the rheostat (dimmer switch) to decrease the illuminator brightness to minimum and then back up to maximum. Note the range of brightnesses available.

Particularly when you are viewing unstained specimens, get in the habit of varying the brightness as you examine the specimen. Detail that's visible with relatively dim illumination may be invisible with very bright illumination, and vice versa.

FOCUSING THE MICROSCOPE

1. Observing from the side, rotate the turret (nosepiece) until the low-power (4X) objective is in position. Note that this objective is the shortest of the objectives in the turret. The 10X objective is longer, the 40X objective longer still, and the 100X objective longest of all.

2. Gently turn the coarse-focus knob to close focus until you reach the focus stop, at which point you can no longer close focus. Note that there is still a significant amount of separation (called *working distance*) between the bottom of the 4X objective lens and the top of the stage.

3. Looking through the eyepiece, use the coarse-focus knob to open focus gradually until the first paper fibers come into reasonably sharp focus. Note that only part of the fiber you've focused on is in focus and that many of the fibers are still completely out of focus, as shown in Figure I-1-3. This occurs because the specimen you are examining is three-dimensional (has depth). Even at low magnification, the image has very shallow *depth of focus*, so you can focus sharply on only small portions of it at a time.

> Some microscopes cannot close focus far enough to bring a sheet of paper into focus unless you place the paper on top of a standard microscope slide.

Figure I-1-3: *Paper fibers showing depth of focus*

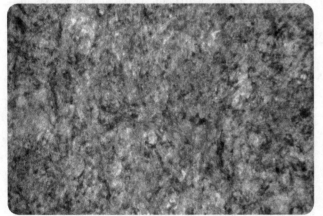

4. Use the fine-focus knob to tweak the focus on the fibers. Note that as one part of the field of view comes into sharp focus, other parts go out of focus. This phenomenon occurs unless the specimen is effectively planar (two-dimensional), with all of it falling within the depth of focus. Because depth of focus decreases with increasing magnification, a specimen that appears planar at 40X may show depth at higher magnifications.

DETERMINING FIELD OF VIEW AND MEASURING SPECIMENS

The *field of view* specifies how much of a specimen is visible in the eyepiece. Field of view and magnification are inversely related: the higher the magnification the narrower the field of view, and vice-versa. For example, at 400X the field of view is one tenth as wide as at 40X. For optical microscopes, field of view is usually specified in millimeters for low magnification objectives and in *micrometers* (μm, or one one-thousandth of a millimeter) for higher magnifications.

> **OPEN WIDE**
>
> A wide field of view makes it much easier to locate objects, which can then be centered in the field and examined at higher magnification. For that reason, the low-magnification objective is sometimes called the *scanning objective*.

Knowing the true field of view at each magnification allows you to make quick estimates of the actual size of objects in the field. For example, if you know that the true field of view is 400 μm (0.4 mm) wide and the length of a paramecium occupies half that field, you know that paramecium is about 200 μm (0.2 mm) long.

1. Place a millimeter-scale ruler on the stage, with the edge centered under the low-power objective.

> Thick or opaque objects are best viewed by reflected rather than transmitted light. You can use a desk lamp or other light source to illuminate such specimens from above. We use a Mighty Bright book light for this purpose, as shown in Figure I-1-4. Its goose neck makes it easy to position the light for best viewing and its dual white LEDs provide plenty of light.

Figure I-1-4: *Using a book light to illuminate an opaque specimen*

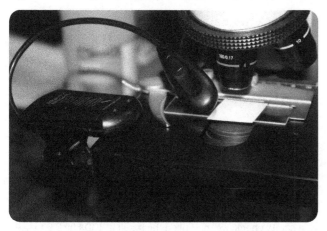

2. Focus carefully on the edge of the ruler and then position the ruler so that one index line is at the edge of the field of view and the edge of the ruler bisects the field. Count the number of divisions visible in the field to determine the true field of view in millimeters. (At 40X you'll probably be able to see about five markings, indicating the field of view is about 5 mm.)

3. Carefully rotate the medium-power objective into place and refocus on the edge of the ruler. At 100X magnification, your true field should be about 2 mm.

4. Carefully rotate the high-and-dry objective into place and refocus on the edge of the ruler. At 400X, your true field should be about 0.5 mm, with only one index line visible.

Note that the 10X objective barrel is longer than the 4X objective barrel, and the 40X objective barrel is longer still. As you rotate higher-power objectives into position, the decrease in working distance is obvious when viewed from the side.

Note further that you can always safely rotate a higher-power objective into position *if the lower-power objective is already focused*. The reason you should always start with the lowest-power objective is that the focus stop on the microscope prevents it from coming into contact with the slide, which is not true of the higher-power objectives. If focus is closed too far on a lower-power objective, rotating the next higher-power objective into position may crash it into the coverslip, damaging the slide or (worse) the objective lens itself.

If you have an *ocular micrometer*, also called a *graduated eyepiece reticle*, you can use it to determine the size of objects with high accuracy. For example, if your reticle is divided into 100 units and you know that those 100 units cover a 400 μm field of view, each reticle unit represents 4 μm. If a paramecium is 46 units long and 11.5 units wide, you can calculate its actual size as 184 x 46 μm.

The obvious problem is that the actual distance spanned by each reticle unit varies with the magnification you're using. Using a millimeter ruler gave us a crude idea of field width, but to calibrate an eyepiece reticle accurately, we need a ruler with much finer graduations. We need something that provides a regular, accurately spaced array of microscopic marks at a known separation. You're holding the solution right now.

Four-color printing, as used to produce the cover of this book, is done by using a screen to put tiny dots of ink on the page. "Four-color" refers to the CMYK inks (Cyan, Magenta, Yellow, and blacK) used to produce full-color images. These dots of ink are not placed randomly within the image. Instead, they're printed on a precise grid, shown in Figure I-1-5. The separation between the dots of that grid is known precisely. That's all we need to calibrate our ocular micrometer. We'll lead you through the process so that you can calibrate your own.

In the dot pattern shown in Figure I-1-5, we'd use the black dots for calibrating because they're much smaller and more sharply defined than the cyan, magenta, and yellow dots.

Figure I-1-5: *A CYMK screen pattern at 40X, shot with the book light*

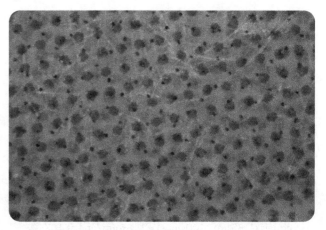

To calibrate your eyepiece reticle, take the following steps:

1. Obtain a screen-printed specimen of known resolution. One obvious source is the cover of this book, which is printed at 300 dpi.

2. Position the specimen under the 4X objective. Use a desk lamp or similar light source to light the specimen from above.

3. Use the mechanical stage to put the center of one dot on the center of the zero mark on the eyepiece reticle. You can use any color dot. The spacing is the same, so choose whichever color has the smallest, sharpest dots. In Figure 1-4, the small black dots (which are actually the black ink) are the clear choice.

4. Orient the eyepiece reticle so that a row of the same color dots aligns with the reticle scale. (Not all eyepiece reticles rotate with the eyepiece; if yours does not, you'll have to realign the specimen on the stage until a row of dots is lined up with the reticle scale.)

5. Look for a dot near one end of the reticle scale, and count the dots between the initial dot and the furthest dot visible near the other end of the scale. Record the number of dots and the number of scale units between them.

6. Based on the known resolution of the screen used to produce the image, the number of dots you counted, and the number of scale units between the first dot and the last, calculate the number of μm represented by each reticle unit.

We did the calculations for our microscope, reticle, and specimen as follows:

- At 40X, 14 dots, center to center, occupied exactly the full length of the reticle scale, 100 units.

- The screen resolution of our print sample was 300 dpi (dots per inch), so those 14 dots represent (14/300) = 0.0467".

- Converting to μm, we multiplied (0.0467" * 25.4 mm/inch * 1000 μm/mm) = 1185 μm.

- Since 1185 μm covers 100 reticle scale units, one unit equals (1185 μm / 100 units) = 11.85 μm/unit at 40X.

We repeated the calibration at 100X, as follows:

- At 100X, five dots, center to center, occupied 89 reticle scale units. (The sixth dot was within the field of view, but off the end of the scale.)

- Again, the screen resolution of our print sample was 300 dpi (dots per inch), so those five dots represent (5/300) = 0.0167".

- Converting to μm, we multiplied (0.0167" * 25.4 mm/inch * 1000 μm/mm) = 423.3 μm.

- Since 423.3 μm covers 89 reticle scale units, one unit equals (423.3 μm / 89 units) = 4.757 μm/unit at 100X.

As a sanity check, we multiplied the result at 100X by 2.5 to see how well it corresponded to the calibration at 40X. That gave us (4.757 * 2.5) = 11.89 μm/unit, very close to the 11.85 μm/unit we calculated at 40X. It's a mistake to expect extreme accuracy from this procedure, so we simply remember that one scale unit corresponds to about 12 μm at 40X, 4.8 μm at 100X, and 1.2 μm at 400X. That's good enough for any work we'll do in these lab sessions.

The calibration values you determine will likely differ from ours because your equipment is probably different from ours. But once you obtain good values at 40X and 100X, write them down where you won't lose them. You'll use them frequently.

POSITIONING AND OBSERVING A SLIDE

1. Turn on the illuminator, open the diaphragm to its widest setting, and rotate the turret to place the 4X objective in position.

2. Position a prepared slide—ideally one of bacteria or diatoms—on the stage, and secure it with the stage clips or the mechanical stage. Make sure the specimen is centered over the hole in the stage.

> If you do not have a prepared slide, you can use a slide you make yourself following the instructions in the next lab session.

3. Focus critically on the specimen. Adjust the diaphragm setting and illumination brightness and observe the differences visible in the specimen. For example, high brightness may reveal maximum detail in the cell wall but make other internal structures invisible, while lower brightness may do the converse. Note that the setting that produces maximum contrast may not reveal the maximum amount of detail, and vice versa.

4. Move the slide slightly right and left and up and down and observe the motion through the eyepiece. Does the object you're observing move in the expected direction or the opposite direction?

5. Carefully rotate the 10X objective into position and then look through the eyepiece. If your microscope is *parfocal* (as it should be) then the specimen should already be focused, or nearly so. If necessary, use the fine focus knob to focus critically. Repeat step 3 to observe the differences in the image at different diaphragm and brightness settings.

6. Repeat step 5 using the 40X (high-and-dry) objective. Make sure that the specimen is centered in the field of view before proceeding to the next step.

7. Rotate the nosepiece (turret) until it is halfway between the 40X and 100X (oil-immersion) objective. Put one drop of immersion oil on the coverslip. Watching from the side, carefully rotate the 100X objective into position, making sure that it does not contact the coverslip and is immersed in the oil drop.

8. Adjust the diaphragm and illuminator brightness to reveal maximum detail in the specimen. Record your observations in your lab notebook.

Richard H. Kessin comments: What will happen here is that they will rotate the turret the wrong way and the oil will get on the 40X objective, making it useless. Either forget oil, which most people do not need, even pathologists, or explain the problem and how to clean off the oil.

RBT responds: Dr. Kessin is absolutely right, and we confess that we frequently use our oil-immersion objective without oil.

The downside to that is that the resolution is lower without the oil. The upside is that using oil (and cleaning up afterward) is time consuming. If you do use immersion oil, when you are finished using it, clean off the oil immediately using a lens tissue and the solvent recommended by your microscope manufacturer.

BUILD A DARKFIELD APPARATUS FROM POCKET CHANGE

So far in this lab session, we've used only bright field illumination, which is all most students and science enthusiasts ever use. *Darkfield illumination* reveals structural details that are invisible with bright field illumination, and is worth exploring. Fortunately, it's quick and cheap to set up nearly any compound microscope to use basic darkfield illumination, at least at low to medium magnification.

Research-grade (very expensive) microscopes include special darkfield condensers and objectives with lower than standard numerical apertures to optimize darkfield image quality. Although they are not equipped with special darkfield condensers or objectives, some student-grade microscopes include the hardware needed to use basic darkfield illumination, or offer it as an inexpensive option. You can also buy a darkfield assembly for $50 or so that can be used with many standard microscopes, but there's really no need for it. You can build your own darkfield apparatus in less than a second for pocket change (literally), and get results about as good as you'd get from the $50 apparatus. Figure I-1-6 shows you how to do it.

Yep, it's that easy. Just place a coin in the center of the ground-glass illuminator diffuser disk. (You might want to use a piece of clear plastic or tissue paper to prevent the coin from scratching the disk.) You can experiment with different size coins to find the best size for your own setup.

Once you've positioned the coin, place a piece of thin paper on the stage. Turn on the illuminator and adjust the condenser focus and diaphragm opening to produce an annular (doughnut-shaped) light spot. Reposition the coin as necessary to center the dark spot in the bright ring. Replace the paper with a slide you want to observe—ideally an unstained thin section or unstained microorganisms—and get your first look at darkfield illumination.

Figure I-1-6: *Assembled dark field apparatus*

With many microscopes, the coin method works pretty well with the 4X or 10X objective, but if you want to use darkfield illumination with your 40X objective you'll need to do a bit more work and spend a few more cents. With a 40X objective, the mask must be be smaller than a coin and closer to the diaphragm and the bottom of the condenser. The filter holder is the obvious location.

You'll need a piece of glass or thin transparent plastic that fits into or on top of your substage filter holder. You can use an ordinary glass slide, if it fits between the top of the filter holder and the bottom of the diaphragm assembly, or a piece of stiff, transparent plastic. (We used a plastic microscope slide, which for adults is about all plastic microscope slides are good for.)

You'll also need an opaque, round mask. How big? We can't say, other than that it must be smaller in diameter than the maximum opening of your microscope's diaphragm—probably much smaller. Ideally, when it's focused you want the dark

central part inside the bright ring to be just slightly larger than the field of view of your 40X objective, which is probably about 0.5 mm. The size of the mask needed to produce a dark area that size at the focal plane depends on the physical and optical characteristics of your condenser and diaphragm.

We suggest you start by placing the slide you used for testing with the 4X and 10X objectives in the stage. Rotate the 40X objective into position, focus on the bright field image, and adjust the diaphragm opening and condenser focus to provide the best image quality. Once that's done, you can start testing to determine the proper mask size.

We started with a mask about 6 mm (a quarter inch) in diameter. We used a small sticky label dot, but you can also use a three-hole punch to punch out circles from heavy paper or thin cardboard. Our sticky label was white paper, so we used a black Sharpie marker to color it in. Stick the mask to the plastic or glass slide and position it in the filter holder. Open the diaphragm to its widest setting. Move the mask around until the dark central portion appears in the field of view and then close the diaphragm to cut off all but a thin annulus around the mask. If the dark spot doesn't cover the entire field of view, you'll need a larger mask. If the dark central spot is much larger than the field of view, you'll need a smaller mask. Ideally, you want the mask to produce a dark central spot that's just slightly larger than the field of view.

Of course, producing a reasonably circular mask of a particular diameter can be difficult. We tried using a black permanent marker to produce tiny black circles, but those circles were neither perfectly round nor completely opaque. Thinking about it, we decided to try using small drops of opaque paint applied with a toothpick, which worked perfectly after a bit of experimentation to produce opaque circles of any size we wanted.

The first thing you'll notice using darkfield illumination at high magnification is that the image is very dim. That's unavoidable, because the only light on the specimen is coming from the periphery. You can improve matters somewhat by removing the ground glass illuminator diffuser disk.

Once you have a mask of the proper size, keep it stored safely so that you don't need to reinvent the mask each time you want to use darkfield illumination at 400X. Unfortunately, it's not possible to build a field-expedient darkfield apparatus for use with the 100X oil-immersion objective. Or, if it is, we can't figure out how to do it.

REVIEW QUESTIONS

Q1: What relationship does physical movement of a slide have to apparent movement while viewing the slide through the microscope?

_____

_____

_____

_____

Q2: Describe the proper procedure for focusing a microscope. Why is it important to use this procedure?

_____

_____

_____

_____

Q3: What is the relationship between magnification and field of view?

_____

_____

_____

_____

Q4: Your microscope provides magnifications of 40X, 100X, and 400X. Do you need immersion oil? Why or why not?

_____

_____

_____

_____

Mounting Specimens

EQUIPMENT AND MATERIALS

You'll need the following items to complete this lab session. (The standard kit for this book, available from *www.thehomescientist.com*, includes the items listed in the first group.)

MATERIALS FROM KIT

- Goggles
- Coverslips
- Forceps
- Glycerol
- Pipettes
- Scalpel
- Slide (well)
- Slides (flat)
- Stain, methylene blue
- Stirring rod (optional)

MATERIALS YOU PROVIDE

- Gloves
- Butane lighter (or other flame source)
- Carrot (raw)
- Microscope and illuminator
- Microtome (purchased or homemade)
- Petroleum jelly
- Specimen: carrot
- Specimen: human hair
- Specimen: pond water
- Toothpicks
- Vegetable oil (olive or similar)
- Water, distilled

BACKGROUND

You must mount a specimen before you can observe it with your microscope. Mounting consists of preparing the specimen (if necessary), placing it on a microscope slide, adding a drop of water or another mounting medium, and covering the specimen and medium with a coverslip. In this lab session, we'll prepare and observe several types of slide mounts.

PROCEDURE I-2-1: MAKING WET MOUNTS

A *wet mount* is the simplest and quickest way to prepare a specimen for observation. To make a wet mount, simply position the specimen in the center of a slide, place a drop of water or other mountant on the specimen, and lower a coverslip into place, making sure to avoid bubbles.

The advantage of a wet mount is that it takes about 15 seconds to make one. The disadvantage is that the mount is only temporary because the mounting fluid will evaporate. In fact, using hot quartz-halogen illumination, water may evaporate after only a minute or two.

> To avoid the problem of wet mounts drying out, you can make a *permanent mount* by substituting a permanent mounting fluid for the water or other temporary mounting fluid. You can purchase permanent mounting fluids—Permount is one popular brand—from science supply vendors, or you can simply use a drop of colorless nail polish. We've used Sally Hansen Hard as Nails successfully. It dries more slowly than Permount, but it does eventually dry to produce a good-quality permanent mount.

MOUNTING A HAIR SPECIMEN

1. Obtain a hair specimen. You can simply snip a bit of your hair, but it's more interesting to examine a specimen that includes the root. (It's also interesting to compare and contrast the appearance of specimens from different people of different ages and hair colors, from the head, beard or mustache, trunk, limbs, pubic, axillary [armpit] regions, from pets, and so on.)

2. Use forceps to position the specimen in the center of a microscope slide. Add a drop of water, carefully position a coverslip at a 45° angle to the slide, and then lower the coverslip into place. If there are any bubbles under the coverslip, use the tip of the forceps to force them out from under the coverslip.

3. Place the slide under the stage clips or in the mechanical stage, and center the specimen under the 4X objective. Adjust the illumination level and diaphragm to obtain the maximum contrast. Readjust them, if necessary to obtain maximum image detail. Note the amount of detail visible in the main body of the hair versus the boundaries.

4. Repeat step 3 to view the hair at 100X and 400X, noting any additional detail that's visible at higher magnifications. Record your observations in your lab notebook. Shoot an image or make a sketch, as shown in Figure I-2-1, to illustrate the major structural elements visible in the hair. If you have an eyepiece reticle, use it to estimate the width of the hair and the size of any structures visible.

Retain this slide for later comparison.

Figure I-2-1: *Wet (water) mount of human facial hair, 40X*

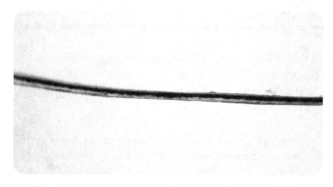

Because you're mounting an entire macroscopic specimen, this method is called a *whole mount*, abbreviated *wm*. The alternative is to slice (*section*) the specimen, which results in a *sectional mount*. The types of sectional mounts are a *cross-section mount* (*cs* or *xs*), a *vertical-section mount* (*vs*), a *radial-section mount* (*rs*), a *longitudinal-section mount* (*ls*), and a *tangential-section mount* (*ts*). Bacteria and other individual cells are usually mounted with a *smear mount* (*sm*).

COMPARING TEMPORARY MOUNTING FLUIDS

The refractive index of the mounting fluid you use can have a major effect on the amount and type of detail visible in a specimen. The RI of keratin, a protein that is the major component of hair, is about 1.52, close to that of the glass used in the microscope slide and coverslip, but far from the 1.33 RI of water.

If a colorless solid object is immersed in a colorless liquid with the same refractive index as the object, it disappears. You can verify this with a glass stirring rod and some water and vegetable oil. Fill a beaker or other container with water and immerse the stirring rod. The rod remains clearly visible because its RI (about 1.5+) is quite different from the 1.33 RI of water. If you replace the water with vegetable oil (RI about 1.5+) the rod becomes invisible in the liquid (or nearly so) because the refractive indices of the liquid and solid object are so similar. If they're identical, the object disappears completely. That's why it can be nearly impossible to see bacteria or other microorganisms in water unless they are stained.

1. Make a wet mount of a hair specimen, but this time use glycerol (or vegetable oil) as the mounting fluid.

2. Place the slide under the stage clips or in the mechanical stage, and center the specimen under the 4X objective. Center the specimen in the field of view and then rotate the 40X objective into place. Focus carefully on the specimen, and then adjust the illumination level and diaphragm to obtain the maximum contrast. Readjust them, if necessary to obtain maximum image detail. Note the amount of detail visible in the main body of the hair versus the boundaries.

3. Switch to the slide you made previously, and compare the amount and type of detail visible using water as the mounting fluid versus using glycerol or vegetable oil. Record your observations and conclusions in your lab notebook. Figure I-2-2 shows human facial hair mounted in glycerol.

Figure I-2-2: *Wet (glycerol) mount of human facial hair, 40X*

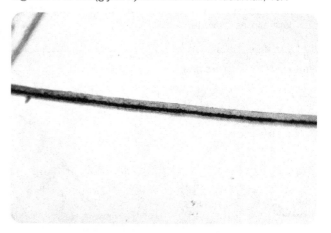

Consider the implications of the mounting fluid you use. In the first instance, you used water, by far the most common mounting fluid used for wet mounts. In the second instance, you used glycerol or vegetable oil, whose RI is close to that of Permount and other permanent mounting fluids. The amount and type of detail differed noticeably between the two fluids. That's why experienced microscopists generally keep a selection of different mounting fluids readily available.

PROCEDURE I-2-2: MAKING SMEAR MOUNTS

A *smear mount* is just what it sounds like. You place a solid or liquid specimen that contains bacteria or other cells on a slide and then smear it to produce a thin layer. If the specimen is liquid, use it as-is. If it's a solid specimen, mix a bit of it with a drop of water or another mounting fluid before doing the smear.

1. Transfer one drop of distilled water to the center of a slide.

2. Use a toothpick to gently scrape a specimen from your teeth along the gum line. Immerse the end of the toothpick in the water drop and stir to mix the solid material from the toothpick with the water.

3. Choose a smearing tool. Some people make smears with glass stirring rod held flat against the surface of the slide. Others use the edge of a second microscope slide or a coverslip held at a 45° angle to the specimen slide and spread the specimen into a thin layer about the size of your coverslip, as shown in Figure I-2-3.

4. Although you can now observe the specimen, it is in a very fragile state because it has not adhered to the slide. *Heat-fix* the slide by passing it several times, specimen side up, through a butane lighter or alcohol lamp flame, as shown in Figure I-2-4. The slide should remain in the flame for a second or two on each pass.

Figure I-2-4: *Heat-fixing a smear*

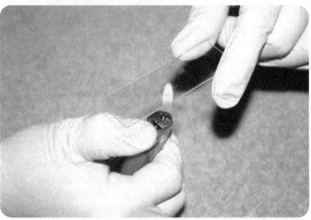

The goal of heat-fixing is to gently evaporate any remaining moisture and cause the specimen to stick to the slide. If you don't heat the slide enough, the specimen will not adhere; if you overheat the slide, the specimen will char or burst.

Figure I-2-3: *Making a smear*

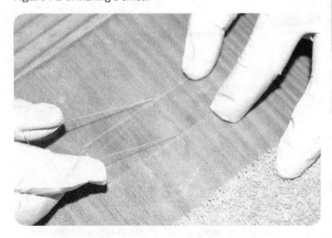

5. Position the slide on the stage, without a coverslip, and use the 4X objective to scan for an interesting area of the smear. Center that area in the field of view, and rotate the 40X objective into position. Record your observations in your lab notebook.

PROCEDURE I-2-3: MAKING HANGING DROP MOUNTS

The *hanging-drop method* is used to observe living microorganisms, typically protozoa, in their natural environment. The advantage to using the hanging-drop method is that you can observe the organisms live and undamaged going about their business, moving around, feeding, and so on. The disadvantage is that some of the little buggers are *fast*. (You can slow them down by adding a drop of methyl cellulose or glycerol to the water.) Because they can move around in three dimensions, it can be challenging to keep them in the field of view and in focus.

You can obtain a specimen to observe anywhere you find standing or slowly moving water. Water from the edge of a pond contains an amazing diversity of microlife—even in cold weather—but if there's no convenient pond you can use water from a puddle or a birdbath. In a pinch, you can even try water from a flower vase.

1. Place a coverslip on a clean, flat surface. Use a toothpick to spread a small amount of petroleum jelly along all four edges of the coverslip. Use just enough to form a seal between the coverslip and the slide.

2. Place one drop of the specimen in the center of the coverslip, as shown in Figure I-2-5.

Figure I-2-5: *Transferring a drop of specimen to a prepared coverslip*

3. Invert a well slide (well down). Center the well over the coverslip and lower the slide onto the coverslip. Use just enough pressure to make sure the coverslip seals to the slide. (Don't press too hard, or excess petroleum jelly will squoosh into the specimen area.)

4. Invert the assembly to put the coverslip on top. Note that the drop is now hanging from the coverslip, suspended in the well.

5. Place the slide on the stage, centering the edge of the drop under the 4X objective.

6. Adjust the diaphragm to its smallest setting and the illuminator to the middle of its range.

7. Focus on the edge of the drop, which should be visible as an irregular line, and reposition the slide if necessary to place the edge of the drop in the center of the field of view.

8. If you're fortunate, there may be a paramecium, amoeba, or other protozoan in the field of view. If so, center it (as best you can) in the field of view and switch to the 10X objective to observe further details. Note that it is motile (moves on its own) and may be difficult to keep focused as it zooms around in the water drop.

9. If there are no protists visible in the field of view, recenter the field on the edge of the drop and rotate the 4X objective into position. Scan the water drop to locate one or more lifeforms, center the area of interest in the field of view, and rotate the 10X objective into position to observe it more closely. Continue scanning at low power to locate additional lifeforms, and identify as many different ones as you can find.

PROCEDURE I-2-4: MAKING SECTIONAL MOUNTS

Making a *sectional mount* allows you to view the internal structure of a specimen that is too thick, large, or opaque to view in its entirety by transmitted light. The goal of sectioning is to produce a very thin slice of the specimen, so thin that it is nearly transparent.

There may be one, two, or three sectional types possible, depending on the nature of the specimen. For illustrative purposes, assume that you are making sections of yourself. Cutting through yourself horizontally (for example, at the waist) produces a *cross-section*, also called an *x-section*, a *transverse section*, or an *axial section*. Cutting through yourself vertically, from head to foot, produces a *longitudinal section*, of which there may be one or two types, depending on the type of specimen.

- Longitudinally asymmetric specimens, such as a person, have left and right halves and front and back halves. A *sagittal section* divides the specimen into left and right halves, while a *coronal section* divides the specimen into front and back halves.

- Longitudinally symmetric specimens, such as a carrot, have no left, right, front, or back, so the longitudinal sectioning angle is arbitrary and only one type of longitudinal section is possible.

Symmetric specimens, such as a potato, can be sectioned only one way, because they have no front, back, left, right, top, or bottom. By convention, such sections are called simply sections.

In this procedure, we'll make and mount cross sections and longitudinal sections of a carrot.

1. Obtain a raw carrot that is thin enough to fit the well of your microtome.

> If you don't have a microtome (purchased or homemade), you can use a scalpel or single-edge razor blade to cut carrot sections freehand, although it's difficult to get the sections thin enough—so thin that they're almost transparent. Keep trying until you have shaved off a usable section. It needn't be the full diameter of the carrot. Even a tiny piece is sufficient, as long as it's thin enough.
>
> **Use extreme care when using a scalpel or razor blade to avoid cutting yourself.**

2. Use the scalpel to cut the carrot crosswise, producing a piece large enough to fill the well of the microtome and extend above the cutting surface.

3. Place the carrot specimen in the microtome, oriented to produce a cross-sectional cut, and use the scalpel, oriented nearly parallel to the cutting surface, to trim the surface of the specimen flush with the cutting surface of the microtome.

4. Turn the microtome micrometer screw a quarter turn or so to raise the surface of the specimen slightly above the cutting surface.

5. Again orient the scalpel almost parallel to the cutting surface, and cut a thin cross-section of the carrot, as shown in Figure I-2-6. (It may take you several tries to get this right.)

Figure I-2-6: *Cutting a section on a microtome*

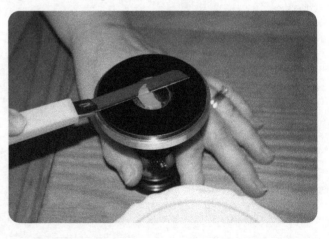

6. Transfer the section from the scalpel blade to a microscope slide. You may find it easier to flush the section onto the slide using a drop or two of water from a pipette.

7. Repeat steps 4 through 6 to produce a second carrot cross-section slide.

8. To the first slide, add one drop of water and carefully lower a coverslip into place, making sure there are no bubbles. (If there are bubbles, press gently on the coverslip to expel them.) To the second slide, add one drop of methylene blue stain, allow it to work for 30 seconds or so, and then use the corner of a tissue or paper towel to draw off the excess stain. Add one drop of water to the specimen and put a coverslip in place.

9. Observe both slides at low, medium, and high magnification. Record your observations in your lab notebook.

10. Repeat steps 2 through 9, using a piece of carrot oriented to produce a longitudinal section. Compare and contrast the stained cross-section and longitudinal-section slides and record your observations in your lab notebook.

REVIEW QUESTIONS

Q1: What are the advantages and drawbacks of wet mounts?

Q2: Why might you use a mounting fluid other than water for making a wet mount?

Q3: What are the advantages and drawbacks of a hanging-drop mount?

Q4: How would you slow down fast-moving live microorganisms to make it easier to view them?

Q5: Why might you make both cross-sectional and longitudinal-sectional mounts of a specimen?

Q6: If a forestry company cuts down a tree, which type of section are they performing? If they cut that tree trunk into long planks, what type of sectioning are they doing?

Staining

Lab I-3

EQUIPMENT AND MATERIALS

You'll need the following items to complete this lab session. (The standard kit for this book, available from *www.thehomescientist.com*, includes the items listed in the first group.)

MATERIALS FROM KIT

- Goggles
- Coverslips
- Pipettes
- Slides (flat)
- Stain, eosin Y
- Stain, Gram's iodine
- Stain, Hucker's crystal violet
- Stain, methylene blue
- Stain, safranin O
- Stirring rod (optional)

MATERIALS YOU PROVIDE

- Gloves
- Butane lighter (or other flame source)
- Ethanol, 70%
- Microscope and illuminator
- Paper towels
- Toothpicks
- Water, distilled

BACKGROUND

As we learned in the last lab session, staining is an essential tool for optical microscopy because it allows viewing structural details that would otherwise be difficult or impossible to discriminate. A *biological stain* (or *biostain*) is selectively absorbed or adsorbed to some parts of a specimen's structure, but not others. The resulting color contrast makes structural details jump out at you.

There are literally thousands of biostains and many *staining protocols*, of which dozens are used frequently and hundreds are used in specific circumstances. *Simple staining*, which we'll use in the first procedure, uses a single stain to selectively stain only some parts of a cell (or some types of cell in a mixed specimen). More complex staining protocols, one of which we'll use in the second procedure, use two or more stains to produce contrasting colors for different structural elements or types of cells.

Dr. Kessin notes: Like dyes that stain cloth, which is where many of these dyes originated.

PROCEDURE I-3-1: SIMPLE STAINING

1. Transfer one drop of distilled water to the center of a slide.

2. Use the flat end of a toothpick to scrape (gently) the inside of your cheek.

3. Immerse the end of the toothpick in the drop of water and stir to transfer the epithelial cells to the water. Position a coverslip over the specimen. If there are bubbles under the coverslip, press gently on the coverslip with your forceps to force the bubbles out from under the coverslip.

4. Position the slide in the stage clips or mechanical stage, rotate the 4X objective into position, turn on the illuminator, adjust the illumination, and focus on the slide.

5. Center an epithelial cell or a group of cells in the field of view and then rotate the 10X objective into position. Again center a cell or group of cells, and then rotate the 400X objective into position and focus critically.

6. Adjust the brightness and diaphragm setting to reveal the most image detail possible. Note the difficulty of resolving nearly transparent cell structures against the bright field, as shown in Figure I-3-1.

7. Place one drop of methylene blue stain at one edge of the coverslip. Touch the corner of a paper towel to the slide at the opposite edge of the coverslip, as shown in Figure I-3-2. The paper towel wicks the water from under the coverslip, drawing the drop of methylene blue stain under the coverslip (and around the edges).

Figure I-3-1: *Unstained squamous epithelial cells, 100X*

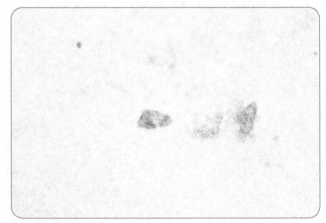

Figure I-3-2: *In-place staining of squamous epithelial cells*

8. Allow the stain to work for 30 seconds or so and then observe the cells. If the cell or group of cells you had centered has drifted out of the field of view, reposition the slide to bring them back into the field.

> If the remaining methylene blue stain is so intense that it makes it difficult to observe cell details, simply put a drop of water at the edge of the coverslip and wick off the stain. If necessary, repeat with a second drop of water to remove sufficient stain.

9. Adjust the brightness and diaphragm to reveal as much detail in the cells as possible. Record your observations and conclusions in your lab notebook, including your estimate of cell size.

> A typical squamous epithelial cell from the buccal mucosa (which is biologist-speak for a cell from your inner cheek) is irregularly shaped and about 60 μm across.

10. Repeat steps 7 through 9, substituting a drop of eosin Y stain. Note which parts of the cells are stained by the methylene blue and which by the eosin Y. Record your observations in your lab notebook.

Figure I-3-3 shows squamous epithelial cells stained with methylene blue (with stain still present under the coverslip), and Figure I-3-4 shows squamous epithelial cells stained with methylene blue and then counterstained with eosin Y (after rinsing the stains by drawing water under the coverslip). Note the different structure elements revealed by the two stains.

Figure I-3-3: *Squamous epithelial cells stained with methylene blue, 100X*

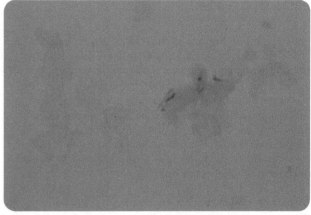

Figure I-3-4: *Squamous epithelial cells stained with methylene blue and eosin Y, 100X*

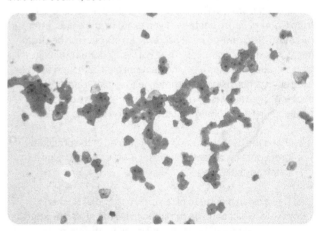

PROCEDURE I-3-2: GRAM STAINING

Gram staining is the mostly frequently used bacteriology staining protocol. When Danish bacteriologist Hans Christian Joachim Gram developed this protocol in 1884, he envisioned it only as a means of making tiny bacteria more visible. He soon realized, however, that his protocol allowed bacteriologists to discriminate among different types of bacteria that otherwise appear identical. This newfound ability had great implications for the diagnosis and treatment of bacterial diseases.

Even today, Gram staining is often the first step in identifying an unknown bacterium, particularly in less-developed countries. Modern instrumental methods provide much more information, but in addition to being cheap, the Gram staining protocol has the inestimable advantage of speed, providing at least some useful information very quickly. A skilled technician can make a smear mount, Gram stain it, and be on the phone to the physician with the results within a few minutes. If the patient has a virulent bacterial infection, these quick results may be the difference between life and death.

Gram staining depends on a difference in the cell walls of different types of bacteria that affects their retention of stains. Bacteria of both types are stained purple by crystal violet during the first step. The second step, Gram's iodine, acts as a *mordant* to fix the violet stain in bacteria of one type, called *Gram-positive bacteria*. The third step, decolorizing with ethanol, does not affect the stain in Gram-positive bacteria, but removes it from *Gram-negative bacteria*. If you examine the smear after this step, the Gram-positive bacteria are purple and the Gram-negative bacteria colorless. The final step, counterstaining with safranin O, stains all of the bacteria pink, although the pink coloration is not visible in the intensely purple Gram-positive bacteria. The result is a smear with Gram-positive bacteria stained purple and Gram-negative bacteria stained pink.

1. Place the heat-fixed slide you produced in the preceding procedure on a clean, flat surface. Use a paper towel to catch any spills.

2. Use a clean pipette to place a drop or two of Hucker's crystal violet stain on the smear. Use the tip of the pipette gently to spread the stain until it covers the entire smear, as shown in Figure I-3-5. Do not touch the smear with the tip of the pipette.

3. Allow the Hucker's crystal violet stain to remain in contact with the smear for one minute.

4. Rinse the slide, smear-side down under a faucet set to provide a trickle of water, as shown in Figure I-3-6. Don't allow the water to fall directly on the smear; instead tilt the slide gently to flood the smear with water. Rinse for at most a second or two.

5. Drain the slide and place it flat on the paper towel.

6. Use a clean pipette to place a drop or two of Gram's iodine stain on the smear. Again, use the tip of the pipette carefully to spread the stain over the entire smear.

7. Allow the Gram's iodine stain to remain in contact with the smear for one minute.

8. Fill a clean pipette with ethanol (drugstore 70% ethanol is fine). Hold the slide at an angle over the sink and gently flood the smear with the ethanol. Continue until the ethanol runs colorless.

Figure I-3-5: *Staining the bacterial smear*

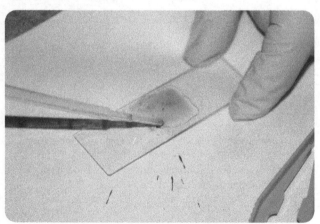

> You can substitute acetone for destaining, but acetone works much faster than ethanol. If you use acetone be careful not to decolorize the smear completely.

9. Repeat step 4 to rinse all of the ethanol from the slide. (It's important to remove all of the ethanol, because the following step won't work if ethanol is still present.) Drain the slide and place it flat on the paper towel.

10. Use a clean pipette to place a drop or two of safranin O stain on the smear. Use the tip of the pipette carefully to spread the stain over the entire smear.

11. Allow the safranin O stain to remain in contact with the smear for one minute.

12. Repeat step 9 to rinse excess safranin O stain from the slide.

Figure I-3-6: *Flooding the smear to remove excess stain*

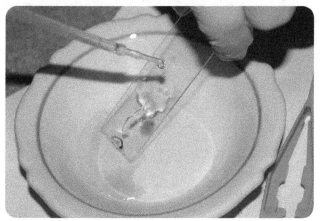

13. Allow the slide to air-dry. If you're in a hurry, you can gently pat the slide dry with a lint-free cloth or tissue. Do not rub the smear area.

14. Position the slide on the stage—you needn't use a coverslip—and use the 4X objective to locate an interesting area of the smear. Change to 400X magnification and observe the smear. If you have an oil-immersion objective, put one drop of immersion oil on that area of the smear, and carefully rotate the 100X objective into position, making sure it comes into contact only with the oil drop. Use the fine-focus knob very carefully to focus critically, and observe the bacteria.

15. Adjust the diaphragm and illuminator brightness to reveal the maximum detail in the bacteria. Gram-negative bacteria appear pink or red, as shown in Figure I-3-7, and Gram-positive bacteria appear violet, as shown in Figure I-3-8.

Figure I-3-7: *Gram-negative bacilli, 400X*

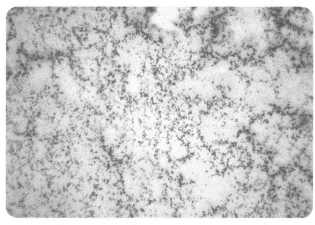

Figure I-3-8: *Gram-positive cocci, 400X*

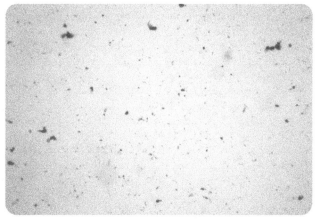

These two figures illustrate why it's important to have an oil-immersion objective if you intend to do much work in microbiology or to pursue AP Biology. Even at 400X (actually, even at 1,000X) there is very little detail visible in prokaryotic cells and other tiny microscopic items. Using the microscope at 400X, we were barely able to identify the bacteria in the first image as bacilli (rods) and those in the second image as cocci (spheres). At 1,000X, it was much easier to identify the shape of the bacteria.

REVIEW QUESTIONS

Q1: Why are stains useful?

_____

_____

_____

_____

Q2: If you could have only three microscope stains for general microscopy, which would you choose and why? (Hint: use Internet resources to choose.)

_____

_____

_____

_____

Building and Observing Microcosms

EQUIPMENT AND MATERIALS

You'll need the following items to complete this lab session. (The standard kit for this book, available from *www.thehomescientist.com*, includes the items listed in the first group.)

MATERIALS FROM KIT

- Goggles

MATERIALS YOU PROVIDE

- Gloves

- Bag, brown paper

- Camera (optional)

- Eggshell and yolk (see text)

- Funnel (or aluminum foil to make your own)

- Jars, wide-mouth (see text)

- Mixing bowl or similar container

- Newspaper

- Pond water/sediment/vegetation (see text)

- Shredder and/or scissors

- Soft drink bottles, 500 mL or 1 L

- Trowel, ladle, or other large scoop

- Water, spring or boiled tap

BACKGROUND

Microcosms are simplified artificial ecosystems that are used to simulate and observe the behavior of natural ecosystems under controlled conditions. Open or closed microcosms provide an experimental area for ecologists to study natural ecological processes. Microcosm studies can be very useful to study the effects of disturbance or to determine the ecological role of key species. A *Winogradsky column* is an example of a microbial microcosm.

Most of the other lab sessions in this book are organized into related groups, may be performed in any order you wish, and many (not all) can be completed over the course of one lab period to at most several lab periods over a week or two. Unfortunately—unlike chemistry, physics, and most other sciences—biology labs can't all be broken down into self-contained single lab sessions. Living things run on their own schedules. Life cycles are what they are, and we're often powerless to speed them up or slow them down.

Building and observing microcosms is an excellent example of one of these on-its-own-schedule activities. There is much to be learned by doing this activity, but pursuing it properly requires frequent and detailed observations over a period of weeks to months. Accordingly, rather than attempt to distribute these activities out over many lab session groups, we decided to consolidate them in this group of lab sessions, which we'll begin at the start of the semester and continue throughout the semester.

Some of the individual sessions in this group will be repeated periodically throughout the semester, beginning early and continuing. For example, we'll observe microscopically the populations of our pond water microcosms throughout the semester, noting changes that occur in the types and mix of different organisms as the microcosm life cycle progresses from juvenile to mature to senescent. Our microcosms will also serve as resources for some later topics. For example, we'll observe succession in the microcosms, which properly belongs in ecology, but requires extensive and repeated observations over the course of many weeks. Similarly, we can use our microcosms as a source of live protists when we do the lab session on protists.

In this lab session we'll build two types of microcosms based on water, sediment, and plant life gathered from a pond or stream. We'll then observe these microcosms periodically over the course of several weeks, noting changes that occur through the life cycles of the various organisms present.

The first type of microcosm resembles a standard aquarium. We'll use wide-mouth jars filled with a relatively thin layer of sediment, with pond water and aquatic plants above it.

> We actually used square nut jars from Costco, but any similar clear, colorless glass or plastic wide-mouth jars of about 1 to 2 liters capacity suffice. Other possibilities include canning jars, pickle jars, large peanut butter jars, and so on.
>
> Before using them, wash the jars and lids thoroughly with dish-washing detergent, rinse thoroughly in tap water, and allow them to dry. Do not use soap, which leaves a film that is toxic to many microorganisms.

These aquaria will provide the opportunity to observe a wide diversity of microlife, especially protists. We'll build several of these microcosms initially, some of which we'll modify later to observe the effects of environment changes on the populations present.

The second type of microcosm, called a *Winogradsky column*, is a tall column that contains a thick layer of pond sediment and pond water. We'll build several Winogradsky columns using 500 mL or 1 L soft drink bottles, with different columns containing different nutrients, and observe the changes in these columns over the course of several months. Although many types of lifeforms will be present initially, the real purpose of the Winogradsky column is to produce a mixed bacterial culture and observe the different types of bacteria that thrive in different microenvironments within the columns.

The Winogradsky column has been used for decades in biology classes to demonstrate the metabolic diversity of prokaryotes (bacteria and archaea). All organisms require a carbon source and an energy source to live and grow. Some organisms, called *autotrophs*, obtain carbon from atmospheric carbon dioxide, while others, called *heterotrophs*, obtain carbon from the organic compounds such as carbohydrates that are produced by autotrophs. Similarly, some organisms, called *phototrophs*, obtain energy directly from sunlight via photosynthesis, while others, called *chemotrophs*, obtain energy from breaking down chemical compounds.

Different organisms use all four possible combinations of these methods to obtain carbon and energy:

- *Photoautotrophs* obtain carbon from carbon dioxide and energy from sunlight.

- *Chemoautotrophs* obtain carbon from carbon dioxide, but cannot obtain energy from sunlight, and so must obtain it by breaking down chemical compounds.

- *Photoheterotrophs* obtain energy from sunlight, but cannot obtain carbon from carbon dioxide, and so must obtain it from carbohydrates or other organic compounds.

- *Chemoheterotrophs* obtain both carbon and energy by breaking down chemical compounds.

> In fact some organisms use different methods depending on the environment they happen to find themselves in. For example, the bacterium *Rhodospirillum rubrum*, which we'll examine in a later lab session, is chemotrophic under aerobic conditions but phototrophic under anaerobic conditions.

All four of these strategies are represented in a typical Winogradsky column, which also provides a visible example of how different organisms occupy different ecological niches, according to how they obtain carbon and energy.

A Winogradsky column, from top to bottom, may include the following ecological micro-niches:

Aerobic water layer

The cellulose present initially causes a microbial bloom, which quickly exhausts the oxygen in the sediment and most of the water column, leaving only the top centimeter or so of the water column aerated. Aerobic species survive and flourish only in this part of the column.

The surface of this layer is populated by sheathed bacteria. The remainder of the aerobic water layer is populated by cyanobacteria (formerly called blue-green algae), which are the only photosynthetic bacteria species. In some columns, cyanobacteria may bloom, oxygenating the entire water column and even the top portion of the sediment column.

Anaerobic water layer

Because oxygen diffuses very slowly in water, most of the water column is anaerobic (oxygen-depleted). Species such as purple nonsulfur bacteria thrive in this layer.

Anaerobic boundary layer

This boundary layer between the anaerobic sediment layer beneath it and the anaerobic water layer above it is populated by green sulfur bacteria just above the sediment and purple sulfur bacteria just above the green sulfur bacteria. These bacteria consume the hydrogen sulfide gas produced in the anaerobic sediment layer beneath, converting sulfide ions to sulfate ions. The bacteria are visible as a thin purplish layer lying just above a thin greenish layer, which in turn lies just above the sediment.

Anaerobic sediment layer

This layer, populated by anaerobic sulfate-reducing species such as *Clostridium* and *Desulfovibrio*, consumes the sulfate ions produced by the sulfur bacteria in the layer above it and releases hydrogen sulfide gas, which in turn feeds the sulfur bacteria.

Each Winogradsky column is unique, even if it started with the same combination of water, sediment, and nutrients, so we'll make several Winogradsky columns and observe and compare them as they mature.

PROCEDURE II-1-1: GATHERING MATERIALS

FIELD TRIP!

The raw materials we need for our microcosms are simple, and can all be obtained from the nearest pond or stream: water, sediment, and aquatic plant life. You can obtain these specimens at any time of year, although it is best if the source is not frozen over.

Even in winter, it's possible in most areas to find accessible pond or stream water. In very cold weather, the populations of the organisms present will usually be much lower than during the other seasons, and the mix of species present may differ. Those populations should rebound quickly once you build your microcosm and allow it to sit at room temperature for a day or two.

The amounts needed depend on the number and size of microcosms you intend to build. For example, if you intend to build six aquarium microcosms in 1.25 L peanut butter jars and four Winogradsky columns in 500 mL soft drink bottles, you will need about 10 liters of water (clean 2 L soft drink bottles make convenient collection vessels), perhaps three liters of sediment, and enough aquatic plant life to provide substantial and diverse plant populations in each of the aquarium microcosms. It's better to get too much raw material than not enough.

WARNING

Use extreme caution when you obtain and handle pond or stream water, sediment, and plant life. Most of the organisms found in typical ponds and streams are harmless to humans, but more than a few are human pathogens. Always wear gloves and goggles when obtaining or handling specimens, and always wash thoroughly with soap and water after doing so. It's also a good idea to spray or rinse the contaminated exteriors of sealed collection containers with disinfectant before taking them home.

When collecting the plants and sediment, try to avoid collecting rocks, sticks, and other large objects. (Well, actually, a snail or two can't hurt, but if you include a snail in one of your aquaria make sure also to include lots of vegetation to make sure the poor snail doesn't suffocate.)

Try to obtain as many types of plants as possible, including those rooted to the bottom, floating in the water, and lying on the bottom. If there is visible pond scum (algae) present, obtain some of it as well. If possible, keep the plants segregated by type, using small jars or zip seal plastic bags to contain them. Collect sufficient examples of each type of plant to provide populations of it in each of your aquarium microcosms. On each collection container, note that plant's environment (for example, free-floating, rooted in the sediment, and so on.)

Build your microcosms as soon as possible after you collect the specimens. Even if you allow the plant and sediment specimens to dry out, most of the types of organisms present will survive, but our goal is to reproduce as closely as possible the original ecosystem present in the pond or stream, so the less delay the better.

Retain some mixed pond or stream sediment and vegetation for later use. Spread the material in a thin layer on newspaper and allow it to dry thoroughly in the shade. (Avoid direct sunlight.) Store the material in a labeled paper bag.

PROCEDURE II-1-2: BUILDING AQUARIUM MICROCOSMS

Insofar as is possible, our goal is to make each of our aquarium microcosms identical, using the same amount of sediment, the same amount of water, and the same amount and mix of aquatic plant life.

1. Put on your gloves and goggles.

2. Arrange all of your wide-mouth jars on your work surface, side by side.

3. Transfer sufficient pond or stream sediment to each jar to fill it about one-quarter full. Try to keep the levels in all jars the same.

4. If your plant specimens include examples that you found rooted to or lying on the bottom, place those plant specimens accordingly in each of your containers, trying to keep the quantities and mix the same in each of your containers.

5. Transfer sufficient pond or stream water to each jar to fill it about 5 cm from the top. Again, try to keep the water levels as similar as possible.

6. Transfer examples of any free-floating plants, algae, and other nonanchored flora to the containers, again trying to keep the populations and mixes the same between containers.

7. Replace the lids, seal them tightly, and then spray or drench the containers thoroughly with disinfectant to kill any organisms on the exteriors of the containers.

8. Transfer all of the containers to an area where they can remain undisturbed and will be exposed all day to bright daylight, but not to direct sunlight. If necessary, you can put the containers under a plant-grow light to ensure they receive adequate light for long periods every day. It's important that the containers be maintained at about room temperature, so avoid areas where they might be exposed to large temperature variations.

9. Remove and discard your gloves and wash your hands thoroughly with soap and water.

PROCEDURE II-1-3: BUILDING WINOGRADSKY COLUMN MICROCOSMS

Just as with the aquarium microcosms, our goal is to make each of our Winogradsky column microcosms as similar as possible, other than differences in added nutrients. With the exception of one control Winogradsky column, which contains only unenriched sediment, each of the other Winogradsky columns begins with sufficient enriched sediment to fill the column roughly half full. To that, we'll add add about half as much unenriched sediment to fill the column to a total of about three quarters. Finally, we'll fill the column with pond or stream water, leaving a small air gap.

We assume that you're making four Winogradsky column microcosms, each in a 1 L soda bottle, which will require a total of about 3 L of sediment and sufficient pond or stream water to nearly fill the bottle. If you're making a different number or size of columns, adjust the quantities accordingly.

1. Put on your gloves and goggles.

2. All but the first Winogradsky column will contain sediment supplemented with shredded newspaper (cellulose) to provide a carbon source. One double-width sheet of newsprint will provide sufficient cellulose for those three columns. Shred or cut the newspaper into small pieces.

3. Transfer enough sediment to half fill three bottles to a mixing bowl or similar container. Stir in the paper bits to mix them throughout the sediment. Retain the remaining sediment for use later.

4. Label one bottle "paper only." Transfer sufficient cellulose-enriched sediment to that bottle to fill it about half full and set the bottle aside.

5. Add a raw egg yolk to the sediment remaining in the mixing bowl. Mix the egg yolk thoroughly into the sediment. Egg yolk provides a sulfur source.

6. Label a second bottle "paper + yolk." Transfer enough of the cellulose and egg yolk enriched sediment to fill that bottle to the same level as the first bottle and set the bottle aside.

7. Crush the egg shell into small pieces and mix them into the sediment remaining in the mixing bowl. Eggshell provides a calcium source.

8. Label a third bottle "paper + yolk + shell." Transfer enough of the cellulose, egg yolk, and egg shell enriched sediment to fill that bottle to the same level as the first and second bottles and set the bottle aside. Discard any remaining enriched sediment.

9. Carefully transfer enough unenriched sediment to the first bottle to fill about half the remaining space. Do so gently to avoid mixing the sediment layers. Your goal is to have distinct layers of sediment, with the enriched sediment on the bottom, covered by a layer of unenriched sediment. Repeat for the remaining two bottles, bringing the sediment to the same level in all three.

10. Label a fourth bottle "unenriched." Transfer sufficient unenriched sediment to that bottle to bring the level to the same as the other three bottles.

11. Disturbing the sediment layers as little as possible, carefully transfer sufficient pond or stream water to each bottle to fill it, leaving a small air gap at the top of the bottle.

12. Replace the lids on all four bottles, seal them tightly, and then spray or drench the containers thoroughly with disinfectant to kill any organisms on the exteriors of the containers.

13. Transfer all of the bottles to an area where they can remain undisturbed and will be exposed all day to bright daylight, but not to direct sunlight. If necessary, you can put the bottles under a plant-grow light to ensure they receive adequate light for long periods every day. It's important that the bottles be maintained at about room temperature, so avoid areas where they might be exposed to large temperature variations.

14. Remove and discard your gloves and wash your hands thoroughly with soap and water.

PROCEDURE II-1-4: OBSERVING WINOGRADSKY COLUMN MICROCOSMS

Observing a Winogradsky column microcosm is a long-term project. Some visible changes occur soon after you build the microcosm, but the major changes take place gradually over weeks and months.

1. For the first week after you build your Winogradsky column microcosms, observe each of them daily. Note any change in the appearance of the layers. If you have a camera, shoot images of each of the microcosms and file those images by date and microcosm makeup.

2. From day 8 until day 30, observe the microcosms every two to three days. After day 30, begin observing the microcosms weekly. Again, note any visible changes and shoot images to record them.

WARNING

The exact contents of these microcosms is unknown, so it's possible they contain pathogenic organisms. When you have finished observing your Winogradsky column microcosms, dispose of them by incineration or by submerging them in a bucket of chlorine bleach solution before removing the cap and mixing the contents of the microcosm with the bleach solution.

Advanced students can use these microcosms as sources of different bacteria species to culture—for example, the *Rhodospirillum* rubrum we use in a later lab session—but it's much less risky simply to buy pure cultures of the species you need rather than attempting to culture them from wild sources. **Do not open any of these microcosms unless you are confident that you are competent to work with possibly pathogenic organisms.**

REVIEW QUESTIONS

Q1: We emphasize safety in collecting and handling pond or stream specimens because of the possible presence of pathogens. Using the Internet or other resources, determine two common pathogens in each of the three classes: viruses, bacteria, and protozoa.

_____

_____

_____

_____

Q2: We built Winogradsky columns with sediment enriched by carbon, carbon+sulfur, and carbon+sulfur+calcium. Using the Internet or other resources, suggest several other elements we might have used to enrich the sediment.

_____

_____

_____

_____

Observing Succession in Aquarium Microcosms

Lab II-2

EQUIPMENT AND MATERIALS

You'll need the following items to complete this lab session. (The standard kit for this book, available from *www.thehomescientist.com*, includes the items listed in the first group.)

MATERIALS FROM KIT

- Goggles

- Coverslips

- Methylcellulose

- pH test paper

- Pipettes

- Slides, flat

- Stain: eosin Y

- Stain: methylene blue

- Thermometer

MATERIALS YOU PROVIDE

- Gloves

- Aquarium microcosms (from preceding lab)

- Particulate masks, N100 (see text)

- Pond life reference material(s) (see text)

- Watch or clock with second hand

BACKGROUND

Succession, a fundamental concept in ecology, is the process by which a community progressively transforms itself from an arbitrary starting point to the steady-state equilibrium of a stable community. The point at which that equilibrium is reached depends on environmental parameters and the types of organisms present at the starting point, although not necessarily their initial numbers.

Our aquarium microcosms represent arbitrary communities. We hope that those communities are very similar, but we cannot know that they are identical. For example, one of the microcosms might contain a few examples of a rare organism, while another contains none of that organism. If that organism is, say, a predator upon another species that is present in both microcosms, the succession we observe in the two microcosms will be very different. In the first microcosm, stable populations of the predator and prey organisms will likely develop, while in the second microcosm the absence of the predator may result in uncontrolled growth of the prey organism. Or a third organism may step in to assume the ecological role of predator.

In short, microcosms like ours are somewhat predictable in terms of succession, but even very small changes in the community makeup may cause dramatic changes in succession. In the extreme case, one microcosm may flourish over a period of weeks to months (if not indefinitely), while another, apparently identical, microcosm may quickly approach senescence and die off. That's one of the reasons we suggest building several of these microcosms. The other reason is that in the next lab session we'll intentionally alter the environmental characteristics of some microcosms and compare their succession to those of a control microcosm.

In terms of subject matter, these microcosm lab sessions actually belong with the ecology group. However, because we'll be observing the microcosms regularly over a period of several weeks to several months, we needed to get the microcosms started early, so we elected to place these lab sessions in their own group early in the book.

In this lab session we'll observe and catalog some of the microscopic life present in our aquarium microcosms.

PROCEDURE II-2-1: OBSERVE SUCCESSION IN MICROCOSMS

With regard to microcosms, succession is just another word for changes, and much of the educational value in microcosms comes from observing and documenting those changes on a regular (and frequent) schedule. Changes can be observed on many levels, from simple observations of gross changes such as the color and turbidity of the water to detailed observations of the types, numbers, and other characteristics of the various microorganisms present in the microcosm and how they change over time.

Obviously, detailed observations require more time than casual observations. Doing detailed observations and counts requires a significant commitment of time and effort, particularly if you have several microcosms. If time is limited, simply do the best you can, keeping in mind that changes are likely to occur more rapidly at first and more slowly later in the life cycle of the microcosms.

Ideally, you should perform this procedure initially on the day you create the microcosms, repeat this procedure for each of your aquarium microcosms every day for the first week, then every two to three days once the microcosms begin to stabilize, and then once a week after full equilibrium has been reached (which may occur at different times for different microcosms).

1. Put on your gloves and goggles.

2. Observe the color and degree of turbidity of the microcosm and record your observations in your lab notebook.

3. Carefully remove the lid from the microcosm, disturbing the contents as little as possible. (It's not uncommon for microcosms to stink to high heaven—particularly after they have incubated for several days or longer—so make sure you have adequate ventilation.)

WARNING

Remember that you don't know exactly what's growing in those microcosms. It's probably not the Andromeda Strain, but there might be a real nasty growing in there, maybe even several nasties.

Use full aseptic precautions each time you open a microcosm. Wear gloves, goggles, and (if you really want to be fully protected) an N100 particle mask. Wash the gloves in soap and water before removing and discarding them, and then wash your hands thoroughly in soap and water. When you remove specimens, sterilize them before discarding them. Always replace the lid of the microcosm immediately when you're not actually obtaining samples from it.

4. Measure the temperature of the water and record it in your lab notebook. (Do not stir the microcosm; simply dip the thermometer into the water and withdraw it once it registers the temperature.)

5. Use the tip of the thermometer to transfer one drop of the microcosm water to a piece of pH test paper. Record the pH value in your lab notebook.

6. Use a clean pipette to withdraw a drop of water from the surface of the microcosm. Transfer it to a flat slide, add a drop of methylcellulose to slow down the fast movers, and put a coverslip in place.

7. Observe the slide at low magnification (40X), and note as many different organisms as possible. Even at low magnification, depending on the state of the microcosm, it's not unusual to observe a dozen or more discrete species, nor is it unusual to see only a few. Scan the full area under the coverslip to make sure you don't miss any species.

8. Using a pond-life reference manual or Internet resources, attempt to identify each species present, or at least the genus.

9. Scan the entire populated area of the slide and record your impressions of the relative numbers of each species present as "very abundant," "abundant," "moderate," "rare," or "very rare."

10. Center a populated area of the slide under the objective and perform an actual count of each species visible in the field of view, beginning with those species that move slowly or not at all. For motile species, do a 15-second count of each species. If an individual swims into the field of view during the 15-second period, increment your count by one; if an individual swims out of the field of view, decrement your count by one. After you complete the count, center another random populated area under the objective and

repeat the count. Do that again, for a total of three counts, and average the number of individuals in each species for the total sample.

11. Repeat your observations at medium (100X) and high-dry (400X) magnification, noting and recording any species that become visible at higher magnification.

> At 400X in particular, a wide variety of bacterial species may be visible. You needn't attempt to determine their species nor even genera, but do note their form (coccus, rod, spiral), anything otherwise special about their appearance, and their relative abundance.

12. To document the original status of the microcosm, label and date a clean microscope slide, transfer another drop of water from the surface of the microcosm to the slide, and spread the drop across the central area of the slide. Flame the slide to kill the microorganisms present and affix them to the slide. Stain the slide with methylene blue and then with eosin Y. Dry the slide and observe it to ensure that all species you noted are present on the slide.

> You needn't use a coverslip for this reference slide unless you want to make a permanent mount.

13. Using clean pipettes, repeat the preceding steps using specimens from near the vegetation and from near the bottom sediment. You will probably find that the relative abundance of species changes significantly. Some species may be present in abundance or entirely absent in different areas of the microcosm.

If you have time, repeat this procedure for each of your other aquarium microcosms to establish baseline population estimates for each of them. Don't make the mistake of focusing on plentiful species and ignoring organisms that are not present in large numbers. Relative population counts will probably change over time, and those changes may (or may not) be dramatic.

Schedule follow-up observations now. If possible, observe them daily for the first week, ideally at the same time of day each time. Follow-up observations needn't be as detailed or time-consuming as this first observation. Just do quick population estimates for as many species as possible, looking for obvious changes and trends. For example, a species that was initially present in small numbers may increase its numbers significantly over the first few days, while the population of another that was plentiful initially may decrease noticeably.

If time is limited, focus your attention on only one of your microcosms, which you can designate as your control microcosm. You should be able to do a follow-up observation on the surface, vegetation, and sediment of your control microcosm in perhaps 15 minutes. If you have more time and particularly if the initial state of one or more of your other microcosms differed noticeably from your control microcosm, it's useful to do follow-up observations on it or them as well.

As time passes, you can change the frequency of your follow-up observations depending on what changes you see occurring, and how quickly. During the first few days, changes are likely to occur quickly. After a week, you can probably reduce the frequency of follow-up observations to once every two or three days without missing much. As the microcosms begin to settle down and near equilibrium, you'll probably find that few changes occur from one week to the next.

Do not be surprised if one or more of your microcosms dies off. That's why we made spares.

REVIEW QUESTIONS

Q1: What gross changes did you observe in your microcosms as they aged?

_____

_____

_____

_____

_____

Q2: Did you notice any change in the relative sizes of particular types of organisms as the microcosm aged?

Q3: As your microcosm progresses toward equilibrium, what changes did you observe in terms of the types of organisms present and their relative populations?

Q4: In terms of the species mix, did you notice any significant changes as the microcosm matured?

Q5: Why is it important to observe aseptic procedures when working with water from a microcosm?

Q6: What safety procedures should you use when working with microcosms?

Observing the Effects of Pollution in Microcosms

Lab II-3

EQUIPMENT AND MATERIALS

You'll need the following items to complete this lab session. (The standard kit for this book, available from *www.thehomescientist.com*, includes the items listed in the first group.)

MATERIALS FROM KIT

- Goggles
- Coverslips
- Centrifuge tubes, 50 mL
- Hydrochloric acid
- Fertilizer concentrate A
- Lead(II) acetate
- Methylcellulose
- pH test paper
- Pipettes
- Slides, flat
- Yeast, live (baker's or brewer's)

MATERIALS YOU PROVIDE

- Gloves
- Foam cup
- Pond sediment/vegetation (from Lab II-1)
- Tablespoon
- Water, spring or boiled tap

BACKGROUND

Any externally applied change to the environment of a community forces changes to occur within the community until a new equilibrium is reached. For example, after a major flood or a forest fire, dramatic changes occur to the community and the populations within that community. Eventually, a new equilibrium is reached, in which the community and populations may or may not be similar to the original equilibrium. Species formerly present may be absent in the new equilibrium, for example, and species not originally present in the original community may be present in large numbers in the new community.

One of the most common causes of such equilibrium changes is pollution, whether of natural or human origin. Floods, hurricanes, forest fires, volcanic activity, and other natural phenomena often cause dramatic changes to the environment, as does pollution of human origin, such as acid rain, drainage from mine tailings, and so on. Nor are all equilibrium changes that occur from natural causes abiotic in origin. For example, algae blooms may cause gigantic die-offs of fish and other aquatic creatures, in some cases severe enough to result in long-term changes to the environment.

In this lab session we'll make up six small microcosms in 50 mL centrifuge tubes. (We call them *nanocosms*, although that's not an official term.) We'll make these microcosms as similar as possible initially, retain one as a control, and then subject the others to various environmental pollutants, as follows:

- Tube A is the control microcosm.

- Tube B treated with fertilizer to simulate phosphate pollution.

- Tube C treated with lead(II) acetate to simulate heavy-metal pollution.

- Tube D treated with hydrochloric acid to simulate acid rain pollution.

- Tube E treated by adding live yeast (foreign microorganisms can shift equilibrium in a microcosm).

- Tube F exposed to direct sunlight (yes, excessive light can be considered a pollutant).

We'll observe succession over a period of several days in each of these microcosms and attempt to draw conclusions about the effects of various pollutants upon the various organisms present.

PROCEDURE II-3-1: BUILD POLLUTED MICROCOSMS

1. Put on your gloves and goggles.

2. Label six 50 mL centrifuge tubes A through F.

3. Obtain about 500 mL of spring water. You can substitute tap water if you boil it for a few minutes and then allow it to cool. (Water treatment chemicals kill the same microorganisms we're trying to grow in our microcosms.)

4. Transfer eight rounded tablespoons of the dried sediment and vegetation you obtained in lab session II-1 to the foam cup. Add sufficient spring water to make a runny mud in the cup.

5. Transfer sufficient mud to each of the centrifuge tubes to bring the sediment level in the bottom of the tube to the 15 mL line. Try to make sure that the sediment in each tube is of similar makeup and that the amount of vegetation is about the same.

6. To tubes A, E, and F, add sufficient spring water to bring the water level to the 40 mL line. Cap the tubes, invert them several times, and then place them aside.

7. To tube B, add sufficient spring water to bring the water level to the 35 mL line. Add 1 mL of the fertilizer concentrate A, which contains phosphates, and then

sufficient spring water to bring the water level to the 40 mL line. Cap the tube, invert it several times, and then place it aside.

8. To tube C, add sufficient spring water to bring the water level to the 35 mL line. Add 1 mL of 0.1 M lead(II) acetate solution, and then sufficient spring water to bring the water level to the 40 mL line. Cap the tube, invert it several times, and then place it aside.

9. To tube D, add sufficient spring water to bring the water level to the 40 mL line. Cap the tube, invert it several times, and allow the sediment to settle. Add one drop of 6 M hydrochloric acid, mix the solution and use the pH test paper to test the pH. Continue adding 6 M hydrochloric acid dropwise, mixing the solution, and retesting the pH until the pH of the solution reaches about 5.0. (The amount of acid you'll need to add varies according to the makeup of your sediment specimen.)

10. To tube E, add a breadcrumb size piece of live yeast. Cap the tube, invert it several times, and then place it aside.

11. Incubate tubes A through E at room temperature in a location where they'll be exposed to daylight but not direct sunlight. (For better consistency you can use a plant-grow light if you have one available.) Incubate tube F in direct sunlight outdoors or on a windowsill. If outdoor temperatures are extremely hot or cold, do the incubation indoors. A tube exposed to direct sunlight will obviously be warmed by it, but our goal is to avoid temperature extremes as much as possible.

The pH of tube D may change over time as the acid reacts with components in the sediment. Our goal is to reach a reasonably stable acid pH in tube D. On the day you build this microcosm, retest the pH every hour or two and then again the morning of the second day. If necessary, adjust the pH by adding more acid dropwise until the pH stabilizes.

PROCEDURE II-3-2:
OBSERVE SUCCESSION IN POLLUTED MICROCOSMS

WARNING

Once again, remember that you don't know exactly what's growing in those microcosms. Use full aseptic precautions each time you open a microcosm. Wear gloves, goggles, and (if you really want to be fully protected) an N100 particle mask. Wash the gloves in soap and water before removing and discarding them, and then wash your hands thoroughly in soap and water. Always replace the lid of the microcosm immediately when you're not actually obtaining samples from it.

Ideally, you should begin this procedure the day after you build the 50 mL microcosms.

1. Put on your protective gear.

2. Open tube A, measure the temperature of the water, and record it in your lab notebook. (Do not stir the microcosm; simply dip the thermometer into the water and withdraw it once it registers the temperature.)

3. Use the tip of the thermometer to transfer one drop of the microcosm water to a piece of pH test paper. Record the pH value in your lab notebook.

4. Use a clean pipette to withdraw a drop of water from the surface of the microcosm. Transfer it to a flat slide, add a drop of methylcellulose to slow down the fast movers, and put a coverslip in place.

5. Observe the slide at low magnification (40X), and note as many different organisms as possible. Scan the full area under the coverslip to make sure you don't miss any species. If necessary, use medium magnification (100X) to verify the identity of small organisms. Using your notes and sketches from the preceding lab session, identify each species present and record it in your lab notebook.

6. Scan the entire populated area of the slide and record your impressions of the relative numbers of each species present as "very abundant," "abundant," "moderate," "rare," or "very rare."

> If you have time, you can also use high-dry magnification (400X) to identify and do population estimates on bacteria and other tiny organisms.

7. Using clean pipettes, repeat the preceding steps using specimens from near the vegetation and from near the bottom sediment.

8. Repeat steps 2 through 7 for each of the other microcosms.

9. Repeat steps 1 through 8 each day at about the same time for several days or until the microcosms reach equilibrium.

> Do not be surprised if one or more of your microcosms dies off within a day or two. We introduced significant levels of pollution into tubes B through E, and it's very likely that those pollutants will suffice to eradicate some or even most of the species present either directly or indirectly (for example, by killing off the food source for a species that was otherwise unaffected or only minimally affected by the pollutant in question). Conversely, one or more of the pollutants may provide a selective advantage for one or more of the species present by killing or inhibiting competitors, allowing some species to flourish at the expense of other species.

REVIEW QUESTIONS

Q1: In detail, what effects did you observe from the differing pollutants in the various microcosms?

Acids, Bases, and Buffers

Lab III-1

EQUIPMENT AND MATERIALS

You'll need the following items to complete this lab session. (The standard kit for this book, available from *www.thehomescientist.com*, includes the items listed in the first group.)

MATERIALS FROM KIT

- Goggles
- Acetic acid solution
- Ammonia solution
- Beaker, 50 mL
- Centrifuge tubes, 15 mL (5)
- Hydrochloric acid solution

- pH test paper
- Pipettes
- Reaction plate, 24-well
- Reaction plate, 96-well
- Sodium hydroxide solution
- Stirring rod

MATERIALS YOU PROVIDE

- Gloves
- Marking pen
- Paper towels

- Scissors
- Specimens: household materials (see III-1-3)
- Water, distilled

BACKGROUND

Acids, bases, and buffers are fundamental components of the chemistry of life. All lifeforms use these chemicals in numerous essential life processes, from respiration to digesting food to maintaining the cellular environments necessary for life, to give just a few examples.

Living organisms use many different acids, bases, and buffers, from simple compounds to very complex organic molecules. For example, the gastric juice in your stomach is largely made up of one of the simplest of all acids, hydrochloric acid, while the DNA present in your cells is actually deoxyribonucleic acid, the most complex of all acids.

In this lab session, we'll learn how biologists quantify these chemicals and examine some of the important characteristics of acids, bases, and buffers.

PROCEDURE III-1-1: PERCENTAGE AND MOLAR CONCENTRATIONS

A *solution* is a mixture of two or more components, each of which may be a solid, liquid, or gas. The *solute* is the material being dissolved, and the *solvent* is the material that dissolves the solute. For example, if you add a teaspoon of sugar to a mug of tea, the sugar is the solute and the tea is the solvent.

> When solids are dissolved in liquids, the solids are always considered the solutes, regardless of the amount being dissolved. Otherwise, by convention, the material or materials present in the smaller amounts are considered to be the solute or solutes, while the material or materials present in the larger amounts are considered to be the solvent or solvents.

The *concentration* of a solution specifies the amount of solute per unit of solvent. There are many ways to specify concentration, but the five most widely used by biologists are *weight-to-volume percentage*, *weight-to-weight percentage*, *volume-to-volume percentage*, *molarity*, and *molality*.

Weight-to-volume percentage

Weight-to-volume percentage, abbreviated *w/v*, is the number of grams of solute per 100 mL of solution. For example, a 1% w/v solution of methylene blue contains 1 gram of methylene blue dissolved in sufficient water to make up 100 mL of solution. W/v percentages are most often used for stains and other bench reagents that comprise a solid dissolved in water or alcohol. In recent texts this method is usually called *mass-to-volume percentage*, abbreviated *m/v*, although the older (and incorrect) weight-to-volume remains very widely used.

Weight-to-weight percentage

Weight-to-weight percentage, abbreviated *w/w* or *m/m*, is the number of grams of solute per per 100 grams (not mL) of solution. For example, a 36% w/w solution of hydrochloric acid contains contains 36 grams of hydrochloric acid gas per 100 grams of solution. Note that the w/w percentage may differ significantly from the w/v percentage because the density of the solution may be greater or less than 1.00 grams/mL. W/w percentages are most often used for gases dissolved in water, such as hydrochloric acid, ammonia, and formaldehyde. This makes it easy to obtain a specific amount of the chemical by simple weighing. For example, if you need 3.6 grams of hydrochloric acid, you can get it simply by weighing out 10.0 grams of the 36% solution.

Volume-to-volume percentage

Volume-to-volume percentage, abbreviated *v/v*, is ordinarily used for solutions that contain water and another miscible liquid such as ethanol, isopropanol, or acetone, but is sometimes used to specify the concentration of an aqueous solution of a gas. V/v percentage is simply the number of mL of solute per 100 mL of solution. For example, a 70% v/v solution of isopropanol contains 70 mL of pure isopropanol mixed with water to make 100 mL of total solution. We phrased that very carefully, because volumes are not necessarily additive and are often greater or less than you might expect. For example, mixing 70 mL of pure isopropanol with 30 mL of pure water does not yield 100 mL of solution.

The preceding three methods of specifying concentration are common, but all of them share a disadvantage. They are based on the mass or volume of solute present in the solution, but do not take into account the fact that different solute molecules have different molecular masses. Molecules react with each other in fixed ratios based on the *number* of molecules rather than their masses or volumes. For example, two molecules of sodium hydroxide, NaOH, react with one molecule of sulfuric acid, H_2SO_4, to form one molecule of sodium sulfate, Na_2SO_4 and two molecules of water, H_2O, by the following balanced equation.

$$2\ NaOH + H_2SO_4 \rightarrow Na_2SO_4 + 2\ H_2O$$

One *mole* of any chemical compound contains the same number of molecules, so it's also correct to say that two *moles* of sodium hydroxide react with one mole of sulfuric acid to produce one mole of sodium sulfate and two moles of water. Because the molecular mass of different compounds varies, the mass of one mole of different compounds also varies.

SIX OF ONE, 60,000,000,000,000,000,000,000 DOZEN OF THE OTHER

Mole is a group noun, like dozen or gross but larger. How much larger? If you have a dozen eggs, you have 12 eggs; if you have a gross of microscope slides, you have 144 slides; if you have a mole of molecules, you have 6.0×10^{23} (6 followed by 23 zeros) or 60 sextillion dozen molecules.

The mass of one mole of sodium hydroxide is 40.00 grams. (You can also say that the *molar mass* of sodium hydroxide is 40.00 g/mol.) The mass of one mole of sulfuric acid is 98.07 grams. So, for any procedure that involves reacting chemicals with each other, it's much more convenient to specify concentrations in terms of moles rather than with mass or volume percentages.

For example, if you want to react a 10% w/v solution of sodium hydroxide with a 10% w/v solution of sulfuric acid to produce pure sodium sulfate without an excess of either reactant, you have some calculating to do to get the mole ratios to come out right. On the other hand, if the concentrations of your solutions are specified in moles, the calculations are trivially simple. That brings up the final two methods of specifying concentration that are widely used by biologists.

Molarity

Molarity, abbreviated *mol/L* or *M*, is the number of moles of solute per per liter of solution. For example, a solution that contains 6 moles of solute per liter of solution is 6 molar, usually abbreviated 6 M.

Chemists usually use fractional molarities to refer to solutions less concentrated than 1 M. For example, a chemist would probably refer to a 0.1 M solution as "tenth-molar" and a 0.01 M solution as "hundredth-molar." Biologists often work with much more dilute solutions than chemists—down into the 0.00001 M range or less—so biologists often specify concentrations in millimoles/L (0.001 mol/L, or *millimolar*, abbreviated *mM*) or micromoles/L (0.000001 mol/L, or *micromolar*, abbreviated *µM*). A 1 M solution is 1,000 mM and 1,000,000 µM.

SI VERSUS THE REAL WORLD

The term molarity is officially deprecated; the official SI unit is *moles of substance per cubic meter*, abbreviated mol/m^3. That value is too large to be practical—a 1 molar solution contains 1,000 moles of solute per cubic meter—but young, pedantic scientists sometimes use the unofficial SI-like unit *moles of substance per cubic decimeter*, abbreviated mol/dm^3, which has the same value as molarity. That is, a 1 molar solution contains 1 mole of solute per cubic decimeter of solution.

In practice, particularly in the United States, most scientists continue to use the old-style terminology. The word molar is simply too useful to give up in favor of the long, awkward official SI terminology.

Molality

Molality, abbreviated *mol/kg* or *m*, is the number of moles of solute per kilogram of solvent. For example, a solution that contains 6 moles of solute per kilogram of solvent is 6 molal, usually abbreviated 6 m. Molality is useful when working with procedures that involve the colligative properties of solutions, such as diffusion and osmosis.

To continue our example from above, a six molar (6 M) solution of sodium hydroxide contains 240.0 grams (6 moles) of sodium hydroxide per liter of solution, and a 1 M solution of sulfuric acid contains 98.07 grams (1 mole) of sulfuric acid per liter of solution. Since we know that two moles of sodium hydroxide react with one mole of sulfuric acid, if you are using these molar solutions you can simply mix one volume of the 6 M sodium hydroxide solution with three volumes of the sulfuric acid solution to produce a pure solution of sodium sulfate, without either reactant in excess.

Similarly, if you need 1 M sodium hydroxide, you can simply dilute one volume of the 6 M sodium hydroxide solution with five volumes of distilled water. As it happens, we need to do exactly that to produce solutions that we'll use later in this lab session.

MAKING UP 1 M SOLUTIONS

For the following procedure, we'll need 1 molar solutions of acetic acid, ammonia, hydrochloric acid, and sodium hydroxide. The following instructions assume you are using the 6 M solutions supplied with the kit. If you are using chemicals from another source, dilute them appropriately. For example, if you are using concentrated (12 M) hydrochloric acid, add one part

of the acid to 11 parts distilled water to make up a 1 M solution. If you are using solid sodium hydroxide, dissolve 4.00 grams of the solid in about 80 mL of water and make up the solution to 100 mL. (**Warning: adding concentrated acids or bases to water produces considerable heat and may cause the solution to boil and spatter. Take care to avoid being burned or splashed.**)

1. If you have not already done so, put on your goggles, gloves, and protective clothing. **Warning: The concentrated solutions you'll be using are extremely corrosive.**

2. Using a clean pipette, transfer 1.25 mL of distilled water to each of four centrifuge tubes or other small containers labeled "1 M acetic acid," "1 M ammonia," "1 M hydrochloric acid," and "1 M NaOH."

3. Using clean pipettes each time, transfer 0.25 mL of the corresponding 6 M solutions to each of the labeled containers. Swirl or stir the containers to mix the solutions, and replace the caps.

Retain all of these solutions for use later in this lab session.

PROCEDURE III-1-2: EFFECT OF CONCENTRATION ON PH

Acids and bases are key components of the chemistry of life. For our purposes, we can consider an *acid* to be any chemical that, when dissolved in water, increases the concentration of *hydrogen ions* (H$^+$) of the solution. (Actually, a free hydrogen ion immediately reacts with a water molecule, H$_2$O, to form a *hydronium ion*, H$_3$O$^+$, but the effect is the same.) A *base* is any chemical that, when dissolved in water, increases the concentration of the *hydroxide ion*, OH$^-$.

In pure water, only a tiny percentage of water molecules dissociate to form hydrogen ions and hydroxide ions. In any aqueous solution, including pure water, the product of the concentrations of the hydrogen and hydroxide ions is 10^{-14}.

$$[H^+] \cdot [OH^-] = 10^{-14}$$

In pure water, the concentrations of the hydrogen and hydroxide ions are equal, at 10^{-7} molar, or one ten-millionth molar.

$$[10^{-7}] \cdot [10^{-7}] = 10^{-14}$$

If you add an acid to the water to increase the hydrogen ion concentration to, say, 10^{-4}, a tiny percentage of those additional hydrogen ions react with hydroxide ions to form water molecules, thereby reducing the hydroxide ion concentration.

$$[H^+] \cdot [OH^-] = 10^{-14}$$

$$[10^{-4}] \cdot [10^{-10}] = 10^{-14}$$

In other words, adding the acid to the water increased the concentration of hydrogen ions from 0.0000001 molar to 0.0001 molar, while at the same time reducing the concentration of hydroxide ions from 0.0000001 molar to 0.0000000001 molar.

Obviously, such tiny numbers are inconvenient to deal with, so chemists use the *pH* scale to quantify the *acidity* or *basicity* of a solution. The pH of a solution is the negative base-10 log of its hydrogen ion concentration. So, for example, if the hydrogen ion concentration of a solution is 10^{-4}, that solution's pH is:

$$-\log_{10}(10^{-4}) = 4$$

A *neutral solution*, one that is neither acidic nor basic, has a pH of 7. Pure water is neutral. An *acid solution* has a pH lower than 7, and a *basic solution* has a pH higher than 7. Each pH unit corresponds to a ten-fold increase or decrease in acidity or basicity. For example, a solution with a pH of 4 is ten times more acidic than a solution with a pH of 5, and a solution with a

pH of 12 is ten times more basic than a solution with a pH of 11. Although the pH scale is often thought of as extending from 0 to 14, it's actually open-ended. A concentrated solution of a strong acid can have a pH lower than 0, and a concentrated solution of a strong base can have a pH higher than 14.

In this procedure, we'll prepare various dilutions of two acids, acetic and hydrochloric, and two bases, ammonia and sodium hydroxide. We'll then use pH test paper to determine the pH of each of these solutions.

POPULATING THE REACTION PLATE

The first step is to populate the reaction plate with various dilutions of the four solutions whose pH we'll be testing. We'll use a procedure called *serial dilution* to produce ten-fold dilutions from each well to the next. For example, well A1 will contain 1.0 M acetic acid, well B1 0.1 M acetic acid, well C1 0.01 M acetic acid, and so on.

1. Fill the clean 50 mL beaker nearly full with distilled water.

2. Using a clean pipette, transfer nine drops of distilled water to each of wells B1 through H1, B3 through H3, B5 through H5, and B7 through H7 of the reaction plate.

3. Use a clean pipette to transfer a few drops of 1 M acetic acid to well A1 and one drop of 1 M acetic acid to well B1. Return any acid still in the pipette to the labeled container.

4. Use the pipette to draw up and expel the contents of well B1 several times to ensure the solution is thoroughly mixed. Well B1 now contains one drop of 1 M acetic acid mixed with nine drops of distilled water, which produces 0.1 M acetic acid.

5. Transfer one drop of the solution from well B1 to well C1. Return any solution remaining in the pipette to well B1, and use the pipette to mix the solution in well C1 thoroughly. Well C1 now contains one drop of 0.1 M acetic acid and nine drops of water, for a concentration of 0.01 M.

6. Repeat the serial dilution for the remaining wells in column 1. When you finish you'll have eight wells populated, A1 through H1, with acetic acid dilutions of 1, 0.1, 0.01, 0.001, 0.0001, 0.00001, 0.000001, and 0.0000001 M, respectively.

7. Using a clean pipette, repeat the serial dilution procedure with ammonia in rows A through H of column 3 to produce final dilutions of 1, 0.1, 0.01, 0.001, 0.0001, 0.00001, 0.000001, and 0.0000001 M, respectively.

8. Using a clean pipette, repeat the serial dilution procedure with hydrochloric acid in rows A through H of column 5 to produce final dilutions of 1, 0.1, 0.01, 0.001, 0.0001, 0.00001, 0.000001, and 0.0000001 M, respectively.

9. Using a clean pipette, repeat the serial dilution procedure with sodium hydroxide in rows A through H of column 7 to produce final dilutions of 1, 0.1, 0.01, 0.001, 0.0001, 0.00001, 0.000001, and 0.0000001 M, respectively.

Retain the beaker of distilled water and the leftover 1 M solutions for use later in this lab session.

DETERMINING THE PH OF THE SOLUTIONS

With the reaction plate populated with acids and bases at various concentrations, the next step is to determine pH values for each of those solutions.

1. Use scissors to quarter eight strips of pH test paper. You'll need 32 pieces of pH test paper, one for each of the occupied wells.

2. Being careful not to slosh solution from one well into another, dip the tip of the stirring rod into well H1 (the most dilute acetic acid solution) and touch the tip of the rod to a piece of pH test paper to transfer the drop of acid to the paper. Compare the color of the test paper to the color chart and estimate the pH of that solution as closely as possible. Record that value in your lab notebook.

3. Dip the tip of the stirring rod into the beaker of distilled water to rinse it. Dry the stirring rod with a clean paper towel.

4. Repeat steps 2 and 3 to test each of the remaining wells in columns 1, 3, 5, and 7.

The actual pH values you observe and record may vary slightly from expected values, particularly at the mid-range pH levels, depending on how accurate your serial dilutions were and how accurately you interpret the colors of the pH test paper.

Hydrochloric acid is a strong acid and sodium hydroxide is a strong base, which means they fully dissociate in solution to form hydrogen ions and hydroxide ions, respectively. The 1 M hydrochloric acid should have a pH value of 0, with each succeeding ten-fold dilution increasing that value by 1. Similarly, the 1 M sodium hydroxide should have a pH value of 14, with each ten-fold dilution decreasing that value by 1.

Acetic acid is a weak acid and ammonia is a weak base, which complicates matters. In concentrated solutions, neither of these chemicals dissociates fully, so the pH of the acetic acid solution is higher than you might expect, and the pH of the ammonia solution lower. Table III-1-1 lists the theoretical pH values for various concentrations of these four chemicals, from 10^0 molar (1 molar) to 10^{-7} molar (0.0000001 molar).

Table III-1-1: *Theoretical pH values*

	10^0 M	10^{-1} M	10^{-2} M	10^{-3} M	10^{-4} M	10^{-5} M	10^{-6} M	10^{-7} M	
Acetic acid	2.4	2.9	3.4	3.9	4.5	5.2	6.0	7.0	1
Ammonia	11.6	11.1	10.6	10.1	9.5	8.9	8.0	7.0	3
Hydrochloric acid	0.0	1.0	2.0	3.0	4.0	5.0	6.0	7.0	5
Sodium hydroxide	14.0	13.0	12.0	11.0	10.0	9.0	8.0	7.0	7
	A	B	C	D	E	F	G	H	

Note that in the most dilute solutions, the pH values for the weak acid and weak base are very close to those of the corresponding strong acid and strong base. In such dilute solutions, the weak acid or base is sufficiently dissociated to provide enough hydronium or hydroxide ions to establish the pH at the expected level. Conversely, as the concentration increases, the pH values begin to diverge. In the highest concentration, 1 M, the weak acetic acid has a pH of only 2.4, versus the 0.0 pH of the strong and therefore fully dissociated hydrochloric acid. Similarly, in a 1 M solution, the weak base ammonia has a pH of only 11.6, versus the 14.0 pH of the strong and therefore fully dissociated sodium hydroxide. This difference in behavior between strong and weak acids and bases has profound implications for the chemistry of life.

PROCEDURE III-1-3: PH OF HOUSEHOLD MATERIALS

To move our study of pH into a more familiar realm, let's take a few minutes to test the pH of common household materials. If the materials are liquids, test them directly. For solids, dissolve a tiny amount in a few drops of water in a reaction plate. Quarter pH test strips, and use a fresh strip for each material.

Here are some materials you might test: water (distilled, tap, and bottled), saliva, urine, table salt, table sugar, baking soda, mouthwash, alcohol or hand sanitizer, washing soda, laundry detergent, chlorine bleach, toilet cleaner, drain opener, window cleaner, soap, shampoo, juices (apple, grape, grapefruit, etc.), vinegar, soft drinks (carbonated and uncarbonated), coffee and tea (with and without milk and/or sugar), and milk.

Record your observations in your lab notebook, along with any conclusions you make concerning the range of pH values found in food and other biological materials versus those intended for cleaning and similar purposes.

PROCEDURE III-1-4: BUFFERS

As we learned in the first procedure, even a tiny amount of acid or base can have a major effect on the pH of a solution, particularly if the pH of that solution is originally neutral or nearly so. For example, adding one drop of 1 M hydrochloric acid to nine drops of distilled water reduces the pH from 7 to 1; adding one drop to 99 drops of water reduces the pH from 7 to 2; and adding one drop to 999 drops of water reduces the pH from 7 to 3.

Such large variations in pH are incompatible with the biological processes of life. For example, many enzymes function only within an extremely narrow pH range, and are denatured (destroyed) if exposed to a pH much outside that range. Living things avoid these extremes in pH by using *buffers*, which resist changes in pH. Human blood plasma is one example of a buffer. It ordinarily maintains a pH of 7.40 ± 0.05. If the pH of someone's blood plasma is outside the range of about 7.3 to 7.5, that person is ill; if the pH is much outside that range, the person is dead.

A buffer solution is a mixture of a weak acid and a strong base or (more rarely) a weak base and a strong acid. For example, acetic acid (a weak acid) reacts in solution with sodium hydroxide (a strong base) to produce sodium acetate. A pure solution of sodium acetate is fully dissociated into sodium ions and acetate ions, and has an alkaline (basic) pH. But if you mix that sodium acetate solution with additional acetic acid—which does not dissociate fully in solution—you produce a solution that contains sodium ions, acetate ions, and molecular (nondissociated) acetic acid.

If you add a small amount of a strong acid to this acetate buffer solution, it absorbs the additional hydrogen ions by forming additional acetic acid molecules from those hydrogen ions and free acetate ions. If you instead add a small amount of a strong base to the acetate buffer solution, it absorbs the additional hydroxide ions by breaking down acetic acid molecules into hydrogen ions and acetate ions. The newly freed hydrogen ions react with the excess hydroxide ions to form neutral water molecules. In either case, the pH of the solution changes only slightly.

In this procedure, we'll prepare a buffer solution and test its ability to resist pH changes as we add small amounts of a strong acid and a strong base.

PREPARING A 0.1 M ACETATE BUFFER SOLUTION

The first step is to prepare a 0.1 M acetate buffer solution. That solution will be 0.1 M with respect to sodium acetate and 0.1 M with respect to acetic acid and will have a pH just under 5.0. We'll use 6 M solutions of acetic acid and sodium hydroxide to make up the buffer.

Because one mole of acetic acid reacts with one mole of sodium hydroxide to form one mole of sodium acetate, we'll need to use twice as much acetic acid as sodium hydroxide. (Half of the acetic acid reacts with sodium hydroxide to form sodium acetate, and the other half of the acetic acid remains in solution as acetic acid.) Because a 6 M solution is 60 times more concentrated than our target 0.1 M concentration, we'll need to dilute the solutions by using one part 6 M sodium hydroxide and two parts 6 M acetic acid to 57 parts of water.

1. If you have not already done so, put on your goggles, gloves, and protective clothing.

2. Label a 15 mL centrifuge tube "acetate buffer" and transfer 10 or 12 mL of distilled water to the tube.

3. Use a clean pipette to transfer 0.25 mL of 6 M sodium hydroxide solution to the tube. Swirl to mix the solution.

4. Use a clean pipette to transfer 0.50 mL of 6 M acetic acid to the tube. Swirl to mix the solution.

5. Fill the tube to the 15 mL line with distilled water. Cap the tube and invert it several times to mix the solution.

TESTING THE BUFFER SOLUTION

With the acetate buffer prepared, the next step is to test its resistance to changes in pH and compare that resistance to that of a nonbuffer solution (ordinary water). We'll use the 1 M solutions of hydrochloric acid and sodium hydroxide we prepared earlier in this lab session.

1. If you have not already done so, put on your goggles, gloves, and protective clothing.

2. Use scissors to quarter 8 or 10 pH test paper strips.

3. Use a clean pipette to transfer 2.0 mL of water to each of wells A1 and D1 of the 24-well reaction plate and 2.0 mL of acetate buffer solution to each of wells B1 and C1 of the reaction plate.

4. Add one drop of 1 M hydrochloric acid to well A1. Stir the contents of the well with the stirring rod, and use the tip of the stirring rod to transfer one drop of the contents to a piece of test paper. Record your results in your lab notebook.

5. Using a fresh piece of test paper each time, repeat step 4 until you have added a total of five drops of 1 M hydrochloric acid to well A1.

6. Repeat steps 4 and 5, but this time adding 1 M sodium hydroxide to well D1.

7. Repeat steps 4 and 5, but this time adding the 1 M hydrochloric acid dropwise to well B1, which contains buffer solution. Continue adding acid to well B1 until the pH starts to show a sharp decrease, or until you run out of acid.

8. Repeat steps 4 and 5, but this time adding the 1 M sodium hydroxide dropwise to well C1, which contains buffer solution. Continue adding base to well C1 until the pH starts to show a sharp increase, or until you run out of base.

Note the key difference between a buffer solution and an unbuffered solution such as water. An unbuffered solution reacts to the addition of small amounts of a strong acid or strong base by showing large changes in pH. A buffer solution absorbs small quantities of a strong acid or strong base without much change in pH. Note also that the capacity of a buffer is limited. Adding larger amounts of a strong acid or strong base eventually overwhelms the ability of the buffer solution to resist changes in pH. Once the buffer solution reaches its capacity, it behaves as an ordinary, unbuffered solution.

REVIEW QUESTIONS

Q1: How would you make up 100 mL of 1 M potassium iodide solution?

Q2: You need to make up 1 L of 1 M sulfuric acid. Your concentrated sulfuric acid is 98% sulfuric acid and has a density as 1.84 g/mL. Describe two methods for making up the 1 M solution.

Q3: You have a hydrochloric acid solution with a known pH of 3.00. How much of that solution would you dilute with distilled water to a final volume of 1 L to produce a 100 μM (micromolar) solution? What would the pH of that solution be? (Assume complete dissociation of the acid.)

Q4: How did the pH values you observed for food and other material intended for human consumption differ from the pH values of household cleaners and similar items?

Carbohydrates and Lipids

Lab III-2

EQUIPMENT AND MATERIALS

You'll need the following items to complete this lab session. (The standard kit for this book, available from *www.thehomescientist.com*, includes the items listed in the first group.)

MATERIALS FROM KIT

- Goggles
- Barfoed's reagent
- Beaker, 250 mL
- Benedict's reagent

- d-glucose (dextrose)
- graduated cylinder, 10 mL
- Gram's iodine stain
- Hydrochloric acid

- Pipettes
- Reaction plate, 24-well
- Seliwanoff's reagent
- Slides (flat) and coverslips

- Sudan III stain
- Test tubes
- Test tube rack

MATERIALS YOU PROVIDE

- Gloves
- Butane lighter (or other flame source)
- Butter
- Diet sweetener
- Fruit juice (unsweetened)

- Hair dryer (optional)
- Honey
- Isopropanol
- Marking pen
- Microscope
- Microwave oven

- Milk (whole)
- Non-dairy creamer
- Onion
- Paper bag (brown)
- Peanut (or cashew, etc.)
- Potato

- Soft drink, colorless (Sprite or similar)
- Sucrose (table sugar)
- Vegetable oil
- Water, distilled

BACKGROUND

In this lab session we'll investigate two classes of *biologically important molecules*, carbohydrates and lipids.

CARBOHYDRATES (SACCHARIDES)

Carbohydrates, also known as *saccharides* (sack'-uh-rides), are a fundamental building block of life. Carbohydrates perform many key biological functions, notably storing and transporting energy and providing physical structure.

Carbohydrates contain only carbon, hydrogen, and oxygen, and have the empirical formula $C_n(H_2O)_m$, where some number of carbon atoms is combined with some number of water molecules. (In fact, the structure of carbohydrates is considerably different than a simple grouping of carbon atoms with water molecules, but they were named carbohydrates because the empiric ratio of atoms in their structures corresponds to hydrated carbon.) Carbohydrates are categorized as members of the following groups.

Monosaccharides

Monosaccharides, also called *simple sugars*, are the smallest and simplest carbohydrates, and are important both for themselves and as the fundamental building blocks of larger, more complex carbohydrates. (You can think of complex carbohydrates as a brick wall in which monosaccharides are the bricks.) *Glucose, fructose, galactose, ribose,* and *xylose* are examples of biologically important monosaccharides.

Monosaccharides are classed by the number of carbon atoms and the functional group they contain. A *triose* contains three carbon atoms, a *tetrose* four, a *pentose* five, a *hexose* six, and a *heptose* seven. Monosaccharides that contain an aldehyde group—a carbonyl group (C=O) bonded to a hydrogen atom and a carbon atom—are *aldoses*, and those that contain a ketone group—a carbonyl group bonded to two other carbon atoms—are *ketoses*.

These two important characteristics may be combined to describe both the number of carbon atoms and the type of functional group in one term. For example, because ribose contains five carbon atoms and an aldehyde functional group, it is an *aldopentose*. Fructose contains six carbon atoms and a ketone functional group, and so is a *ketohexose*.

CHIRALITY

Chirality is the property of handedness. For example, the structures of a right glove and a left glove are identical but the gloves are chiral because each is the mirror image of the other. Many organic molecules, including monosaccharides, are chiral. The *dextrorotary* form of a chiral molecule rotates the plane of polarized light to the right, and the *levorotary* form to the left.

The naturally occurring form of glucose rotates polarized light to the right, and is called *D-glucose* or *dextrose*. L-glucose is useless biologically because hexokinase, the enzyme that metabolizes D-glucose, is itself chiral and cannot operate on L-glucose.

Disaccharides

Disaccharides, also called *compound sugars*, comprise two bound monosaccharides, which may be the same or different. For example, the disaccharide *lactose* contains glucose bound to galactose, *sucrose* contains glucose bound to fructose, and *maltose* contains two bound glucose molecules.

Disaccharides can be *hydrolyzed* (split) into their component monosaccharides using enzymes, heat, or an acid or base catalyst. For example, sucrose can be hydrolyzed into its component monosaccharides, glucose and fructose, by heating it in a dilute hydrochloric acid solution.

Oligosaccharides

An *oligosaccharide* is a polysaccharide polymer made up of more than 2 but fewer than 10 monosaccharide units. Many oligosaccharides are biologically important molecules. Oligosaccharides are components of many proteins, glycoproteins, and glycolipids, where they often function as chemical markers. For example, blood types A and B contain different oligosaccharides; blood type AB contains both of those oligosaccharides; blood type O contains neither.

Polysaccharides

A *polysaccharide* is a *macromolecule* (large molecule) made up of 10 or more (often *many* more) monosaccharide and/or disaccharide units. Two familiar polysaccharides are *cellulose*, which plants use as a structural element, and *starch*, which plants use to store energy. (Animals store energy using *glycogen*, which is structurally similar to starch.)

Some polysaccharides are easily hydrolyzed into their component mono- or disaccharides. For example, the glucose (dextrose) sold in drugstores and health-food stores is produced by hydrolyzing starch. Other polysaccharides are very difficult to hydrolyze. For example, while it is possible in principle to hydrolyze cellulose into glucose—which would immediately make biofuels cheap and universally available—in practice it has so far proven impossible to do so economically.

> ## REDUCING SUGARS
>
> A *reducing sugar* (more properly, *reducing saccharide*) is one that, when in solution, exposes a reactive ketone or aldehyde group. All monosaccharides (ketoses and aldoses) are, by definition, reducing sugars. Many di-, oligo-, and polysaccharides are reducing saccharides. Sucrose is one example of a nonreducing disaccharide. The "reducing" part of the name refers to the ability of these saccharides to react with mild oxidants such as Fehling's, Benedict's, or DNSA reagent to yield a color change or precipitate that identifies the presence of a reducing sugar.

Biologists frequently use four color-test reagents to discriminate among types of sugars. All of these reagents are used in the same way: transfer a small amount of the reagent to a test tube, add a few drops of the specimen, and place the tube in a boiling water bath for a few minutes.

Barfoed's reagent

Barfoed's reagent is used to discriminate monosaccharides. A brick-red precipitate forms within five minutes if a monosaccharide is present. Disaccharides generally cause no precipitate even after 10 to 15 minutes of heating.

Benedict's reagent

Benedict's reagent is used to discriminate reducing sugars. A precipitate forms within five minutes if a reducing sugar is present. The amount and color of the precipitate vary with the amount of reducing sugar present. With increasing concentration of the reducing sugar, the precipitate color varies from green to yellow to orange to brick-red, with the brick-red color generally occurring at concentrations of 1% or higher.

Seliwanoff's reagent

Seliwanoff's reagent is used to discriminate ketoses from aldoses. An orange to red color forms within five minutes if a ketose is present. If the ketose is present in low concentration, the color may be anything from a straw yellow to yellowish-orange. Aldoses yield no color change even after 10 to 15 minutes of heating.

If you're using the kit, you can make up Seliwanoff's reagent by adding one part 6 M hydrochloric acid to one part of Seliwanoff's Reagent A solution. For example, to make up 3 mL of Seliwanoff's Reagent, use a graduated pipette to transfer 1.5 mL of 6 M hydrochloric acid to 1.5 mL of Seliwanoff's Reagent A.

> Bial's reagent, which we won't use in this session, is used to discriminate pentoses from hexoses. Bial's reagent forms a green, greenish-blue, or blue color in the presence of pentoses, but shows no color change for hexoses.

Starches can be detected using a dilute aqueous solution of iodine, such as Gram's stain. The brown iodine solution reacts with starch to form an intense blue complex. Adding a drop of iodine solution to a few mL of even extremely dilute starch solution causes a blue coloration to appear; with more concentrated starch solutions, the blue is so intense it may appear black. Iodine is also used to stain cells for microscopic examination. Any starch present in the cell is stained black.

> ## THE MOLISCH GENERAL TEST FOR CARBOHYDRATES
>
> In a biology lab, a quick test is sometimes needed to determine if carbohydrates of any type are present in a specimen. Molisch's reagent is widely used for this purpose.

LIPIDS

Lipids are members of a large group of biomolecules that includes *oils*, *fats*, *waxes*, *sterols*, *fatty acids*, and other classes. Lipids perform many key biological functions. Like carbohydrates, lipids are widely used for storing energy and providing physical structure in cell membranes. Many important signaling molecules are lipids, as are several vitamins.

Some lipids are *hydrophobic* (water-hating), which means they do not mix with water or aqueous solutions. For example, vegetable oil contains hydrophobic lipids. If you mix vegetable oil with water and agitate the liquid, it initially forms a suspension of tiny globules of vegetable oil suspended in the water. If you allow the liquid to sit undisturbed, it eventually separates into two layers, with the denser water forming the bottom layer and the vegetable oil the top layer.

Other lipids are *amphiphilic*, which means they mix well with both water and other lipids. The molecular structure of these lipids has a hydrophobic group on one end of the molecule and a *hydrophilic* (water-loving) group on the other end. Amphiphilic lipids called *phospholipids* form the structure of the *bi-layer membranes* found in many cells. These phospholipids arrange themselves into a double layer with their hydrophobic groups in the center of the layer and their hydrophilic groups facing outward toward the aqueous solutions on both sides of the membrane, thereby isolating the aqueous areas with a hydrophobic lipid layer that is impermeable to water.

Lipophilic dyes (fat-loving dyes) such as Sudan III are the best general test for the presence of lipids. These dyes are nearly insoluble in water but readily soluble in lipids (such as oils and fats). When a lipophilic dye is applied to a specimen that contains lipids, it is selectively attracted to the lipids, staining them and leaving parts of the specimen that do not contain lipids unstained. This property is useful for discriminating lipids both macroscopically and microscopically.

We will attempt to resolve the following:

- The type or types of sugars present in the various specimens.

- Whether it is possible to hydrolyze sucrose into its component monosaccharide(s) and, if so, how closely you might categorize the component monosaccharide(s).

- The presence or absence of starch in the various specimens.

- Whether it is possible to hydrolyze starch into its component monosaccharide(s) and, if so, how closely you might categorize the component monosaccharide(s).

- Whether it is possible to identify the presence of starch as a component of plant cells and, if so, how.

- The presence or absence of lipids in the various specimens.

- Whether it is possible to identify the presence of lipids as a component of plant cells and, if so, how.

PROCEDURE III-2-1: INVESTIGATING SUGARS

1. If you have not already done so, put on your goggles, gloves, and protective clothing.

2. Prepare a hot water bath by filling the 250 mL beaker about one third full of tap water and heating it in the microwave until it comes to a gentle boil. Alternatively, simply bring a pot of water to a gentle boil on the stove, and use that pot as your source of boiling water during this procedure.

WARNING: USE EXTREME CAUTION

A microwave oven can actually heat water above its boiling point without causing the water to boil, a phenomenon called superheating. The slightest disturbance can cause superheated water to boil violently, expelling it from the container.

The water needn't be actually boiling, as long as its temperature is close to 100 °C. As the hot water bath cools during these procedures, periodically replace the water with freshly boiled water to maintain the temperature near 100 °C.

3. Label a test tube for each of your first six specimens, including one tube for distilled water, and place the tubes in the rack.

4. Prepare your specimens as described in the footnotes of Table III-2-1. Any specimens that do not include preparation instructions can be used as-is.

5. Transfer about 0.5 mL of Barfoed's reagent to each test tube.

6. Add three drops of each specimen to the corresponding tube and swirl the tubes to mix the contents.

7. Transfer the first six tubes to the boiling water bath and allow them to remain for five minutes.

8. Record your observations in your lab notebook.

9. After the tubes have cooled, dispose of their contents by flushing them down the drain with plenty of water. Rinse and then wash the tubes.

10. Repeat steps 5 through 9 for your other six specimens.

11. Repeat steps 5 through 10 using Benedict's reagent and Seliwanoff's reagent. (We'll fill out the Gram column in the next procedure.)

Retain the prepared specimens in the reaction plate and the (clean) labeled test tubes for the following procedure.

Table III-2-1: *Detecting and Classifying Saccharides—experimental observations*

#	Specimen	Barfoed	Benedict	Seliwanoff	Gram
1	distilled water				
2	diet sweetener[a]				
3	fruit juice				
4	glucose solution				
5	honey[b]				
6	milk (whole)				
7	nondairy creamer[c]				
8	onion water[d]				
9	potato water[e]				
10	Sprite soft drink				
11	sucrose[f] solution				
12	hydrolyzed sucrose[g]				

a. Dissolve a pinch of diet sweetener in about 2 mL of distilled water in a reaction plate well.

b. Dissolve two or three drops of honey in about 2 mL of distilled water in a reaction plate well.

c. Dissolve a pinch of non-dairy creamer in about 2 mL of distilled water in a reaction plate well.

d. Mash bits of onion with a stirring rod in about 2 mL of distilled water in a reaction plate well.

e. Mash bits of potato with a stirring rod in about 2 mL of distilled water in a reaction plate well.

f. Dissolve a pinch of table sugar in about 2 mL of distilled water in a reaction plate well.

g. Dissolve a pinch of sugar in about 2 mL of distilled water in a test tube and add two drops of hydrochloric acid. Heat for 10 minutes in the hot water bath, allow to cool, and transfer to a reaction plate well.

If you don't use them normally, you can obtain diet sweetener and nondairy creamer in small packets from any restaurant.

PROCEDURE III-2-2: INVESTIGATING STARCHES

This procedure has two parts. In the first, we'll test the specimens we used in the last procedure for the presence of starch. In the second part, we'll test for starch at the cellular level of some solid specimens.

TESTING SPECIMENS FOR THE PRESENCE OF STARCH

1. If you have not already done so, put on your goggles, gloves, and protective clothing.

2. Transfer about 0.5 mL of each of the first six specimens listed in Table III-2-1 to the corresponding test tubes and place the tubes in the rack.

3. Add one drop of Gram's stain to each of the test tubes.

4. Record your observations in your lab notebook and in Table III-2-1.

5. Dispose of the contents of the tubes by flushing them down the drain with plenty of water. Rinse and then wash the tubes.

6. Repeat steps 2 through 5 for each of the remaining six specimens.

Retain the prepared specimens in the reaction plate and the (clean) labeled test tubes for the following procedure.

TESTING FOR STARCH AT THE CELLULAR LEVEL

1. If you have not already done so, put on your goggles, gloves, and protective clothing.

2. Transfer a tiny amount of solid potato to a slide, add one drop of distilled water, and use the stirring rod to crush and spread the potato to make a smear mount. Heat-fix the smear mount.

3. Repeat step 2 using a tiny amount of solid onion.

4. Repeat step 2 using a tiny amount of solid peanut.

WARNING

Skip step 4 if you are or anyone else in the vicinity is allergic to peanuts.

5. Place one drop of Gram's iodine stain on the smear area of each slide. Allow the stain to work for 30 seconds and then rinse gently with distilled water to remove excess stain. Allow the slides to dry naturally or use a hair dryer set to low to speed drying.

6. Examine each slide at low magnification to locate a cluster of cells. Examine those cells at medium and high magnifications. Record your observations, including sketches, in your lab notebook. Retain the slides for use in the next procedure.

PROCEDURE III-2-3: INVESTIGATING LIPIDS

This procedure has four parts. In the first, we'll test the solubility of lipids in water and isopropanol. In the second, we'll use the grease-spot test to detect lipids in our various liquid specimens. In the third, we'll investigate the effect of Sudan III stain (a lipophilic dye) on lipids. In the fourth, we'll use Sudan III stain to stain the slides we made in the previous procedure to determine if we can detect lipids at the cellular level.

SOLUBILITY OF LIPIDS

1. If you have not already done so, put on your goggles, gloves, and protective clothing.

2. Transfer 5.0 mL of water to one test tube and 5.0 mL of isopropanol to a second test tube. Place the tubes in the rack.

3. Use a pipette to add 0.5 mL of vegetable oil to each of the test tubes. Agitate the tubes and replace them in the rack.

4. Allow the tubes to remain undisturbed for a minute or so, and then record your observations in your lab notebook.

Retain the contents of the isopropanol test tube for the following sections.

THE GREASE-SPOT TEST FOR LIPIDS

1. If you have not already done so, put on your goggles, gloves, and protective clothing.

2. Draw a 4x4 grid of 5 cm squares on a brown paper bag. Label the squares for each of the 12 solutions you tested earlier, plus a thirteenth square for the mixture of isopropanol and vegetable oil.

3. Transfer one drop of each of the 13 liquids to the corresponding square. Allow the paper to dry naturally, or use a hair dryer set on low to speed drying.

4. Hold the paper up to a window or bright lamp, and examine it for translucent grease spots, which indicate the presence of lipids. Record your observations in your lab notebook.

THE EFFECT OF SUDAN III STAIN ON LIPIDS

1. If you have not already done so, put on your goggles, gloves, and protective clothing.

2. Label four test tubes. Transfer 1.0 mL of distilled water to the first tube, 1.0 mL of honey to the second, 1.0 mL of vegetable oil to the third, and 0.5 mL of vegetable oil and 0.5 mL of distilled water to the fourth.

3. Add two drops of Sudan III stain to each of the test tubes, and agitate the tubes to mix their contents. Replace the tubes in the rack and allow them to sit undisturbed for a minute or so. Record your observations in your lab notebook.

OBSERVING LIPIDS AT THE CELLULAR LEVEL

1. If you have not already done so, put on your goggles, gloves, and protective clothing.

2. Add one drop of Sudan III stain to the smear areas of the slides you prepared earlier and stained with Gram's iodine. Allow the Sudan III stain to work for 30 seconds and then rinse gently with distilled water to remove excess stain. Allow the slides to dry naturally or use a hair dryer set to low to speed drying.

3. Examine each slide at low magnification to locate a cluster of cells. Examine those cells at medium and high magnifications. Record your observations, including sketches, in your lab notebook.

REVIEW QUESTIONS

Q1: In procedure III-2-1, what is the purpose of the test tube that contains only distilled water?

Q2: If you obtain a positive result with one of the reagents when testing distilled water, what can you conclude? What action would you take?

Q3: Which of the specimens you tested in procedure III-2-1 contained a monosaccharide? How do you know?

Q4: Which of the specimens you tested in procedure III-2-1 contained a reducing sugar? How do you know?

Q5: Which of the specimens you tested in procedure III-2-1 contained a ketose? How do you know?

Q6: Based on your tests of hydrolyzed sucrose in procedure III-2-1, what can you conclude about the monosaccharide hydrolyzation products present?

_____

_____

_____

Q7: Based on your tests in procedure III-2-2, which of your specimens contained starch?

_____

_____

_____

Q8: Based on your microscopic examinations in procedure III-2-2, which of the three specimens contained starch and in what amounts? Was the starch evenly distributed throughout the cells or localized? How can you tell?

_____

_____

_____

Q9: Based on your tests in procedure III-2-3, are the lipids in vegetable oil more soluble in water or isopropanol? Which of your 13 specimens contained lipids?

_____

_____

_____

Q10: Based on your microscopic examinations in procedure III-2-3, which of the three specimens contained lipids and in what amounts? Were the lipids evenly distributed throughout the cells or localized? How can you tell?

_____

_____

_____

Q11: What do you conclude about the similarities and differences in how potatoes, onions, and peanuts store food energy?

_____

_____

_____

_____

Proteins, Enzymes, and Vitamins

Lab III-3

Other than collecting household items, no advance preparation is required for any of the procedures in this lab session. Procedure 2 requires at least one urine specimen, which you should obtain immediately before you begin the session. If there will be a significant delay between obtaining the specimen and doing the lab session, refrigerate the specimen in a full, tightly capped container until shortly before you begin the session.

WARNING

We'll assume that you're healthy and using your own urine, which presents no biohazard to you. In fact, urine obtained from any healthy person is sterile unless it is contaminated during collection. If you obtain urine specimens from others, wash your hands with soap and water thoroughly after handling the containers. In any event, wash the outsides of the containers after collecting the specimens.

EQUIPMENT AND MATERIALS

You'll need the following items to complete this lab session. (The standard kit for this book, available from *www.thehomescientist.com*, includes the items listed in the first group.)

MATERIALS FROM KIT

- Goggles
- Ascorbic acid tablet
- Beaker, 100 mL
- Beaker, 250 mL
- Biuret reagent
- Centrifuge tubes, 15 mL

- Gelatin (unflavored)
- Gram's iodine stain
- Graduated cylinder, 10 mL
- Graduated cylinder, 100 mL
- Hydrochloric acid
- Lead(II) acetate

- l-Glutamine
- pH test paper
- Pipettes
- Reaction plate, 24-well
- Reaction plate, 96-well
- Sodium hydroxide

- Spatula
- Stirring rod
- Test tubes
- Test tube clamp
- Test tube rack

MATERIALS YOU PROVIDE

- Gloves
- Blood from uncooked meat
- Desk lamp or other strong light source

- Egg white, raw
- Freezer
- Hydrogen peroxide, 3%
- Isopropanol, 99%

- Marking pen
- Microwave oven
- Paper or cloth, black
- Paper towels

- Starch water (see text)
- Urine specimen(s) (see text)
- Water, distilled

BACKGROUND

In this lab session we'll investigate three more classes of biologically important molecules, proteins, enzymes, and vitamins.

PROTEINS

A *protein* is a member of a class of organic chemical compounds that are made up of *amino acid* building blocks. An amino acid is an organic chemical compound that includes a *carboxyl group* (COOH) and an *amino group* (NH_2). In proteins, amino acids link together in chains, with the carboxyl group from one amino acid in the chain connected to the amino group of the next amino acid by a *peptide bond*.

There are a huge number of different amino acids, but only 20 of them (22 in some lifeforms) are used as protein building blocks. Simple proteins may include just a few amino acids in the chain, while complex proteins may include hundreds or thousands of individual amino acid building blocks. The sequence (order) of the amino acids in a protein is defined by the gene that is responsible for producing that particular protein. Small changes in that sequence can result in huge differences in the characteristics and functioning of the resulting proteins.

Although proteins can be thought of as simple chains of amino acids, the physical reality is different. Rather than existing as long linear chains, during production most actual proteins fold in upon themselves to produce molecular structures that have the form of globes or fibers. These structures, called *conformations*, may be relatively rigid and fixed or they may be flexible and therefore subject to conformational changes. Proteins of the first sort may be found as structural elements in cells and other applications where their fixed conformations are essential; those of the second sort are often used for signaling and are important elements of the *cell cycle*.

Most proteins can be *denatured* by heat, acids or bases, heavy metals, or other *denaturants*. For example, when you cook liquid egg white, the albumin protein present in the egg white is denatured and precipitates as a white solid.

The process of *denaturation* unfolds the conformation of the protein, disrupting its three-dimensional structure and inactivating it. Depending on the protein and the denaturant used, denaturation may or may not be reversible.

The term denatured is used in two distinct ways. For example, changing the pH of a solution slightly may inactivate a particular protein, but returning the pH of that solution to the range in which that protein is active may reactivate the protein. Conversely, making a large change of pH in that solution may inactivate the protein irreversibly. Some people use the term denaturation to refer to either reversible or irreversible inactivation; others use it to refer only to irreversible inactivation.

Like carbohydrates and lipids, proteins play a key role in life processes, participating in nearly every cellular process. Some proteins fill structural roles at the cell and organ levels. Others are enzymes that catalyze biochemical processes, including metabolism. Still others play key roles in cell signaling, immune system responses, and most other critical life processes.

Biologists use numerous chemical and instrumental tests to detect, identify, and quantify the types and amounts of amino acids and proteins present in a specimen. The oldest of these is the *biuret test*, which is a general test for proteins and can be run qualitatively or, with a colorimeter or spectrophotometer, quantitatively. Other common tests include the *Bradford protein assay*, which is more sensitive than the biuret test but does not respond to all proteins, the *cystine test*, which detects the presence of cysteine residues in proteins, and the *xanthoproteic acid test*, which is specific to the two amino acids tyrosine and tryptophan. (We'll use the biuret test and the cystine test in this lab session.)

ENZYMES

An *enzyme* is a protein that functions as a *catalyst*. (A catalyst is a chemical that increases the rate of a chemical reaction but is not consumed during the reaction.) In the presence of a catalyst, a reaction may proceed millions of times faster than it would in the absence of that catalyst. Nearly all of the biochemical reactions that occur in cells require enzymes to proceed at useful rates. About 4,000 different biochemical reactions are known to be catalyzed by enzymes.

Most enzymes are specific to one reaction. For example, the enzyme *invertase* catalyzes the hydrolysis (splitting) of the disaccharide sucrose (table sugar) into the monosaccharides glucose and fructose, but has no effect on DNA, while the enzyme *DNA polymerase* plays an essential role in DNA replication, but has no effect on sucrose. Because most enzymes are so selective, the types of enzymes made in a cell determine the types of biochemical reactions (*metabolic pathways*) that occur in that cell.

Enzyme activity is affected by many factors, including concentration, pH, and temperature. It can also be increased or decreased by the presence of other organic molecules. Molecules that increase enzyme activity, including some vitamins, are called *enzyme activators*. Molecules that decrease enzyme activity, including many drugs and poisons, are called *enzyme inhibitors*. As proteins, most enzymes can be denatured, reversibly or irreversibly.

In this lab session, we'll investigate some of the properties of *peroxidase*, an enzyme that speeds up the decomposition of hydrogen peroxide millions of times. (That's why hydrogen peroxide foams when you pour it on a cut; the peroxidase enzyme in your blood catalyzes the decomposition of hydrogen peroxide into water and oxygen gas, which causes the foaming.) Hydrogen peroxide occurs naturally, and is a strong oxidizing agent that damages cells. In the absence of a catalyst like peroxidase, hydrogen peroxide decomposes very slowly, and continues damaging cells until it has completely decomposed. In the presence of peroxidase, hydrogen peroxide breaks down very quickly into harmless water and oxygen gas.

VITAMINS

A *vitamin* is an organic compound or group of related compounds that is a required *trace nutrient*, but either is not produced by or is produced in insufficient quantity by an organism and so must be obtained from the diet. (Many dietary minerals—such as iron, copper, zinc, and iodine—are also required trace nutrients, but no organism can produce these from simpler precursors, so by convention they are considered essential trace minerals rather than vitamins.)

For example, vitamin C (ascorbic acid) is required by all forms of plant and animal life. Nearly all species produce sufficient ascorbic acid for their own needs, so in these species ascorbic acid is not considered a vitamin. Conversely, a few species—including bats, guinea pigs, some birds and fish, and some primates, including humans—produce little or no ascorbic acid, and so must obtain it from their diets. For these species, ascorbic acid is a vitamin.

As is true of some proteins, some vitamins must be obtained from the diet only in some circumstances. For example, humans require vitamin D, but we also make it ourselves when our skin is exposed to the ultraviolet wavelengths in sunlight. Under some conditions, we make sufficient vitamin D to meet our needs, but in other conditions we may need supplemental vitamin D from our diets.

Skin color and sunlight intensity have significant effects on vitamin D production. The ancestors of light-skinned people originated at high latitudes, where sunlight is weak, while the ancestors of dark-skinned people originated at equatorial latitudes, where sunlight is intense. Dark skin contains large amounts of the pigment melanin, which blocks UV light, preventing skin damage and melanomas but at the same time reducing production of vitamin D. Light skin contains little melanin, which maximizes production of vitamin D, but does not protect against damage from intense sunlight. Accordingly, someone with light skin exposed to the strong equatorial sunlight may overproduce vitamin D (not to mention suffering a bad sunburn), while someone with dark skin exposed only to the weak sunlight at high latitudes may produce insufficient vitamin D.

Unlike the other biologically important molecules we've looked at so far, vitamins are classified by their biochemical activity rather than by their structures. The biochemical functions of vitamins are diverse. Some regulate metabolism or cell and tissue growth. Others function as free-radical scavengers (anti-oxidants). But most function as *enzyme cofactors*, which assist enzymes in their catalytic activities.

What constitutes a "trace" also differs from vitamin to vitamin. For example, a multivitamin tablet might contain 100,000 micrograms (µg) of vitamin C, but only 750 µg of vitamin A and only 15 µg of vitamin D3.

In this lab session, we'll use iodometric titration to determine the concentration of vitamin C in urine, how that concentration varies from person to person and over the course of time, and the importance of timeliness in assaying concentrations of some organic compounds in living organisms.

We will attempt to resolve the following:

- The presence or absence of proteins in the various specimens.

- Whether or not general tests for proteins yield positive results for amino acids.

- The denaturation of proteins by various mechanisms.

- Whether an enzyme is denatured by extreme temperatures or heavy-metal ions.

- The effect of pH on the activity of an enzyme and whether that effect is reversible.

- The presence and concentration of vitamin C in a urine specimen.

- Whether the concentration of vitamin C in a urine specimen decreases spontaneously over time.

PROCEDURE III-3-1: INVESTIGATING PROTEINS

Before you begin, prepare a hot water bath by filling the 250 mL beaker about half full of tap water and heating it in the microwave until it comes to a gentle boil. Alternatively, simply bring a pot of water to a gentle boil on the stove, and use that pot as your source of boiling water during this procedure.

HANDLE WITH CARE

A microwave oven can actually heat water above its boiling point without causing the water to boil, a phenomenon called superheating. The slightest disturbance can cause superheated water to boil violently, expelling it from the container.

The water needn't be actually boiling, as long as its temperature is close to 100 °C. As the hot water bath cools during these procedures, periodically replace the water with boiling water to maintain the temperature near 100 °C.

PREPARING SPECIMENS

1. If you have not already done so, put on your goggles, gloves, and protective clothing.

2. Label a centrifuge tube for each of your specimens.

3. Transfer about 2 mL of raw egg white and 10 mL of distilled water to a labeled centrifuge tube. Cap the tube and agitate the contents to mix. This solution contains about a 2% concentration of albumin, the primary protein present in egg white.

4. Transfer one rounded microspatula spoon of unflavored gelatin to a labeled centrifuge tube that contains about 10 mL of cold distilled water. Cap the tube and agitate the contents to mix. Remove the cap and place the tube in the hot water bath. Allow the tube to heat until the gelatin dissolves. Remove the tube, cap it, place it in the rack, and allow it to cool to room temperature.

5. Transfer one rounded microspatula spoon of L-glutamine to a labeled centrifuge tube that contains about 10 mL of cold distilled water. Cap the tube and agitate the contents to mix.

DETECTING PROTEINS WITH THE BIURET TEST

The biuret test is a general color test for the presence of proteins. If a specimen contains a protein or proteins, adding a few drops of the specimen to a small amount of biuret reagent causes the color of the reagent to change from blue to violet. The test is specific for proteins, as opposed to the amino acids that are the building blocks of proteins. Testing a pure amino acid with biuret reagent usually results in no obvious color change or at most a slight darkening of the light-blue color of the reagent itself.

1. If you have not already done so, put on your goggles, gloves, and protective clothing.

2. Place the 96-well reaction plate on a sheet of white paper under a strong light.

3. Transfer eight drops of biuret reagent to each of wells A1, A3, A5, and A7 of the reaction plate.

4. Add two drops of distilled water to well A1, two drops of the egg-white solution to well A3, two drops of the gelatin solution to well A5, and two drops of the L-glutamine solution to well A7. Note any color changes, and record your observations in your lab notebook.

> The time required for the biuret reagent to react varies with the concentration of the protein(s) present in the specimen and other factors. If no immediate reaction occurs, place the reaction plate aside and recheck it after 10 to 15 minutes. You can be working on other parts of this lab session while you wait.

TESTING FOR THE PRESENCE OF CYSTINE

Cystine is a dimeric amino acid that is formed by the oxidation of two cysteine residues linked via a disulfide bond. Cystine is significant because the disulfide bonds it forms within and between protein molecules are major determinants in the conformation of most proteins. At basic pH, cystine reacts with lead(II) ions to form an insoluble black precipitate, which is the basis of the cystine test.

1. If you have not already done so, put on your goggles, gloves, and protective clothing.

2. Label two test tubes A and B.

3. Transfer about 1.0 mL of albumin solution to tube A and 1.0 mL of gelatin solution to tube B.

4. Transfer about 0.25 mL of 6 M sodium hydroxide to each of the two tubes and swirl the tubes to mix the solutions.

5. Transfer about 0.25 mL of lead(II) acetate solution to each of the two tubes and swirl the tubes to mix the solutions.

6. Using the test tube clamp, place the tubes in the hot water bath and allow them to remain for about five minutes.

7. Carefully remove the tubes from the hot water bath and place them in the rack.

8. Observe the tubes for the presence of a black precipitate, which indicates the presence of cystine in the specimen. Record your observations in your lab notebook.

DENATURING PROTEINS WITH ALCOHOL, ACIDS, BASES, AND HEAVY-METAL IONS

Denaturing a protein changes it into a different conformation (physical form). We've already seen that heat is a denaturant for some proteins. For example, when you fry an egg the albumin protein in the egg white is denatured, changing from a colorless liquid to a white gel. There's no way to un-fry the egg and convert the denatured albumin back into native albumin. But proteins can be denatured by many other mechanisms, including strong acids or bases, heavy metals, and alcohols. Some proteins are more resistant than others to being denatured by any particular means, and a means that is very effective for denaturing one protein may be less effective or ineffective in denaturing a different protein.

One sure sign that a protein has been denatured is that it coagulates (clumps) and precipitates from solution. The converse is not true; *denaturation*, sometimes called *denaturization*, may occur without coagulation and precipitation also occurring.

1. If you have not already done so, put on your goggles, gloves, and protective clothing.

2. Place the 24-well reaction plate on a black cloth or sheet of paper and illuminate it with a desk lamp or other strong light source.

3. Transfer about 0.5 mL of the albumin solution to each of wells A1 through A6.

4. Transfer about 0.5 mL of the gelatin solution to each of wells C1 through C6.

5. Transfer five drops of 6 M hydrochloric acid to each of wells A1 and C1.

6. Transfer five drops of 6 M sodium hydroxide to each of wells A3 and C3.

7. Transfer five drops of 0.1 M lead(II) acetate to each of wells A4 and C4.

8. Transfer five drops of isopropanol to each of wells A6 and C6.

9. After 30 seconds have passed, observe each of the wells carefully for any visual indication that a reaction has occurred. Record your observations in your lab notebook.

10. If a visible change occurred in wells A1 and/or C1 (the wells to which you added acid), the protein in question is denatured at low pH. Determine if that denaturation is reversible by transferring five drops of 6 M sodium hydroxide to each well, A1 and/or C1, in which a reaction occurred. Record your observations in your lab notebook.

11. If a visible change occurred in wells A3 and/or C3 (the wells to which you added base), the protein in question is denatured at high pH. Determine if that denaturation is reversible by transferring five drops of 6 M hydrochloric acid to each well, A3 and/or C3, in which a reaction occurred. Record your observations in your lab notebook.

PROCEDURE III-3-2: INVESTIGATING ENZYME CATALYSIS

In this procedure, we'll investigate the phenomenon of enzyme catalysis and examine the effects of heat, pH, and heavy-metal ions on the activity of an enzyme catalyst. We'll observe the catalyzed breakdown of hydrogen peroxide solution into water and oxygen, which is evident as bubbles in the solution. Our catalyst is the enzyme peroxidase, which is present in human and animal blood.

One molecule of peroxidase catalyzes the breakdown of millions of peroxide molecules per second, so it's important to use a peroxidase solution of the appropriate concentration. It there is too little peroxidase present, the catalytic breakdown will proceed too slowly; too much, and the peroxide solution will foam uncontrollably. We'll use uncooked meat juice as a source of peroxidase enzyme and adjust the concentration of the solution to provide a usable level of activity.

PREPARE A STANDARD PEROXIDASE SOLUTION

1. Transfer about 1 mL of uncooked meat juice to a 50 mL centrifuge tube and fill the tube nearly full with tap water. Cap the tube and invert it several times to mix the solution thoroughly.

2. Transfer about 5 mL of drugstore 3% hydrogen peroxide to a test tube, add one drop of your peroxidase solution, and swirl the tube to mix the solutions. Observe the intensity of the bubbling that occurs.

3. If the concentration is correct, numerous individual bubbles will form rapidly in the peroxide solution. If the peroxidase concentration is too high, rapid foaming occurs, and part of the solution may actually be ejected from the tube. If the peroxidase concentration is too low, bubbles form slowly, if at all.

4. If your solution is too concentrated, empty half the contents of the centrifuge tube and add tap water to bring up the volume to about 50 mL. Retest and repeat the dilution if necessary until you have a peroxidase solution with a usable activity level. If your solution is too dilute, add additional uncooked meat juice and repeat the testing until the peroxidase concentration is usable.

PREPARE MODIFIED PEROXIDASE SOLUTIONS

1. Transfer about 5 mL of the standard peroxidase solution to each of six 15 mL centrifuge tubes labeled A through F.

2. Tube A is the unmodified peroxidase solution.

3. Cap tube B and place it in the freezer until the solution freezes. When it has frozen, remove the tube and allow the peroxidase solution to melt and come to room temperature.

4. Place tube C in a hot water bath at boiling or nearly so, and allow it to remain in the bath for 10 minutes or so. Remove the tube and allow the contents to come to room temperature.

5. Add five drops of 0.1 M lead(II) acetate solution to tube D. Cap the tube and invert it several times to mix the solutions.

6. Add five drops of 6 M sodium hydroxide solution to Tube E. Cap the tube and invert it several times to mix the solutions.

7. Add five drops of 6 M hydrochloric acid solution to tube F. Cap the tube and invert it several times to mix the solutions.

PREPARE A NEUTRAL SOLUTION OF HYDROGEN PEROXIDE

The pH of hydrogen peroxide solutions varies significantly, depending on concentration and the type of stabilizer present. We want a hydrogen peroxide solution that is at neutral pH or nearly so for our tests.

1. If you have not already done so, put on your goggles, gloves, and protective clothing.

2. Transfer about 80 mL of 3% hydrogen peroxide solution to the 100 mL beaker.

3. Use the stirring rod to transfer one drop of the solution to a piece of pH test paper. After a few seconds, compare the color of the test paper to the color key.

4. If the solution is acidic, add 6 M sodium hydroxide solution dropwise, with stirring, and continue testing the pH until it is approximately neutral (approximately pH 7). If the solution is basic, use the same procedure, but substitute 6 M hydrochloric acid solution.

> Work quickly. Adding acid or base to the hydrogen peroxide solution destroys the stabilizer present in the solution. Unstabilized peroxide quickly breaks down into its components (water and oxygen).
>
> Complete this procedure as quickly as possible after you prepare the neutral peroxide solution. As a final step, retest some of the peroxide solution with unmodified peroxidase solution to verify that oxygen bubbles are still produced. If they are not, that peroxide solution has already broken down, invalidating your results.

TEST THE ACTIVITY OF PEROXIDASE SOLUTIONS

1. If you have not already done so, put on your goggles, gloves, and protective clothing.

2. Label six test tubes, A through F, transfer about 5 mL of the freshly prepared hydrogen peroxide solution to each tube, and place them in the rack.

3. Transfer one drop of the unmodified peroxidase solution from centrifuge tube A to test tube A, and return any unused peroxidase solution to the centrifuge tube. Swirl the test tube, and observe the formation of bubbles. Record your observations in your lab notebook.

4. Using a clean (or well-rinsed pipette), transfer one drop of the modified peroxidase solution from centrifuge tube B to test tube B, return any unused peroxidase solution to the centrifuge tube, swirl the test tube, and observe the formation of bubbles. Record your observations in your lab notebook.

5. Repeat step 4 for for the remaining four peroxidase solutions, C, D, E, and F.

6. If the peroxidase solutions in centrifuge tubes E (strongly basic) and F (strongly acidic) were about as active as the unmodified peroxidase solution, further adjust the pH of those tubes to even more strongly basic and acidic by adding 10 more drops of 6 M sodium hydroxide solution to tube E and 10 more drops of 6 M hydrochloric acid solution to tube F. Swirl the tubes and then retest their contents with fresh hydrogen peroxide solution. Record your observations in your lab notebook.

7. If the peroxidase solutions in centrifuge tubes E (strongly basic) and F (strongly acidic)—either originally or after adding additional base and acid—were inactive or showed noticeably lower activity than the unmodified peroxidase solution, adjust the pH of those tubes to near neutral by adding 5 (or 15) drops of 6 M hydrochloric acid solution to tube E and 5 (or 15) drops of 6 M sodium hydroxide solution to tube F. Swirl the tubes and then retest their contents with fresh hydrogen peroxide solution. Record your observations in your lab notebook.

PROCEDURE III-3-3: ASSAYING VITAMIN C CONCENTRATION IN URINE

Vitamin C, also called *ascorbic acid*, is an essential nutrient, the lack of which causes the horrible disease scurvy. Most mammals produce sufficient vitamin C for their needs; primates (including humans) and guinea pigs do not.

Humans are doubly unfortunate. Not only do we not produce as much vitamin C as we need; we don't store it, either. We must obtain the necessary amount of vitamin C from our diets, and we waste most of the vitamin C we consume. Very little vitamin C is metabolized in the human body; most is excreted unchanged. About 3% of excreted vitamin C is found in the feces, with the remainder excreted in the urine.

The concentration of vitamin C in human urine can vary dramatically, from less than 10 milligrams (mg) per liter (mg/L) to several thousand mg/L. The concentration varies from person to person and from hour to hour for the same person, depending on the amount of vitamin C consumed, frequency and volume of urination, time of day, state of health, and so on. For healthy people, the normal concentration of vitamin C in fresh urine ranges from about 100 mg/L to about 1,000 mg/L.

Vitamin C is a strong reducing agent. We'll use this fact to do a quantitative assay of vitamin C in urine specimens using a procedure called *iodometric titration*. An aqueous or alcoholic solution of iodine is brown. Vitamin C reacts quickly and quantitatively with iodine, reducing the brown elemental iodine to colorless iodide ions and oxidizing the vitamin C to dehydroascorbic acid, which is also colorless. We'll start with an iodine solution and slowly add urine until all of the iodine has been decolorized. By measuring how much urine is required to reach that point, and comparing that value with the amount of a vitamin C solution of known concentration needed to decolorize the same volume of the iodine solution, we can determine how much vitamin C is present in a known volume of that urine specimen.

One molecule of vitamin C reacts with one molecule of iodine, producing one molecule of dehydroascorbic acid and two iodide ions. The gram molecular weight of vitamin C, $C_6H_8O_6$, is 176.126 g/mol, and that of iodine, I_2, is 253.809 g/mol. Because we know that one molecule of vitamin C reacts with one molecule of iodine, we also know that 176.126 grams of vitamin C reacts with 253.809 grams of iodine. Simplifying this ratio, 176.126:253.809, tells us that this reaction consumes about 0.6939 milligrams (mg) of vitamin C per mg of iodine. Or, another way of looking at it, 1.4411 mg of iodine reacts with 1.0000 mg of vitamin C. We'll titrate a known volume of iodine solution with a solution of urine of unknown vitamin C concentration, and use this ratio to calculate the concentration of vitamin C in the urine specimen.

MAKE UP A STANDARDIZED VITAMIN C SOLUTION

In order to determine the unknown concentration of vitamin C in a urine specimen, we need a comparison standard with a known concentration of vitamin C. The easiest way to obtain such a standard is to dissolve a vitamin C tablet (which has a known mass of vitamin C) in a known volume of water.

Recall that the normal range for vitamin C in human urine is about 100 mg/L to 1,000 mg/L. If we make up our standard solution to a concentration near the center of this range, say 500 mg/L, we can expect that it will contain a roughly similar concentration of vitamin C as our urine sample and that roughly similar volumes of the standard solution and the urine should be needed to neutralize a specific volume of iodine solution. For example, if a given volume of iodine solution is neutralized by 20 drops of our 500 mg/L standard solution, we'd expect that same volume of iodine solution to be neutralized by anything from 10 drops of urine that contains 1,000 mg/L to 100 drops of urine that contains 100 mg/L.

> Try to obtain a vitamin C tablet that includes only vitamin C as an active ingredient. If your tablet contains an amount of vitamin C other than 500 mg, such as 100 mg or 1,000 mg, adjust the volume of water accordingly.

1. If you have not already done so, put on your goggles, gloves, and protective clothing.

2. Transfer a 500 mg vitamin C tablet to a 100 mL beaker that contains about 50 mL of distilled water. Swirl or stir the contents to dissolve the vitamin C tablet. (Don't worry if some solids remain undissolved; the tablet may contain starch or other insoluble binders.)

3. Pour the contents of the beaker into the 100 mL graduated cylinder. Rinse the beaker two or three times with a few mL of distilled water and add this to the cylinder to ensure that all of the vitamin C present in the beaker is transferred to the graduated cylinder. (This is called doing a *quantitative transfer*.)

4. Fill the graduated cylinder to the 100.0 mL mark with distilled water and stir to mix the contents thoroughly. At this point, the solution contains 500 mg of vitamin C per 100 mL, or 5,000 mg/L, which is 10 times more concentrated than we want.

5. Allow any solids in the graduated cylinder to settle, and then carefully pour 10.0 mL of the solution into the 10 mL graduated cylinder.

6. Rinse the 100 mL graduated cylinder thoroughly, and then transfer the 10.0 mL of solution from the 10 mL graduated cylinder to the 100 mL graduated cylinder. Rinse the 10 mL graduated cylinder two or three times with distilled water, and transfer the rinse water to the 100 mL graduated cylinder.

7. Fill the 100 mL graduated cylinder to the 100.0 mL mark with distilled water, and stir to mix the solution. At this point, the solution contains 50 mg of vitamin C per 100 mL, or 500 mg/L, which is the concentration we want for our standard solution.

8. Fill a 15 mL centrifuge tube with the standard vitamin C solution and cap it. (Vitamin C in solution quickly breaks down, particularly when exposed to air and light, so try to minimize such exposure over the course of this experiment.) Discard the vitamin C solution remaining in the 100 mL graduated cylinder.

PREPARING STARCH WATER

Even tiny amounts of free iodine react with a starch solution to produce an intensely colored blue-black complex. This phenomenon is used as an indicator for the presence of iodine (or starch). You can use any starch solution for this purpose. (We used some water that we'd cooked pasta in for dinner the previous evening; it keeps for a day or so in the refrigerator.) To prepare starch solution on the fly, simply boil a small amount of potato, rice, or pasta briefly in half a test tube of water.

ASSAY VITAMIN C CONCENTRATIONS

1. If you have not already done so, put on your goggles, gloves, and protective clothing.

2. Transfer 2.0 mL of water to each of two clean test tubes, and place those tubes in the rack.

3. Add 10 drops of Gram's iodine to each of the two test tubes and swirl to mix the contents.

4. Working as quickly as possible, draw up a full pipette of the standard vitamin C solution and recap the centrifuge tube.

5. Add the vitamin C one drop at a time to the first test tube while swirling the tube, keeping track of the number of drops you've added. As you continue adding the vitamin C solution, the color of the contents of the tube gradually fades from brownish to a pale yellow.

6. Once the tube contents appear pale yellow, add a few drops of starch water to the tube. The solution should immediately turn dark, indicating that some free iodine is still present.

7. Continue adding standard vitamin C solution with swirling until the solution in the test tube turns colorless (or very light blue). This may require only one or two more drops of solution.

8. Record the total number of drops of standard vitamin C solution required to neutralize the 10 drops of Gram's iodine in your lab notebook.

9. Repeat steps 4 through 8 using the urine specimen. Given the known volume and concentration of the standard vitamin C solution needed to neutralize 10 drops of Gram's iodine and the known volume of urine needed to neutralize the same amount of Gram's iodine, calculate the concentration of vitamin C in the urine specimen in mg/L and record that value in your lab notebook.

Urine may be anything from nearly colorless to a fairly deep yellow. Urine voided early in the morning is often darker in color and urine voided during the day is usually paler. Yellow urine presents an obvious problem. As you near the endpoint of the titration, the iodine solution fades to a pale yellow color, which can easily be masked by the color of the urine itself. In this situation, the best option is to add the starch indicator solution before you begin the titration. The intense blue color of the complex formed by iodine and starch causes the yellow urine to appear dark blue black. As the color changes to blue-green and then to green, the titration is very near the endpoint. You'll need to add only a drop or two more titrant to complete the titration.

So why not always add the starch indicator at the beginning of the titration? Because the color change from dark blue to colorless is extremely sharp. One drop of titrant may suffice to turn the solution from dark blue to colorless, so you have to add titrant very slowly and carefully to avoid missing the endpoint. If the urine is colorless, you can watch the color of the solution gradually change from dark brown-yellow to pale yellow, so you can titrate much faster until the endpoint approaches.

REVIEW QUESTIONS

Q1: When you tested the reaction of biuret reagent with the various specimens, what was the purpose of adding only distilled water to well A1? What color changes did you observe with the biuret reagent, and from those observations what do you conclude about the makeup of the various specimens?

_____

_____

Q2: In the denaturation tests, what purpose did wells A2, A5, C2, and C5 serve?

_____

_____

Q3: Why did we adjust the pH of the drugstore 3% hydrogen peroxide to near neutral?

_____

_____

Q4: What activities did you detect with the standard and modified peroxidase solutions? What do you conclude about each modification?

_____

_____

Q5: What key assumption do we make in using iodometric titration to determine the concentration of vitamin C in urine?

Q6: Many texts recommend using a dilute solution of sodium thiosulfate to remove iodine stains. What alternative can you propose?

Q7: What implication does the high concentration of vitamin C typically present in urine have for the human diet?

Q8: We may have been mistaken about the stability of vitamin C in solution. Design an experiment to determine if vitamin C in urine in fact degrades when exposed to air and light and, if so, how quickly. Or perhaps the vitamin C concentration in urine decreases over time, regardless of exposure to air or light. State your hypothesis, the results you expect, and your proposed experimental procedure.

Q9: Chemicals that are excreted in the urine (whether unchanged or as metabolites) are said to have a biological half-time (BHT). That is, after one BHT has passed, half of the material has been excreted, after two BHT's have passed, three-quarters has been excreted (the original half plus half of a half, or a quarter), and so on. This is important for many pharmaceuticals. For example, in order to maintain effective serum levels of an antibiotic, the rate at which that antibiotic is excreted must be known and taken into account. An antibiotic with a BHT of 24 hours might be administered every twelve hours, while one with a BHT of twelve hours might be administered every six hours, twice as often. Design an experiment to determine an approximate BHT for vitamin C.

Coacervates

EQUIPMENT AND MATERIALS

You'll need the following items to complete this lab session. (The standard kit for this book, available from *www.thehomescientist.com*, includes the items listed in the first group.)

MATERIALS FROM KIT

- Goggles
- Beaker, 50 mL or 100 mL
- Beaker, 250 mL
- Centrifuge tube, 15 mL
- Centrifuge tubes, 50 mL
- Coverslips

- Gelatin
- Graduated cylinder, 10 mL
- Graduated cylinder, 100 mL
- Gum arabic
- Hydrochloric acid

- pH test paper
- Pipettes
- Reaction plate, 96-well
- Slides (flat)
- Stain, methylene blue
- Stain, Sudan III

- Test tubes
- Thermometer
- Spatula
- Stirring rod
- Stopper (to fit test tube)

MATERIALS YOU PROVIDE

- Gloves
- Balance (optional)
- Microscope

- Microwave oven
- Water, distilled

BACKGROUND

Scientists believe that life originated from simple molecules, a process called *abiogenesis*, when those simple molecules combined into more complex molecules that eventually became complex enough to be self-replicating. We will never know for certain exactly how this occurred. Even if (or when) scientists eventually create life in the lab, the most they will be able to assert is that their method is *a* way that life on Earth may have originated, not that it is *the* way.

One of the pioneers in proposing possible mechanisms for abiogenesis was a Russian scientist named Alexander Oparin, who studied *coacervates*, which are colloidal droplets of hydrophobic molecules in an aqueous medium. These droplets are typically about the size of a cell—1 to 100 micrometers (μm) in diameter—and at first glance under a microscope resemble a cluster of cells. Coacervates form spontaneously in mixtures of proteins and carbohydrates at specific pH values.

Coacervates are not alive, but they present a simulacrum of life as they form what appear to be cell membranes, ingest materials from their environment, grow, and reproduce. Although most current abiogeneticists assign higher importance to information transfer (e.g., DNA and RNA), coacervates may have had a key role in forming the nonliving precursors to the first living cells.

In this lab session, we'll prepare and observe a coacervate, using gelatin as the protein and gum arabic as the carbohydrate.

> The gelatin is best prepared immediately before use. Allow it to cool to room temperature before proceeding. If necessary, you can prepare the gelatin the day before and store it in the refrigerator. When you remove it from the refrigerator, immerse the container in a bath of hot water to remelt the gelatin.
>
> If you don't have gelatin and/or gum arabic, or if you'd like to prepare a coacervate using other types of protein and carbohydrate, you can substitute a level teaspoon of table sugar dissolved in about 400 mL of water as the carbohydrate and/or about 10 mL of uncooked egg white diluted to about 125 mL with water as the protein. Always add the carbohydrate solution to the protein solution when making coacervates.

PROCEDURE III-4-1: PREPARE AND OBSERVE A COACERVATE

If you have not done so before the session, prepare 1% gelatin and gum arabic solutions as follows:

1. Transfer about 40 mL of distilled water to a labeled 50 mL centrifuge tube, and heat it in the microwave until it reaches about 60 °C. (**Careful: HOT!**)

2. Transfer about 10 mL of cold distilled water to a small beaker, and gradually stir in 0.5 g (one rounded spatula spoon) of gelatin powder. Stir thoroughly, making sure there are no lumps and the powder is completely dispersed.

3. Pour the gelatin slurry slowly and with constant stirring into the hot water in the 50 mL centrifuge tube. Stir until all the gelatin has dissolved and the solution is clear. If necessary, you can reheat the contents of the tube in the microwave and continue stirring until the gelatin solution is clear. Allow the tube to cool to room temperature.

4. Repeat steps 1 through 3 with a second tube, substituting 0.5 g (one rounded spatula spoon) of gum arabic powder.

If you have not done so before the session, prepare a dilute solution of hydrochloric acid as follows:

1. If you have not already done so, put on your goggles, gloves, and protective clothing.

2. Label a 15 mL centrifuge tube "dilute HCl" and transfer about 12.5 mL of distilled water to it.

3. Using a plastic pipette, draw up 0.25 mL of 6 M hydrochloric acid and transfer the acid to the water in the centrifuge tube. (The graduation lines on the pipette stem are at 0.25, 0.50, 0.75, and 1.00 mL.)

4. Replace the cap on the tube, invert the tube several times to mix the solution, and set it aside for now.

To make the coacervate, proceed as follows:

1. If you have not already done so, put on your goggles, gloves, and protective clothing.

2. Transfer 5 mL of the room-temperature gelatin solution to a test tube, followed by 3 mL of the gum arabic solution. Stopper the tube and invert it *gently* several times to mix the solutions. Do not shake or otherwise strongly agitate the contents of the tube, which will hinder the formation of the coacervate.

3. Observe the contents of the test tube against a strong light. Record the appearance of the contents, particularly with regard to clarity, in your lab notebook as "Trial 1."

4. Use the stirring rod to transfer one drop of the liquid to a small piece of pH paper. Compare the color of the paper against the scale provided with the pH paper to determine the approximate pH of the liquid. Record this value in your lab notebook.

5. Label a microscope slide "Trial 1." Transfer one drop of the liquid to the slide, put a coverslip in place, and observe the slide at low magnification. If anything interesting appears, center the object(s) and observe them at higher magnification. Record your observations in your lab notebook. Retain the slide for comparison with later trials.

6. Use a plastic pipette to transfer 0.25 mL of the solution to well A1 of the reaction plate.

If you're unable to find any coacervates, don't despair. The formation of coacervates is strongly dependent on pH. In subsequent trials, we'll adjust the pH of the liquid by adding dilute hydrochloric acid until coacervates begin to appear.

7. For trial 2, transfer three drops of the dilute hydrochloric acid to the test tube. Stopper the tube and invert it gently several times to mix the solutions. Repeat steps 3 through 6, using well A2 of the reaction plate to store 0.25 mL of the solution from trial 2.

8. Repeat step 7 for additional trials until the pH of the liquid in the test tube reaches 3 or lower. After each addition of hydrochloric acid to the liquid, transfer a 0.25 mL sample to the corresponding well in the reaction plate.

Once you have determined experimentally the pH level that is optimum for the formation of coacervates, the next step is to observe the effect of two biostains on the coacervates to determine if coacervates, like living cells, have selectively permeable membranes. The first, methylene blue, is a hydrophilic (water-loving) stain. The second, Sudan III, is a lipophilic (fat-loving) stain.

9. Place one drop of the optimum-pH coacervate liquid on a microscope slide, add one drop of methylene blue stain, and position a coverslip over the specimen. Observe the slide at low, medium, and high magnifications. Note the effect of the stain on the parts of the coacervate and record your observation in your lab notebook.

10. Repeat step 9, using Sudan III stain.

REVIEW QUESTIONS

Q1: What visual evidence suggests that coacervates are forming?

Q2: Based on your observation of the coacervate slides, what pH level is optimum for the formation of coacervates?

Q3: What correlation did you observe between the clarity of the liquid and the population of coacervates?

Q4: In what ways do coacervates resemble cells?

Q5: What do you conclude from your experiments with the methylene blue and Sudan III stains?

Extracting, Isolating, and Visualizing DNA

Lab III-5

EQUIPMENT AND MATERIALS

You'll need the following items to complete this lab session. (The standard kit for this book, available from *www.thehomescientist.com*, includes the items listed in the first group.)

MATERIALS FROM KIT

- Goggles
- Centrifuge tubes, 50 mL
- Coverslips
- Eosin Y stain
- Funnel
- Graduated cylinder, 10 mL
- Methylene blue stain
- Microscope slides (flat)
- Pipettes
- Reaction plate, 96-well
- Scalpel
- Sodium dodecyl sulfate, 10%
- Spatula
- Stirring rod
- Test tubes
- Test tube rack
- Yeast (optional)

MATERIALS YOU PROVIDE

- Gloves
- Balance (optional)
- Beef or pork liver, raw (or yeast)
- Cheesecloth (or muslin, etc.)
- Isopropanol (see text)
- Freezer
- Microscope
- Paper towels
- Saucer
- Table salt
- Teaspoon
- Toothpick
- Water, distilled

BACKGROUND

All organisms are made up of cells, from the tiniest species such as single-cell bacteria and protozoa to the largest animals and plants, which are made up of trillions of individual cells. Every cell contains DNA, which is the hereditary genetic material that allows cells and organisms to function and to reproduce themselves.

With very few exceptions (research *chimera* on the Internet; the genetic kind, not the mythological kind) DNA taken from any cell of a particular individual of any species is identical to that taken from any other cell from that individual, and is unique to that individual. DNA found in unrelated members of the same species—for example, you and your best friend (assuming your best friend isn't a dog or a diamond...) or two rosebushes—is nearly (but not quite) identical. DNA found in different but closely related species—for example, lions and tigers or wolves and coyotes—has greater differences, but is still extremely similar. DNA found in different, unrelated species—for example, a human and a cucumber—has still greater differences, but remains closely similar.

But what does DNA actually look like? Let's find out. In this lab session, we'll extract liver-cell DNA into a solution, isolate that DNA, and stain it to visualize it.

PROCEDURE III-5-1: EXTRACTING AND VISUALIZING DNA

1. If you have not already done so, put on your goggles, gloves, and protective clothing.

2. Transfer five drops of methylene blue stain and five drops of eosin Y stain to one well of the reaction plate. Use a toothpick to mix the solutions thoroughly, and place the reaction plate aside for later use.

3. Make up 50 mL of *normal saline* by dissolving 0.45 g (one slightly rounded spatula spoon) of table salt in 50 mL of distilled water in a 50 mL centrifuge tube, and place the tube aside for later use.

4. Transfer about 40 mL of isopropanol to a second 50 mL centrifuge tube. Cap the tube and place it in the freezer to cool.

> 99% isopropanol yields the best results, but 91% or even 70% isopropanol is usable. You can also substitute ethanol for isopropanol.

5. Obtain a teaspoon-size specimen of fresh beef or pork liver.

> It's best to use fresh liver from a butcher or fresh market. Supermarket liver can be used but may yield inferior results, depending on its age. Also, you may substitute dried baker's or brewer's yeast for the liver, although *lysis*—breaking down the cell membranes and releasing the contents of the cells—is more difficult to observe with yeast.

6. Use the scalpel to cut the liver into very small pieces, and transfer the pieces to the saucer.

7. Add about 10 mL of normal saline solution to the saucer, and use the teaspoon to grind and mash the bits of liver to produce a suspension of liver cells in the saline solution.

8. Place a clean test tube in the rack, with the funnel atop it. Fold and refold a small piece of cheesecloth to provide four layers. Place the cheesecloth in the funnel and pour the solution from the saucer through the cheesecloth into the tube to filter out most of the solid material.

9. Use the stirring rod to transfer one drop of the liver cell suspension to a microscope slide and position a coverslip over the specimen. Observe the specimen at low, medium, and high-dry magnifications. Record your observations in your lab notebook. Include a sketch of a representative liver cell and its nucleus.

10. Add one drop of the mixed stains at the edge of the coverslip, and use the corner of a paper towel to draw the stain under the coverslip. Observe the specimen at low, medium, and high-dry magnification. Record your observations in your lab notebook. Include a sketch of a representative liver cell and its nucleus.

11. Use a pipette to transfer 0.5 mL of 10% sodium dodecyl sulfate (SDS) solution to the liver-cell suspension tube, and swirl gently to to mix the solutions.

12. Repeat steps 9 and 10 to stain and view the liver cells in the presence of the SDS solution.

13. Transfer another 0.5 mL of the 10% SDS solution to the liver-cell suspension and swirl gently to mix the solutions.

14. Repeat steps 9 and 10 to stain and view the liver cells in the presence of the SDS solution.

15. Repeat steps 13 and 14 several times (about five repetitions should do it) and observe the liver cells in the presence of increasing concentrations of SDS. As the cell membranes break down (are *lysed*) and no longer surround the nucleus, the integrity of the cells is destroyed and the nucleus and other cell contents are exposed.

16. Using a pipette, transfer about 4 mL of the liver-cell suspension to a second test tube. Retain the excess liver-cell suspension in the first test tube until you complete the next step successfully.

17. Holding the second tube at a slight angle, very slowly and carefully trickle cold isopropanol into the tube until the liquid level reaches 2 to 3 cm from the top of the tube. The goal is to cause the isopropanol to form a separate layer on top of the liver-cell suspension. If you pour too fast, the layers will mix and you'll have to start over.

18. Place the tube in the rack and observe the interface between the liver-cell suspension layer and the isopropanol layer. You'll see a cloudy whiteness begin to form at the interface. That whiteness is DNA precipitating out of solution.

19. Carefully and gently dip the stirring rod into the test tube until it penetrates both layers. Slowly spin the stirring rod between your thumb and finger to spool the DNA precipitate onto the stirring rod. Continue until no more DNA is deposited on the stirring rod. Withdraw the rod from the test tube.

20. Place one drop of distilled water on a microscope slide. Use a toothpick to remove a tiny amount of the DNA from the stirring rod. Transfer the DNA to the water drop and position a coverslip over the specimen. Add one drop of methylene blue at the edge of the coverslip, and use the corner of a paper towel to draw the stain under the coverslip. Observe the DNA at low, medium, and high-dry magnification. Record your observations in your lab notebook, including a sketch of the stained DNA.

If you want to retain the DNA you just isolated for future use, allow it to dry and then store it in a paper bag or similar porous container.

REVIEW QUESTIONS

Q1: What was the result of adding the mixed stains to the liver-cell suspension?

_____

_____

_____

Q2: Methylene blue stains acidic cell structures but leaves basic cell structures unstained. Based on that information, what do you conclude about cell nuclei?

_____

_____

_____

Q3: What did you observe as you increased the amount of SDS solution you added to the liver cell suspension?

_____

_____

_____

Q4: How did the appearance of the DNA differ before and after staining with methylene blue? What do you conclude about the acidity or basicity of DNA?

_____

_____

_____

Q5: What do you conclude about the relative solubilities of DNA in water and isopropanol? Propose a reason why we chilled the isopropanol.

_____

_____

_____

Build a Gel Electrophoresis Apparatus

Lab III-6

EQUIPMENT AND MATERIALS

You'll need the following items to complete this lab session. (The standard kit for this book, available from *www.thehomescientist.com*, includes the items listed in the first group.)

MATERIALS FROM KIT

- Leads, alligator clip (2)
- Ruler

MATERIALS YOU PROVIDE

- Aluminum foil
- Batteries, 9V transistor (5, 7, or 9)
- Gel casting comb materials (see text)
- Marking pen (Sharpie or similar)
- Plastic containers (see text)
- Scissors
- Tape (electrical or masking)

BACKGROUND

There's a time-honored custom in science: we scientists build our own apparatus. Not always, of course, nor even most of the time. Nowadays, anyway. Sometimes it's cheaper, easier, and faster just to buy what we need. But when we need something to complete an experiment and that something isn't available or we don't have the budget to buy it, we make do. If that involves designing and building a piece of equipment, so be it. For the following lab session, we needed a gel electrophoresis apparatus and power supply, and we didn't feel like spending $300 or $400 to buy one. So we designed and built our own, at a total cost of about $10.

Like paper chromatography, gel electrophoresis is used to separate mixtures of chemical compounds. (In biology, gel electrophoresis is often used to separate proteins or DNA fragments.) The similarities do not end there. In paper chromatography, the fixed phase (matrix) is a strip of paper; in gel electrophoresis, the matrix is a bed or column of gel. In both cases, the matrices provide resistance to the movement of molecules. In paper chromatography, capillary action (wicking) provides the force required to move the mobile phase (solvent) and analyte(s) through the matrix; in gel electrophoresis, the force is supplied by an external electrical current, with positively charged molecules attracted to the negative electrode and vice versa.

Although gel electrophoresis is conceptually simple, commercial gel electrophoresis setups sell for $300 and up, and the gels, buffers, and other supplies they require are not cheap. If you can afford a commercial apparatus and supplies, there's no question you'll find it easier and faster to run gels. However, most homeschoolers and hobbyists can't justify spending several hundred dollars on a gel electrophoresis apparatus and supplies. (Check with your homeschool group or co-op, which may have such an apparatus available to borrow.) Fortunately, with a few minutes' work, you can assemble your own apparatus and supplies for only a few dollars, using items from the kit and some common household items. Here's what you'll need:

Shallow plastic container

You'll need a shallow, easily cut plastic container to serve as the gel casting bed. We used the base of a one-quart Gladware container, trimmed to about 1 cm tall. You can also use a soap dish or similar container. If you want to cast multiple gels in one pass, make several of these gel bed containers.

Deeper container

You'll need another plastic container that's larger and deeper than the gel bed. We used a sandwich container, but you can substitute any suitable container. This container needs to be at least as long and wide as the gel bed container—although, ideally, not all that much longer and wider—and at least a centimeter or so taller.

Tape

We'll cut the ends of the gel bed container off, but we need to be able to seal them temporarily back in place while casting gels. We used plastic electrical tape, but you can substitute ordinary masking tape. Once the gel solidifies, you simply peel off the tape to expose both ends of the gel. Tape may also be useful in making the comb that's used to produce wells in the gel.

Comb

No, not the kind you use on your hair. A gel electrophoresis comb is a comb-like assembly with teeth that are placed in the gel casting container to produce small wells in the gel as it solidifies. You can buy a commercial comb for an outrageous price, or you can improvise your own with common household materials.

We've made combs from (a) cardboard and aluminum foil or disposable coffee stirrers, (b) wooden paint stirrers and nails or disposable chopsticks, and (c) Lego blocks. The only important things are that the prongs be relatively smooth (so as not to tear up the solidified gel when they're removed), roughly 3 mm to 4 mm in diameter (or the equivalent area if the teeth are square or rectangular), long enough to reach the bottom of the gel casting container (or nearly so), and spaced at intervals of roughly 1 cm across the width of the gel casting container. Use your imagination and materials at hand.

Aluminum foil

The apparatus requires an electrode at each end of the larger container. We'll use two pieces of aluminum foil to make these electrodes. You may also use the aluminum foil to make the teeth in the comb used to produce wells in the gel.

9V transistor batteries

The apparatus requires a DC power supply, which can be anything from 30 or 40 VDC to 150 VDC or more. The lower the voltage, or the larger the gels, the longer they take to run.

Five 9V transistor batteries connected in series provide 45 VDC, which is enough to run small gels in a reasonable time. Using seven 9V batteries for 63 VDC or nine 9V batteries for 81 VDC lets you run gels faster, or larger gels in the same amount of time.

Why the odd numbers? Because we'll connect the batteries in series by snapping the positive connector on one battery to the negative connector on the next battery, as shown in Figure III-6-1.

> For reasonably fast run times, use at least 3V per centimeter of gel length. For example, if your gel bed is 15 cm long, your power supply should be at least 45 VDC. (With these parameters, running a gel may require two or three hours or more, depending on the type and concentration of the gel.)

WARNING

Be careful with 9V batteries connected in series. Depending on your skin resistance and other factors, even 45 VDC may produce a painful and potentially dangerous electrical shock. **Never** touch the power leads of the battery assembly or touch the liquid in the gel electrophoresis apparatus while power is connected to it.

Never close the circuit (that is, once they are connected to the battery stack, never allow the free ends of the positive (red) and negative (black) leads to contact each other. That short circuit can rapidly weld the connection and cause the batteries to overheat and explode. (It's safe to connect the positive terminal of one battery to the negative terminal of the next because the circuit has not been closed unless you allow the final open positive and negative terminals to contact each other, such as by allowing the free end of the connected red lead to contact the free end of the connected black lead.)

If you run many gels, the cost of 9V batteries starts to add up fast, so it's cheaper in the long run to buy a DC power supply. For smaller gels, any DC power supply that provides fixed-, multiple-, or variable-output DC voltages in the range of 50 to 70 VDC or higher at a few hundred mA will suffice. We've seen suitable single-voltage units on eBay for $30 to $90, depending on voltage.

Figure III-6-1: *A 9V transistor battery stack*

PROCEDURE III-6-1: MAKING THE GEL CASTING CONTAINER AND COMB

1. To begin, place the marking pen flat on a table and measure the distance between the table surface and the point of the pen. If it is less than 1 cm, place magazines or other spacers between the table surface and the pen until the tip of the pen is about 1 cm above the table surface.

2. Holding the pen firmly in place, place the smaller plastic container flat of the table, and bring its surface into contact with the tip of the pen. Move the plastic container to draw a line around its exterior that is 1 cm above its bottom, as shown in Figure III-6-2.

Figure III-6-2: *Marking the gel casting tray for depth*

3. Use scissors to trim the excess from the plastic container, converting it to a shallow plastic tray.

4. Use the scissors to remove most of both ends of the tray, as shown in Figure III-6-3. Retain these cut ends for future use.

Figure III-6-3: *Removing the ends from the gel casting tray*

5. Hold the ends of the tray in position, and tape them securely back in place, as shown in Figure III-6-4. These taped ends will hold the liquid gel in the tray until it's solidified. Before running the gel, you'll remove the ends to allow current to flow through the gel.

Figure III-6-4: *Taping the ends of the gel casting tray into place*

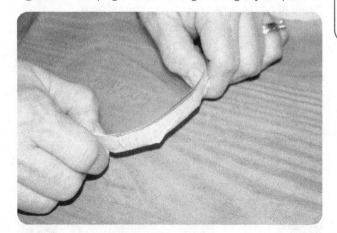

Place the gel tray aside for now. If you want to cast multiple gels in one pass, make several of these trays before continuing. The next step is to make the comb. Here's one way to do it (but feel free to experiment with other methods...).

6. Cut a strip of stiff cardboard about 3 cm wide and 2 cm longer than the width of your gel tray. Draw a centered line the length of the strip. Place tick marks on that line, beginning 1.5 cm from each end of the strip and then evenly spaced at about 1 to 1.5 cm intervals across the width of the gel bed container.

7. Twist pieces of aluminum foil to form stiff cylinders about 2.5 cm long by 3 mm in diameter. (The number you need varies by the width of your gel tray; make one of these cylinders for each of the tick marks across the tray width.) Try to make the surface of the teeth as smooth as possible.

8. Make a 90° bend in each cylinder 1 cm from one end, forming an "L" shape, with one leg 1 cm long and the other 1.5 cm long.

9. Place each cylinder with the bend on a tick mark, and press the longer leg flat against the cardboard. Securely tape the longer leg of each cylinder to the cardboard, with the 1 cm long leg projecting vertically from the surface of the cardboard.

Calculate the volume of the tray by measuring and multiplying its three dimensions. For example, if your tray is 7 cm wide by 10 cm long by 1 cm deep, its total volume is (7 * 10 * 1) = 70 cm^3, which is the same as 70 mL. When you cast gels, you'll fill the tray to about 0.6 cm (6 mm) to 0.8 cm deep, so each gel tray requires something between 42 mL (0.6 * 70 mL) and 56 mL of gel.

PROCEDURE III-6-2: ASSEMBLE THE APPARATUS

1. Fold aluminum foil over each end of the larger container and connect one end of an alligator clip lead to each of the foil electrodes. (Don't connect the other ends of the alligator clip leads to the battery stack until you're actually ready to run a gel.)

2. Place the gel bed container inside the larger container, centered within the larger container. Verify that there is at least a few millimeters of clearance between each end of the gel bed container and the corresponding electrode on the larger container. Also verify that all sides of the larger container are a centimeter or so higher than the sides of the gel bed container.

At this point, you're ready to cast and run gels, which you'll do in the next lab. Incidentally, believe it or not, the apparatus you've just built is fully capable of running gels just as good as those produced by expensive professional apparatus. It's clumsier to use, certainly, but if you use the same agarose and stains that are used on professional apparatus, the resulting gels will be indistinguishable.

Simulated DNA Separation by Gel Electrophoresis

EQUIPMENT AND MATERIALS

You'll need the following items to complete this lab session. (The standard kit for this book, available from *www.thehomescientist.com*, includes the items listed in the first group.)

MATERIALS FROM KIT

- Goggles
- Agar
- Beaker, 250 mL
- Graduated cylinder, 100 mL
- Glycerol
- Pipettes
- Reaction plate, 96-well

- Ruler
- Stain: Hucker's crystal violet
- Stain: Methylene blue
- Stain: Safranin O
- Stirring rod
- Thermometer

MATERIALS YOU PROVIDE

- Gel electrophoresis apparatus (from III-6)
- Measuring spoons
- Microwave oven
- Soda bottle (2 liter, clean and empty)

- Sodium chloride (table salt)
- Sodium bicarbonate (baking soda)
- Toothpicks
- Water (distilled or tap; see text)

BACKGROUND

We originally designed this lab session to use real DNA, but we decided instead to simulate DNA fragments of different sizes by using dyes with different molecular weights. Dyes yield reasonably sharp separations even with ordinary agar and homemade buffer solutions, but achieving sharp separations with actual DNA requires using expensive agarose, restriction enzymes, and special buffers.

> If you would like to delve further into gel electrophoresis and DNA analysis, one of the best sources of equipment and supplies is Edvotek (*http://www.edvotek.com*). Edvotek packages many different kits designed to illustrate various aspects of DNA analysis in a classroom setting.
>
> If you would like to use the gel electrophoresis apparatus you built in the preceding lab session to separate real DNA fragments (from ground beef) but don't want to buy agarose, restriction enzymes, and so on, download the original version of this lab session from *www.thehomescientist.com/biology/lab-3-7a.pdf*. Completing that alternate version of the lab session takes longer and yields blurrier separations than are achieved with the dyes used in this simulated separation, but you will be separating actual DNA.

Gel electrophoresis is conceptually similar to chromatography, but with a slightly different goal. Ordinarily, we use chromatography to separate different compounds from a mixture. With DNA gel electrophoresis, the goal is to separate DNA fragments of different sizes and masses, which are produced by treating a DNA specimen with *restriction enzymes* to cleave it at known locations into fragments.

To imagine the shape and structure of DNA, think of a standard ladder, with two sides and rungs joining them. Twist the tops of the sides of the ladder to form two counter-rotating, interlocked spirals, with the rungs still joining them. In DNA, the sides of the ladder are made up of molecules of a sugar called 2-deoxyribose, with those sugar molecules bonded together by phosphate groups. The rungs, called *base pairs*, are formed by bonded pairs of four amino acid bases: called *adenine* (abbreviated A), *cytosine* (C), *guanine* (G), and *thymine* (T). Adenine bonds only with thymine, forming an *AT base pair*, and cytosine only with guanine, forming a *CG base pair*.

To visualize the complexity of DNA, imagine that your ladder, rather than having only a few rungs, has millions or billions of rungs. Genomic human DNA, for example, has about 3.2 billion base pairs. Untwisted and stretched out, a single strand of genomic human DNA would form a "ladder" about 2.4 µm wide and 2 meters long, with 3.2 billion rungs.

The number of base pairs in genomic DNA is unrelated to the complexity of the organism. For example, while honeybee DNA has only about half as many base pairs as human DNA (1.77 billion versus 3.2 billion), the fruit fly has about 130 million, E. coli bacteria about 4.6 million, and the marbled lungfish has about 130 billion base pairs—more than 40 times the human count—and DNA from one species of amoeba has 670 billion base pairs.

Restriction enzymes function like tiny scissors, cutting the DNA into fragments for subsequent separation by gel electrophoresis. In effect, a restriction enzyme "looks for" a specific sequence of base pairs, such as 3 AGG'CCT. When the restriction enzyme "finds" that specific sequence, it cuts the DNA at that point.

Because that specific short base sequence inevitably occurs many times in any DNA molecule, treatment with that one restriction enzyme cuts that DNA molecule into many fragments. Because repeated occurrences of that base sequence may be relatively close together or far apart on the DNA molecule, the fragments are of differing lengths, typically from hundreds to thousands of base pairs.

The specific size distribution of the DNA fragments obtained by treating DNA with a particular restriction enzyme varies according to the positions of the target base sequence in that particular DNA specimen. So, for example, treating your DNA with a particular restriction enzyme results in a different distribution of fragment sizes than treating your lab partner's DNA with the same restriction enzyme. The distribution of fragment sizes in your DNA specimen is unique to you, just as everyone else's distribution is unique to them.

Once the DNA has been cut into fragments by the restriction enzyme, gel electrophoresis is used to analyze the DNA by producing a map of the fragment sizes present in the specimen. A sample of the fragmented DNA is placed in a small well in a gel, which is then immersed in a buffer solution and subjected to DC electric current. Because DNA fragments are negatively charged, they are attracted to the positive electrode, which is positioned at the far end of the gel from the wells that contain the DNA solution.

The gel selectively retards the migration of the DNA fragments toward the positive electrode. Small DNA fragments pass through the gel relatively unhindered, and so move toward the positive electrode quickly. Larger fragments move proportionally more slowly, because the gel provides more resistance to their progress. If the current is applied until the smallest fragments just reach the end of the gel nearest the positive electrode, larger fragments are strung out along the length of the gel, with the largest fragments barely clear of the well where they originated. The positions of the various fragments provide a graphical map of the fragment size distribution in the specimen.

But DNA fragments are colorless, so it's impossible to track the progress of the electrophoresis visually. For that reason, a *marker dye* is added to the DNA specimen before the electrophoresis run. The marker dye is chosen on the basis of how fast it migrates through the gel. By using a marker dye that moves about the same speed as the smallest DNA fragments of interest, electrophoresis can simply be discontinued when the visible marker dye approaches the positive electrode, at which point the smallest DNA fragments have made about the same amount of progress through the gel.

At this point, the only visible change to the gel is a band of marker dye near the positive electrode end. To visualize the bands of colorless DNA fragments, they're stained using dyes that are selectively attracted to the fragments. The most common stain used in professional laboratories is ethidium bromide, which bonds to the DNA fragments and fluoresces under ultraviolet light to reveal the DNA fragments as bright bands against the dark background of the gel.

PROCEDURE III-7-1: PREPARE RUNNING BUFFER

Running buffer is used to make up the gel initially, and later to immerse the gel while doing the electrophoresis run. The amount of running buffer you need depends on the size of your electrophoresis apparatus, the size of the gels you make, and the number of runs. Fortunately, running buffer is very cheap to make up, so you can make it in excess and discard any that's left over when you complete this lab session. Actually, for our simulation we'll use a very simple running buffer that wouldn't work very well with real DNA but is suitable for separating dyes.

1. Rinse out the 2 L soda bottle thoroughly and fill it to the 2 L level with water.

2. Add 1/8 teaspoon of table salt and a heaping tablespoon of baking soda to the soda bottle, cap it, and invert it several times to dissolve the solids. Invert the bottle every few minutes for half an hour or so to make sure the solution is homogeneous.

> For a real DNA separation, we'd use distilled or deionized water to make up the running buffer and we'd use different salts but for this simulation most tap water will work fine. If your tap water is extremely hard or has much iron in it, use distilled water. Otherwise, tap water should work.

PROCEDURE III-7-2: PREPARE AND CAST THE GEL

1. If you have not done so already, calculate the volume of agar gel you need to cast your gel. Multiply the length and width of the gel chamber in centimeters by the desired thickness of the gel to determine how many cubic centimeters (mL) of gel you'll need. For example, if your gel chamber is 7 cm by 10 cm, and you want a 1 cm thick gel, you'll need (7 * 10 * 1) = 70 cubic centimeters = 70 mL of gel.

2. Transfer that volume of room-temperature running buffer to the beaker.

3. Stir in about 2 g (1/2 to 5/8 teaspoon) of agar powder per 100 mL of running buffer. (The exact amount of agar is not critical.)

> The kit contains 10 grams of agar powder, which is sufficient to make up 500 mL of 2% agar gel. Agar gel is used both in this lab and in Lab VII-2. If you need more agar, you can use agar that is sold by specialty grocery stores and some Chinese and Japanese restaurants. If you use agar from the grocery store, we recommend using a 3% to 4% concentration as a starting point.
>
> You can substitute unflavored gelatin from the grocery store **for dye separations only**. (Using gelatin for DNA separations results in a horrible mess.) Use double the amount of gelatin suggested on the package, otherwise following the procedure described for agar. Run gelatin gels at no more than 63 VDC and keep a careful eye on progress. If the gelatin overheats it will melt and may cause a short in the apparatus.

4. Carefully heat the liquid in the microwave just until it begins to foam slightly. Keep a very close eye on the beaker as you heat it.

WARNING

The first time we made up 100 mL of agar in the microwave, we figured that one minute would be about right to get the stuff hot but not yet boiling. We intended to heat it for that one minute and then blip it for 5 or 10 seconds more at a time until it boiled.

At about the 50 second mark, the contents of the beaker erupted volcanically, with agar foam flowing up out of the beaker and down its side. We opened the door immediately, but we still ended up with 25 mL of our agar on the microwave tray.

The moral here is that if you use the microwave to heat your agar, do so in very short bursts. You can keep an eye on the actual temperature by using the thermometer (carefully) as a stirring rod. It's not actually necessary for the agar solution to boil. Getting it up near the boiling point but not actually boiling the solution is a good way to prevent messy accidents.

5. Using a towel or oven mitt to protect your hand from the heat, remove the agar liquid from the microwave and set it aside to cool. While the agar cools, set up your gel electrophoresis apparatus on a level surface. Verify that the tips of the comb do not contact the bottom of the gel casting tray.

6. Once the agar has cooled to the point where it feels very warm or hot to the touch (about 50 to 60 °C), pour the agar liquid into the gel casting tray to a depth of roughly 1 cm. If there are any bubbles, use the stirring rod to eliminate them. While the agar is still hot and flowing freely, place the comb near one end of the casting tray with its teeth immersed in the liquid agar.

7. Allow the agar to cool and gel completely. Once it has set up, carefully remove the comb. Try to avoid tearing the agar gel when you do so. The goal is to have a solid, flat gel with a series of small neat holes near one end. Remove the ends from the gel casting tray.

PROCEDURE III-7-3: LOAD AND RUN THE DYE SPECIMENS

If you want sharper separations with agar gels, you can prerun the gel. That is, run the gel normally, but without loading specimens in any of the wells. After 15 minutes or so, disconnect the power, remove the gel tray from the running buffer, load the specimens, and rerun the gel. The prerun clears out a lot of the gunk that's present in agar (but not agarose) and allows the agar to function more like actual agarose.

A prerun is not necessary for this lab session. If you do a prerun, you'll probably find that you get much sharper separation of the dyes, but for our purposes, simply measuring from the well to the center of the dye cloud is good enough.

1. Transfer nine drops of Hucker's crystal violet stain to one well of the 96-well reaction plate, nine drops of methylene blue to a second well, and nine drops of safranin O to a third well. To a fourth well, transfer three drops each of these three stains.

> We'll use the reaction plate and plastic pipettes rather than attempting to transfer the stains directly from their bottles to the wells in the gel because the former method is much more precise.

2. Transfer one drop of glycerol to each of the four wells containing dyes. Use a clean toothpick to stir each well until the dye solution and glycerol is thoroughly mixed.

3. Use a clean pipette to transfer the Hucker's crystal violet dye and glycerol mixture to one well of the gel to fill the well about half full. (Depending on the size of your wells, one drop may suffice.)

4. Repeat the preceding step to transfer the methylene blue, safranin O, and mixed stain mixtures to three vacant wells of the gel.

> If you have a fifth well in your gel, you can place the individual stains in wells 1, 3, and 5 and the mixed stains in both wells 2 and 4 to make it easier to compare migration distances.

5. Carefully place the gel tray centered in the outer container (electrophoresis tank). Gently pour sufficient running buffer into the electrophoresis tank to fill it to a level just above the top of the gel tray. The goal is to immerse the gel tray completely in running buffer.

6. Verify that the aluminum foil electrodes are in place on the electrophoresis tank. Connect the red alligator clip lead to the electrode nearest the well end of the gel tray and the black alligator clip lead to the other electrode.

7. Connect the free end of the red alligator clip lead to the exposed positive terminal on your 9V battery stack.

8. Carefully (**shock hazard!**) connect the free end of the black alligator clip lead to the exposed negative terminal on your 9V battery stack.

HANDLE WITH CARE

Do not allow the red and black leads to contact each other directly!

A short circuit is very bad news. Depending on the voltage of the battery stack, the best you can hope for is a show of sparks and destruction of your battery stack. It's possible that the clips will weld together, that the wires will overheat and vaporize, and that the insulation will catch fire.

9. Observe the gel until some migration of the dyes is visible. If all is well, all dyes should be migrating toward the far end of the gel. If instead they are migrating toward the near end of the gel, the polarity is reversed. Switch the positions of the black and red leads.

WARNING

Depending on the voltage you use, there may be enough voltage exposed on the running buffer surface to give you a severe shock. Never touch the apparatus (and particularly the running buffer) while power is connected to the apparatus. You can tell it's working because bubbles appear at the positive electrode.

10. Continue observing the gel until one of the dyes has migrated nearly to the far end of the gel. Depending on the size, type, and concentration of the gel and the voltage of the battery stack, this may require anything from a few minutes to a couple of hours. When the farthest advanced dye has nearly reached the far end of the gel, disconnect both leads from the electrodes and the battery stack. Allow the gel to cool to room temperature before handling it.

11. For each well, measure the distance from the middle of the well to the middle of the corresponding dye streak. Record these values in your lab notebook.

12. Calculate the distances each dye migrated as ratios to the dye that advanced farthest. For example, if dye #1 migrated 10.0 cm, dye #2 7.8 cm, and dye #3 5.3 cm, record the ratios for dye #1 as 1.00, dye #2 as 0.78, and dye #3 as 0.53.

> This calculation is similar conceptually to the R_f value used in paper chromatography. The difference is that in paper chromatography the baseline is the distance that the solvent front advanced. With gel electrophoresis, there is no solvent front, so we use the distance that the farthest-advanced dye migrated as the baseline value.

REVIEW QUESTIONS

Q1: The kit also contains eosin Y stain and Sudan III stain. Why did we not use these stains in our gel run? (Hint: research acidic, basic, and neutral dyes. Second hint: as applied to stains, these terms refer to ionization rather than pH.)

_____

_____

_____

_____

Q2: Could any of the three stains we used be used as marker dyes if we were doing actual DNA electrophoresis? Why or why not? If not, what common dyes, stains, and pH indicators could be used as marker dyes?

_____

_____

_____

_____

Q3: The gram molecular masses of crystal violet, methylene blue, and safranin O are 407.979, 319.85, and 350.84 g/mol, respectively. Based on those molecular masses, what ratios of migration distances would you expect?

_____

_____

_____

_____

_____

Chlorophyll and Photosynthesis

Lab IV-1

EQUIPMENT AND MATERIALS

You'll need the following items to complete this lab session. (The standard kit for this book, available from *www.thehomescientist.com*, includes the items listed in the first group.)

MATERIALS FROM KIT

- Goggles
- Beaker, 250 mL
- Bromothymol blue
- Centrifuge tubes, 15 mL
- Centrifuge tubes, 50 mL
- Chromatography paper
- Forceps
- Gram's iodine stain
- Hydrochloric acid
- Pipettes
- Ruler (mm scale)
- Slides (flat) and coverslips
- Test tube clamp
- Test tube rack
- Thermometer

MATERIALS YOU PROVIDE

- Gloves
- Coin with milled edge (optional)
- Cotton balls
- Elodea (water weed; see text)
- Isopropanol, 70%
- Leaves, various (see text)
- Light source (see text)
- Meter stick or measuring tape
- Microscope (with reflection illumination)
- Microwave oven
- Saucer
- Scissors
- Soda straw
- Toothpicks (plastic)
- UV light source (optional)
- Watch or clock with second hand
- Water, distilled

BACKGROUND

In this lab session we'll investigate *photosynthesis*, the conversion of solar energy to stored energy in the form of saccharides, which is the basis of nearly all life on Earth. (Only nearly all, because a few deep-sea organisms use energy directly from thermal vents on the sea bottom.) Photosynthesis depends largely on *chlorophylls*, a closely related group of organic chemical compounds that contain a central magnesium atom.

In its most common form, photosynthesis is a chemical reaction that converts carbon dioxide and water, in the presence of light as an energy source and chlorophyll as a catalyst, to glucose (or another saccharide) and oxygen.

$$6\,CO_2 + 6\,H_2O \rightarrow C_6H_{12}O_6 + 6\,O_2$$

The products of this reaction, saccharides and oxygen, are the basis of all animal life. Without the saccharides produced by plants, we'd starve; without the oxygen, we'd suffocate. Note the balance. We animals consume the saccharides and inhale the oxygen produced by plants and exhale the carbon dioxide that plants in turn use to produce more saccharides and oxygen.

> Plants, algae, and cyanobacteria (formerly called blue-green algae) are *autotrophs* or self-feeders. They use photosynthesis to produce food—saccharides and other organic carbon compounds—from carbon dioxide and water. Autotrophs comprise about 5% of all species, and are the producers in the food chain. The other 95% of species—animals and most other bacteria and fungi—are *heterotrophs*, which are the consumers in the food chain.
>
> Some species of sea slugs, notably *Elysia chlorotica*, use chlorophyll from plants they eat to perform photosynthesis for themselves; no other animal has been found to have this ability.

PROCEDURE IV-1-1: OBSERVING CARBON DIOXIDE UPTAKE

In this procedure, we'll observe the uptake of carbon dioxide (carbon fixation) and the release of oxygen by plants during photosynthesis. We'll do this by establishing various environments in which the three key elements—sunlight, carbon dioxide, and chlorophyll—are present or absent in different combinations.

During photosynthesis, carbon dioxide is consumed and oxygen is produced. We'll use a simple test for the presence of carbon dioxide. When carbon dioxide is dissolved in water, it produces a slightly acidic solution. In slightly acidic solutions (< pH 6.0) the pH indicator bromothymol blue is yellow. In slightly basic solutions (> pH 7.6) it is blue. Between pH 6.0 and pH 7.6, it transitions through various greenish shades.

We'll produce a saturated solution of carbon dioxide by the simple expedient of exhaling through a soda straw into water that has been tinted with bromothymol blue. As the carbon dioxide from our breath dissolves in the water, the indicator turns yellow. If a chemical reaction occurs in that solution that consumes carbon dioxide, we expect the solution color to shift toward blue (or at least greenish) as the carbon dioxide is removed and the pH of the solution increases.

We have no simple chemical test for the presence of oxygen, so we'll depend on the fact that oxygen gas is much less soluble in water than carbon dioxide gas. If in fact a reaction occurs that consumes carbon dioxide and produces oxygen, we would expect that oxygen to be visible as tiny bubbles in the reaction vessel. In this procedure, we'll observe the uptake of carbon dioxide by elodea (a water plant) and determine which combination of three factors—the presence of carbon dioxide, the presence of light, and the presence of chlorophyll—are necessary for photosynthesis to occur.

We used Elodea (water weed) as our plant. You can purchase Elodea (also called Anacharis) at a pet store or aquarium shop, or you can simply gather it from a lake, pond, or in shallow water near the banks of a stream.

1. If you have not already done so, put on your goggles, gloves, and protective clothing.

2. Label six 15 mL centrifuge tubes as follows:

   A. Light – Elodea – CO_2

   B. Light – Elodea

   C. Light – CO_2

   D. Dark – Elodea – CO_2

   E. Dark – Elodea

   F. Dark – CO_2

3. Place 7 cm to 9 cm lengths of elodea sprig into each of tubes A, B, D, and E.

4. Transfer about 175 mL of distilled water to the 250 mL beaker. Add sufficient bromothymol blue indicator solution to the water to tint it a distinct blue or blue-green color and stir to mix the solution.

5. Fill tubes B and E nearly full of the blue solution you produced in step 4.

If the solution in the tubes is not distinctly colored, return the contents of the tubes to the beaker and add more bromothymol blue. It's necessary that the color of the solution in the tubes be intense enough that any color changes are readily visible.

6. Add one drop of hydrochloric acid to each of tubes B and E, and swirl them to mix the contents. If the solution does not turn yellow, continue adding hydrochloric acid dropwise, with swirling, just until the solution turns yellow. Cap the tubes. Note the appearance of the tubes—color, presence or absence of bubbles, and so on—and record your observations in your lab notebook. Place tube B on a windowsill or other location where it is exposed to direct sunlight. Place tube E in a drawer or other dark location.

The tubes will stand on their caps if you invert them. Just make sure the cap is tight.

7. Use the soda straw to blow bubbles (gently) into the solution in the beaker until it assumes a distinctly yellow color. (Obviously, don't suck any of the solution into your mouth.)

8. Fill tubes A, C, D, and F nearly full with the yellow solution you produced in step 7. Cap the tubes. Note the appearance of the tubes and record your observations in your lab notebook. Place tubes A and C on where they are exposed to direct sunlight. Place tubes D and F in a dark location. Retain the excess solution in the beaker for use in the next procedure.

9. Observe each of the six tubes every few minutes and record your observations in your lab notebook.

PROCEDURE IV-1-2: DETERMINING THE EFFECT OF LIGHT INTENSITY ON PHOTOSYNTHESIS

In the preceding procedure, we established that photosynthesis requires the presence of chlorophyll, carbon dioxide, and light. In this procedure, we'll determine the effect of light intensity on the rate of photosynthesis. We'll measure photosynthesis rate by determining the amount of time needed for water saturated with carbon dioxide and tinted by bromothymol blue to change color when exposed to light at differing intensities, indicating that photosynthesis has consumed the carbon dioxide present.

To do that, we'll need an intense light source, ideally one that is full-spectrum and as close as possible to a point source. We used a 600W quartz-halogen work light, which among common light sources comes closest to meeting those criteria. If you don't have a quartz-halogen light, substitute any bright light source.

> The amount of time necessary for a change to occur depends on the intensity of the light acting on the chlorophyll in the Elodea, which in turn depends on both the actual intensity of the light source and its distance from each tube. With our 600W quartz-halogen light, visible changes occurred very quickly; with a less-intense light source, it may require several minutes or longer for changes to become evident.

1. Label five 15 mL centrifuge tubes: two as "0.25 m," and one each as "0.5 m," "1.0 m," and "2.0 m."

2. Cut four sprigs of Elodea. It's important to make the sprigs as similar as possible in terms of leaf area.

3. Place each sprig in one of the centrifuge tubes, leaving one of the 0.25 meter tubes without a spring. Fill all five tubes nearly to the brim with the yellow solution from the beaker, and cap the tubes.

4. Set up your light source, but leave it off for now. Use the meter stick to place each tube at the corresponding distance from the light, inverted and sitting on its cap. Position each tube so that it will be fully exposed to the light once you switch it on.

> The intensity of a point light source decreases with the square of the distance. For example, light intensity at 2 meters is one-quarter that at one meter, and one-sixteenth that at 0.5 meter. If the rate of photosynthesis varies directly with light intensity, we would expect the nearest tube to change color quickly, with tubes at greater distances taking correspondingly longer to change.

5. Note the time to the second and turn on the light.

6. Watch the 0.25 meter tube with the sprig carefully, and note the elapsed time when you first notice a change in the color of the liquid in the tube. Do the same for the 0.5 meter tube, the 1.0 meter tube, and finally the 2.0 meter tube.

7. Record all of your observations in your lab notebook.

PROCEDURE IV-1-3: CHROMATOGRAPHY OF PLANT PIGMENTS AND LOCATING CELLULAR ENERGY RESERVES

Although chlorophylls are the primary pigment in most vegetation, there are others. Yellow and orange carotenoids are present in leaves year-round, but during the spring and summer their colors are ordinarily masked by the intense green of chlorophyll. Red and purple anthocyanins are not present in spring and summer foliage, but are actually produced during the autumn color change.

In this procedure, we'll extract chlorophyll and other plant pigments from green leaves and use paper chromatography to separate those pigments. In doing so, we'll discover how many pigments are present in our leaf specimens. (The number and types of pigments varies from one plant to another and with the time of year, so your results may differ from ours.)

Some of the pigments are relatively soluble in alcohol but nearly insoluble in water, and vice versa. To extract as many pigments as possible from our specimens, we'll use 70% isopropanol (drugstore rubbing alcohol), which we expect to be a reasonably good solvent for all of the pigments present.

We'll also examine a decolorized leaf microscopically. Plants store the glucose produced by photosynthesis in the form of starches. These cellular energy reserves are ordinarily colorless and therefore difficult or impossible to discriminate visually. We'll therefore stain a decolorized leaf with Gram's iodine solution, which reacts with colorless starches to form an intensely blue-black complex that is readily visible if present.

You can use any type(s) of green leaves available. If you're performing this procedure at a time of year when the leaves have changed color or fallen, you can use evergreen needles or supermarket produce such as spinach. Ideally, you should obtain deciduous leaves during the spring or summer while they're green, and then obtain specimens from the same trees during the autumn leaf color season.

> We obtained spring/summer and autumn leaf specimens from two species of maple. For both species, the spring/ summer leaves are the intense green of chlorophyll. In autumn, the red maple leaves turn red, due to the presence of anthocyanins, which are absent during the spring and summer. The golden maple leaves in autumn are a brilliant yellow color, due to the presence of carotenoid pigments, which are present in the leaves year-round, but with their yellow color screened by chlorophyll during the spring and summer. We expected anthocyanins to be visible in chromatograms of red maple leaves only in autumn specimens, while we expected carotenoids to be visible in chromatograms of golden maple leaves from any time of year.

1. If you have not already done so, put on your goggles, gloves, and protective clothing.

2. Prepare each of your leaf specimens by cutting the leaves into small pieces, the smaller the better. Rough squares about 5 mm on a side are adequate, but even smaller pieces allow more of the plant pigments to be extracted quickly. Cut enough leaf pieces to fill a 15 mL centrifuge tube to at least the 3 mL or 4 mL line loosely packed.

3. Transfer the pieces of each specimen to a labeled 15 mL centrifuge tube and add enough 70% isopropanol to the tube to cover the leaf fragments completely. A small excess of alcohol is fine, but we want the resulting extract to be as concentrated as possible, so don't use a large excess.

> For quick-and-dirty chromatograms, you can skip this extraction process by transferring plant pigments directly to the chromatography paper. To do so, press a leaf against the paper and rub the edge of a milled coin (such as a US quarter) against the leaf to produce a line of plant pigment on the paper. Develop the chromatogram as usual. The results may not be as good as those using the alcoholic extract, but the process is fast enough to allow you to run many more chromatograms if your time is limited.

4. Cap the tube and allow the leaf pieces to soak in the isopropanol for at least several days, inverting the tubes occasionally. It does no harm to allow the extraction process to continue for weeks or even months. For example, we obtained summer leaves and allowed them to continue soaking in the alcohol until autumn leaves were available.

> If you haven't prepared the extracts ahead of time, you can use hot isopropanol to speed the extraction. Plug the mouths of the 15 mL centrifuge tubes loosely with cotton balls and place the tubes in the 250 mL beaker full of tap water heated in the microwave to about 80 °C (**Caution! HOT!**). Much higher than 80 °C will cause the alcohol to boil too vigorously; much lower will slow extraction. If the alcohol boils too vigorously, just add some cold tap water to cool the water bath. If the water cools too much, add a small amount of boiling water to the bath.
>
> Allow the extraction to continue at or near the boiling point of the alcohol until most of the color is extracted from the leaves, which may require several minutes or more, depending on the type of leaf and the size of the pieces. When extraction is complete, use the test tube clamp to transfer the centrifuge tubes to the test tube rack and allow them to cool.

5. Use the forceps to remove one piece of leaf from the first centrifuge tube and transfer it to a microscope slide. Center the specimen on the stage and observe it with low magnification. Depending on the thickness and level of decolorization of the leaf, you may or may not be able to see significant detail using transmitted light. If not, illuminate the specimen from above with a high-intensity desk lamp or similar external illuminator. Use medium and/ or high magnification to reveal additional details. Note your observations in your lab book, and shoot an image if you are equipped to do so.

6. Transfer one drop of Gram's iodine stain to the specimen, allow it to work for 30 seconds or so, and then draw off any excess stain using the corner of a paper towel. Repeat your observations and note any differences. If starch is present in the leaf, it will be visible as intensely blue-black clumps. Once again, record your observations in your lab notebook and shoot an image if possible.

7. Repeat the preceding two steps for each of your other leaf specimens.

The next step is to use paper chromatography to separate the leaf pigments. Depending on the type of leaf, time of year, and other factors, your developed chromatograms may show that anything from two or three to half a dozen or more different pigments are present in an extract.

Some or all of the following pigments may be visible in your chromatograms after you develop them:

- Anthocyanins (red and violet pigments)

- Carotenes (yellow to yellow-orange carotenoid pigments)

- Chlorophyll A (bluish-green pigment)

- Chlorophyll B (yellow to yellowish-green pigment)

- Pheophytin A (bluish-gray pigment)

- Pheophytin B (bluish-gray pigment; often visible only under UV light)

- Xanthophylls (yellowish carotenoid pigments)

1. Prepare as many chromatography vessels as you have extract solutions by filling 50 mL centrifuge tubes to about the 5 mL line with 70% isopropanol and capping them. (If you have more extracts than tubes, don't worry; you can simply develop the chromatograms in two passes.)

Handle the chromatography paper with gloves. Skin oils can interfere with the chromatography process.

2. Cut one chromatography strip about 2.5 x 9 cm for each of your extract solutions. About 1 cm from one end of each strip, draw a light pencil line across the width of the strip. At the far end of the strip, label it, e.g., "extract A."

3. Spot each strip with the corresponding extract. To do so, place the strip on a clean flat surface. Dip the tip of a clean plastic toothpick into the extract solution and transfer the tiny drop on the tip of the toothpick to the center of the pencil line, ideally with the pencil line bisecting the spot. The goal is to make the smallest possible spot on the paper. Allow the spot to dry.

4. Repeat the preceding step several times, transferring more of the extract to the same spot on the paper. The goal is to reinforce that spot, making a small spot with a high concentration of the pigments present in the extract.

5. Repeat the spotting procedure for each of your other extract solutions. Allow all of the spotted paper strips to dry completely.

6. Uncap all of your chromatography vessels and quickly transfer each spotted paper strip to one of the vessels. Immediately recap the vessels.

7. Keep an eye on the development progress, which may take 15 minutes or more. As development progresses, the isopropanol is drawn up the paper strip by capillary action and the pigments begin to separate. When the solvent front reaches nearly the top of the strips, use forceps to remove the strips from the chromatography vessels. Place the strips on a paper towel and immediately make a pencil mark to indicate the furthest extent of the solvent front (which will be invisible after it dries).

8. For each strip, measure the distance from the index line to the solvent front line and record it in your lab notebook. (These values should be nearly identical for all strips, since they were placed into and removed from the chromatography vessels nearly simultaneously.)

9. Measure the distance from the index line to the center of each pigment spot on each chromatogram and record those values in your lab notebook.

If you have a UV (black light) source available, use it to view the chromatograms in a darkened room. Some pigment spots that are invisible under white light may be quite prominent under UV light. (UV LED flashlights are readily available for $10 or less, and are useful not just for biology but for chemistry and forensic science studies.)

As long as you're at it, use the UV lamp to illuminate one of the tubes of pigment extract. Note your observations in your lab notebook.

10. Calculate the R_f value for each pigment spot by dividing the distance from the index line to the pigment spot by the distance from the index line to the solvent front line. For example, if a pigment spot migrated 2.5 cm from the index line and the solvent front line is 7.5 cm from the index line, you'd calculate the R_f value for that pigment as (2.5 / 7.5) = 0.33. Record these values in your lab notebook.

R_f value is a dimensionless number that can be compared to R_f values for other chromatograms only if the combination of substrate (such as this particular chromatography paper) and solvent is identical. For example, if you've run several chromatograms in this procedure, the R_f value for the intense blue-green spot of chlorophyll A should be identical (or nearly so) for each of the extracts. If we ran chromatograms using a different solvent (or a different substrate), the R_f value for chlorophyll A might be very different.

REVIEW QUESTIONS

Q1: In procedure IV-1-1, why did we use a combination of tubes with and without carbon dioxide, Elodea sprigs, and exposure to sunlight?

Q2: In procedure IV-1-1, why did we add a small amount of hydrochloric acid to tubes B and E?

Q3: In procedure IV-1-1, based on your observations of the six tubes, what do you conclude?

Q4: In procedure IV-1-1, propose an explanation for the changes you observed in one or more of the tubes. Why did these changes not occur in all tubes?

_____

_____

_____

_____

Q5: In procedure IV-1-1, what would you expect to happen in tube A if you continued to observe the tube over a period of days, weeks, or months? Why? Would you expect a different result if you also introduced a snail into that tube?

_____

_____

_____

_____

Q6: In procedure IV-1-2, why did we use two tubes at 0.25 meters from the light source, one with and one without a sprig of Elodea?

_____

_____

_____

_____

Q7: In procedure IV-1-2, what did your observed data indicate about the effect of light intensity on photosynthesis rate?

_____

_____

_____

_____

Q8: In procedure IV-1-2, would you expect the (approximate) linearity in photosynthesis rates to remain true at light intensities much, much higher or lower than those you tested?

_____

_____

_____

_____

_____

Q9: The light source we used also produces considerable heat, as do most intense light sources. It is possible that the tubes nearer the light source were warmed by the light source, and that that higher temperature increased the rate of photosynthesis. Design an experiment to determine what effect, if any, different temperatures have on photosynthesis rate.

Q10: In procedure IV-1-3, we implied that the extracts obtained in cold isopropanol and hot isopropanol were for practical purposes identical. What assumption did we make and how might you test that assumption? If that assumption is false, what evidence would you expect to see?

Q11: If you used the UV light in procedure IV-1-3, what results did you observe when you viewed the extract solution? Using Internet resources, research this phenomenon and propose an explanation.

Q12: Using the materials and resources from this group of procedures, design an experiment to determine if photosynthesis requires only chlorophyll, carbon dioxide, and light, or if it can occur only within a plant.

Investigating Osmosis

Lab IV-2

EQUIPMENT AND MATERIALS

You'll need the following items to complete this lab session. (The standard kit for this book, available from *www.thehomescientist.com*, includes the items listed in the first group.)

MATERIALS FROM KIT

- None

MATERIALS YOU PROVIDE

- Gloves
- Balance
- Eggs, uncooked (2)
- Foam cups
- Graph paper, calculator, or software
- Marking pen

- Paper towels
- Syrup (corn, maple, pancake, waffle, etc.)
- Tablespoons, plastic or metal (2)
- Vinegar, distilled white
- Watch or clock with second hand

BACKGROUND

In this lab session we'll investigate *osmosis*, the diffusion of water across a *semipermeable membrane* to equalize solute concentrations on both sides of the membrane. A semipermeable membrane allows the free passage of small molecules (such as water) through the membrane, while blocking the passage of larger molecules (such as salts, sugars, and other organic compounds).

If two aqueous solutions of different concentrations are divided by a semipermeable membrane, a phenomenon called *osmotic pressure* causes water molecules to migrate across the membrane from the more dilute side to the more concentrated side until the concentration is the same (in equilibrium) on both sides of the membrane.

For example, assume that you make up two sugar solutions, one with one teaspoon of sugar dissolved in 100 mL of water, and the other with three teaspoons of sugar dissolved in 100 mL of water. You then place these two solutions in a two-part container with a semipermeable membrane dividing them. Because the first solution is more dilute, water molecules will pass from it through the membrane and into the more concentrated solution, thereby increasing the concentration of the first solution and decreasing the concentration of the second.

The two solutions reach equilibrium when the sugar concentration is the same in both. Because sugar molecules cannot pass through the membrane, that means the volumes of the two solutions must change to effect that change in concentration. Because the second solution contained three times as much sugar as the first, its final volume must also be three times as much as the first solution. This is achieved when 50 mL of water has migrated from the first to the second solution, leaving the first solution containing one teaspoon of sugar in 50 mL of water and the second solution containing three teaspoons of sugar in 150 mL of water, which of course means the concentrations of the two solutions are identical.

> With respect to two solutions divided by a semipermeable membrane, scientists refer to the more concentrated solution as *hypertonic* and the less concentrated as *hypotonic*. If the two solutions have reached equilibrium (have the same concentration), they are referred to as *isotonic*. Note that these three terms are meaningless except when comparing two solutions against each other.

Osmosis is an important phenomenon in biological systems. Semipermeable cell membranes are typically permeable to molecules that are small and/or nonpolar—including oxygen, carbon dioxide, water, as well as lipids—while being impermeable to larger and/or more polar molecules, including polysaccharides, proteins, and other large organic molecules, as well as ions. Osmosis and related diffusive processes plays a role in many biological functions, from water transport into and out of cells to elimination of wastes to cellular respiration.

In this lab session, we'll prepare two decalcified eggs by dissolving their shells in vinegar. We'll then determine the mass of each egg before immersing it for measured times in either water—which, because it contains essential no solutes, is hypotonic with respect to the egg—or syrup, which because it contains a high concentration of dissolved sugars, is hypertonic with respect to the egg. We'll determine the mass change over time for each egg and use that data to determine the effects of osmosis in our controlled environment.

> You may have heard the term *reverse osmosis*, which is a technique used in some industrial processes, particularly water desalinization. Reverse osmosis forces water from the more concentrated solution to the less concentrated by applying physical pressure or other means. As applied to desalinization, for example, sea water (which contains a high concentration of salts) is placed under pressure, which forces pure water through a semipermeable membrane, leaving more concentrated sea water on the pressurized side of the membrane and producing pure drinking water on the other side.

PROCEDURE IV-2-1: OBSERVING OSMOSIS IN CHICKEN EGGS

Eggshells are primarily calcium carbonate, a chemical compound that is also familiar in the form of chalk, limestone, and marble. Calcium carbonate reacts readily with acids to form calcium salts and carbon dioxide gas. In this procedure, we'll immerse two raw eggs in an acidic solution and allow the reaction to consume the eggshells, leaving only the raw eggs within the membrane that surrounds the white and yolk.

1. Place two raw eggs in a large foam cup or other suitable container.

2. Fill the cup with distilled white vinegar, ensuring that both eggs are fully submerged, and cover the cup loosely to allow carbon dioxide gas to escape. (It's okay for the

eggs to be in contact.) Keeping the cup in the refrigerator may slow down the reaction somewhat, but it also keeps the eggs from spoiling. That won't matter unless you accidentally break the egg membrane, in which case it's better if the eggs aren't spoiled.

3. Observe the eggs periodically. Over a period of three to four days, the eggshells gradually react with the vinegar until the eggshells disappear entirely.

WARNING

Raw eggs may be contaminated with salmonella or other pathogens. Wear gloves when handling the eggs and wash your hands thoroughly with soap and water after you finish the experiment. Discard both the eggs and the syrup when the experiment concludes.

4. When the eggshells are completely decalcified, carefully pour off the waste solution and use a spoon (gently!) to remove each egg and place it on a paper towel. Carefully blot each egg to remove as much of the excess liquid as possible.

Be extremely gentle handling the decalcified eggs. With the shells gone, it's very easy to break the membrane and ruin the experiment.

5. Weigh two foam cups and record the mass on each cup.

6. Carefully transfer one egg to each of the cups. Reweigh the cups and subtract the mass of the cup from the mass of the cup+egg to determine the mass of the egg alone.

7. Note the time and fill one of the cups with syrup just sufficient to cover the egg with a centimeter or so to spare. Record the time in your lab notebook.

8. After a couple of minutes, repeat step 7 for the second egg, immersing it in water.

9. After about 10 minutes have elapsed, note the time and use one of the spoons to remove the egg from the syrup. Rinse the egg gently for a moment in a trickle of tap water to remove as much syrup as possible, and then very gently blot the egg dry with a paper towel. Reweigh the egg and record its current mass and the elapsed time in your lab notebook.

10. Return the egg to the syrup, again recording the start time in your lab notebook.

11. By the time you complete steps 9 and 10, the egg in water should be nearing the 10 minute mark. Note the time and use the other spoon to remove the egg from the water. Very gently blot the egg dry with a paper towel. Reweigh the egg and record its current mass and the elapsed time in your lab notebook.

12. Return the egg to the water, again recording the start time in your lab notebook.

13. Repeat steps 9 through 12 every 10 minutes for at least one hour. If possible, extend the experiment for several hours, recording data every 10 minutes until you run out of time or until the mass of one or both eggs remains constant between cycles.

14. After you complete the final mass determination for both eggs, fill two additional cups with water and syrup to the same level as the original cups. Compare the appearances of both unused liquids to the appearances of the liquids used during the experiment, and record your observations in your lab notebook.

15. Using your data, calculate the percentage mass change at each weighing for each egg and graph the results.

REVIEW QUESTIONS

Q1: Which of the eggs gained mass? Which lost mass? Propose an explanation for these mass changes.

Q2: What, if any, changes did you note in the appearance of the water and syrup?

Q3: Using the terms hypotonic and hypertonic, explain what occurred with each of the eggs.

Q4: Can isotonicity (isotonic equilibrium) be reached for the egg immersed in syrup? For the egg immersed in water?

Q5: Explain in terms of osmosis what occurs if you pour table salt on a slug.

Q6: Using Internet resources, find a practical application of osmosis in food preservation. In terms of osmosis, explain how this process works.

_____

_____

_____

_____

Q7: What percentage of mass gain or loss did you observe for each of the eggs?

_____

_____

_____

_____

Investigating Cell Division

Lab IV-3

EQUIPMENT AND MATERIALS

You'll need the following items to complete this lab session. (The standard kit for this book, available from *www.thehomescientist.com*, includes the items listed in the first group.)

MATERIALS FROM KIT

- None

MATERIALS YOU PROVIDE

- Microscope (oil-immersion if available)
- Prepared slide: plant mitosis (onion tip)
- Prepared slide: animal mitosis (optional)

BACKGROUND

The *cell division cycle* (or *cell cycle*) is a fundamental biological process. To understand genetics and reproduction, you must first understand cell division. Prokaryotes use simple binary fission, during which one cell splits in two to form two identical cells. The cell cycle in eukaryotes, which we investigate in this lab session, is considerably more complex.

The eukaryotic cell cycle encompasses three major phases. During *interphase*, shown on the left in Figure IV-3-1, the cell grows, accumulates the nutrients needed for the next phase, and replicates its DNA. Interphase typically occupies about 90% of the cell cycle. During *mitosis*, shown in the center in Figure IV-3-1, the cell nucleus divides, separating its

chromosomes into two identical sets and forming two separate nuclei from these chromosomes. Mitosis typically occupies about 10% of the cell cycle. During *cytokinesis*, the cell splits into two distinct *daughter cells*. The cycle then continues as these daughter cells individually enter interphase and continue to divide.

Figure IV-3-1: *Major events in the eukaryotic cell cycle*

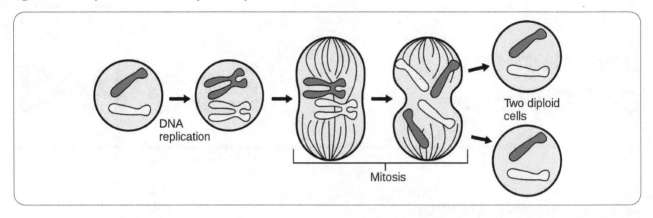

DNA replication

Mitosis

Two diploid cells

Although cytokinesis is often thought of as the final step in the mitotic stage of the cell cycle, it is actually a separate and distinct stage of the cell cycle, occurring after the final stage of mitosis (telophase) and before the cell returns to interphase.

Interphase comprises three phases. During G_1 *Phase* or *Gap One Phase*, the cell grows and carries on its routine activities. During *S Phase*, the cell duplicates its chromosomes, which as the cell enters this phase are dispersed and invisible with a standard microscope. Initially, each chromosome consists of only one *chromatid*. After DNA synthesis, each chromosome comprises two genetically identical *sister chromatids*, attached at the *centromere*. During G_2 *Phase* or *Gap Two Phase*, the cell prepares itself for mitosis.

The exact sequence and timing of mitosis steps varies between cell types, but mitosis is conventionally divided into four or five phases. During the first phase, called *prophase*, the cell exiting the interphase second gap (G_2) stage enters prophase and begins the mitotic sequence.

In early prophase, the *chromatin* condenses (coils) into chromosomes, which become visible with a standard microscope. As prophase continues, the two *centrosomes* (in animal cells, each made up of two *centrioles*) separate to opposite sides of the nucleus, the microtubules begin to lengthen and form the mitotic spindle, and the nucleolus disappears. Figure IV-3-2 is a graphic illustration of prophase, with the centrosomes shown in orange and the microtubules in blue. Figure IV-3-3 is an image of an actual cell in prophase.

Figure IV-3-2: *Prophase of mitosis*

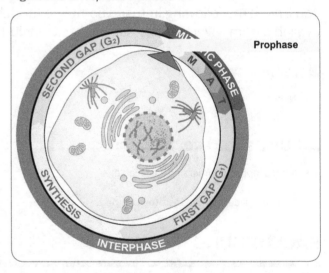

During the second phase of mitosis, called *prometaphase*, the nuclear membrane disintegrates, chromosomes form *kinetochores* at the *centromere*, and the microtubules that make up the mitotic spindle attach to the chromosomes. Figure IV-3-4 is a graphic illustration of prometaphase. Figure IV-3-5 is an image of an actual cell in prometaphase, with the mitotic spindle in green, the condensed chromosomes as a cloud of blue, and the kinetochores and centromeres in magenta.

Figure IV-3-3: *Prophase of mitosis*

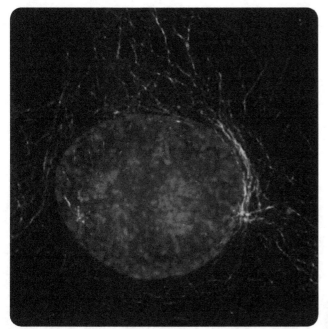

Figure IV-3-4: *Prometaphase of mitosis*

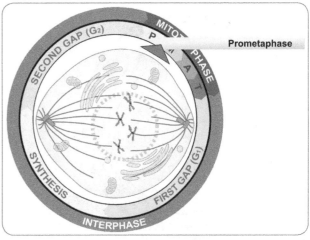

Figure IV-3-5: *Prometaphase of mitosis*

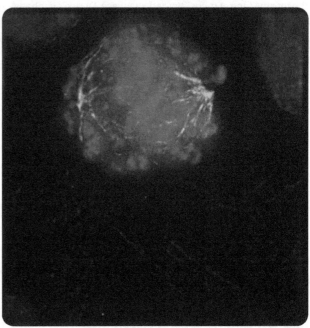

Some texts describe mitosis as occurring in four phases: prophase, metaphase, anaphase, and telophase. Other texts place an additional phase, prometaphase, between prophase and metaphase. Prometaphase incorporates what four-phase texts include in late prophase and early metaphase. The reality is that the divisions between phases are somewhat arbitrary, and it doesn't really matter if you think of mitosis as having four or five phases.

During the third phase of mitosis, called *metaphase*, the *centromeres* of the condensed and coiled chromosomes align in an imaginary plane called the *metaphase plate* (also called the *equatorial plate*) located at the equator of the cell, with the *telomeres* (chromosome ends) drifting away from the equator toward the centrosomes at the poles. This even alignment physically organizes the cell in preparation for later division into two daughter cells. At this point, and before entering the next mitotic phase, the cell does a literal "self-test" to ensure that every chromosome is properly aligned at the metaphase plate and that every kinetochore is properly attached to the mitotic spindle. Figure IV-3-6 is a graphic illustration of metaphase.

Figure IV-3-6: *Metaphase of mitosis*

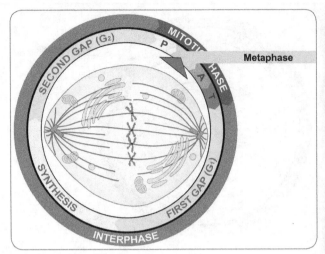

If and only if the cell passes its metaphase self-test, it enters the fourth stage of mitosis, called *anaphase*. During anaphase, the microtubules that make up the mitotic spindle attach to the centromere of each chromosome and literally pull the sister chromatids of each chromosome apart toward the centrosomes at each pole, with the telomeres of each chromosome pointing toward the equatorial plate. Those microtubules that are not attached to kinetochores act to stretch the cell lengthwise into an oval form. Figure IV-3-7 is a graphic illustration of anaphase. Figure IV-3-8 is an image of an actual cell in anaphase.

Figure IV-3-7: *Anaphase of mitosis*

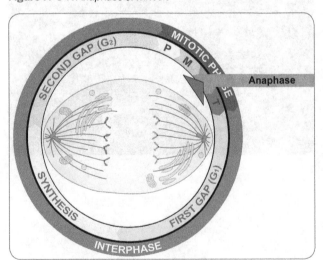

Figure IV-3-8: *Anaphase of mitosis*

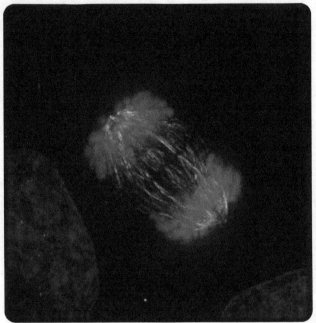

The final phase of mitosis, *telophase*, begins when anaphase is complete. Essentially, telophase is the "clean-up" phase. The cell continues to elongate. Chromosomes uncoil and disperse as chromatin. The mitotic spindle breaks down and microtubules disintegrate. Two daughter nuclei form, one on either side of the ovular cell, with their nuclear membranes forming from the fragments of the original nuclear membrane. As the new nuclear membranes form around each chromatid pair, new nucleoli form. In animal cells, a contractile ring of fibers called a *cleavage furrow* develops. which during cytokinesis closes to pinch off the binucleate cell into two individual daughter cells. In plant cells, cell wall materials coalesce into a structure called the *cell plate*, which develops into a full cell wall, separating the two nuclei. Figure IV-3-9 is a graphic illustration of telophase. Figure IV-3-10 is an image of an actual cell in telophase.

Figure IV-3-9: *Telophase of mitosis*

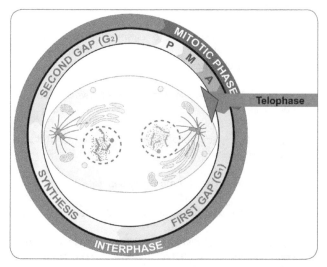

Figure IV-3-11: *Cytokinesis*

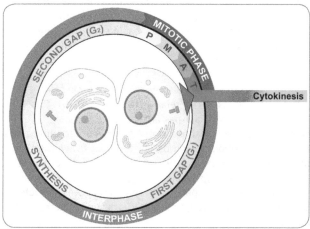

Figure IV-3-10: *Telophase of mitosis*

Finally, *cytokinesis*, shown graphically in Figure IV-3-11, occurs when the cell that has undergone mitosis splits into two new individual daughter cells, each of which then enters interphase to start the cycle anew.

Meiosis is a specialized type of cell division that occurs in eukaryotes to produce the *gametes* (sperm cells and egg cells, produced by all animals and some protists) or *spores* (in all plants and fungi and some protists) required for sexual reproduction. Because of the similarity of names, students sometimes confuse mitosis and meiosis. Although there are similarities in the mechanisms and the names of the phases, the two processes are distinctly different in several respects:

- Mitosis is the asexual reproduction of diploid cells; meiosis is the production of haploid cells (gametes or spores) in preparation for sexual reproduction.

- Mitosis is a part of the cell cycle, a circular process with no beginning or end. New cells produced at the end of the cell cycle restart the cycle to produce still more new cells. Meiosis is a process with a starting point and an ending point. In effect, mitosis runs in a circle, while meiosis runs irreversibly in a straight line from point A to point B.

- Mitosis produces daughter cells that are genetically identical to the mother cell, with no reshuffling of genetic material. In meiosis, the chromosomes recombine, shuffling the genes to produce different genetic combinations in each gamete.

- Mitosis produces two genetically identical diploid cells; meiosis produces four genetically distinct haploid cells.

Meiosis starts with one *diploid cell* that contains two copies of each chromosome, one from each parent of the organism. (For example, your cells contain 23 pairs of chromosomes, with one chromosome of each of the 23 pairs originating from your mother and one from your father.) Meiosis produces four haploid cells (gametes), each of which has one copy of each chromosome. Each of those chromosomes is a unique mixture of the paternal and maternal DNA from the organism. This mixing of genetic material ensures that offspring are genetically distinct from both parents, thereby ensuring diversity in any population that reproduces sexually.

Chromosomes in ordinary eukaryotic cells occur in pairs. The nuclei of such cells are referred to as *diploid* (*2N*). The two chromosomes of each pair are called *homologous chromosomes*. Each of those chromosomes has the same genes at the same positions, called *loci* (singular *locus*), although each of the homologous genes may have different *alleles* (different variants of the same gene) at the corresponding homologous loci. Chromosomes in specialized reproductive cells (gametes and spores) are unpaired, and the nuclei of such cells are called *haploid* (*1N*).

In preparation for meiosis, a cell undergoes a meiotic interphase stage that is similar to the mitotic G_1, S, and G_2 stages. The cell's chromosomes are duplicated via DNA replication, producing the homologous chromosomes in paternal and maternal versions, each of which comprises two exact copies, called *sister chromatids*, attached at the centromere.

Like the cell cycle (including mitosis), meiosis is a seamless and continuous process, but is for convenience normally treated as occurring in phases. The names of those phases are the same as those of the cell cycle—prophase, metaphase, anaphase, telophase, and cytokinesis—and those phases are similar to the corresponding cell-cycle phases in mechanism and appearance. The major difference is that while the cell cycle involves only one division, meiosis involves two divisions, called *Meiosis First Division* (or *Meiosis I*) and *Meiosis Second Division* (or *Meiosis II*). During Meiosis I, the parent cell divides into two diploid daughter cells. During Meiosis II, the two diploid daughter cells divide into four haploid daughter cells The meiotic phases are as follows:

- Prophase I – homologous chromosomes condense and form pairs, with crossing over.

- Metaphase I – the microtubular structure forms, with the microtubules attaching to the chromosomes.

- Anaphase I – homologous pairs of chromosomes separate and the cell elongates.

- Telophase I – one set of paired chromosomes completes its migration to each pole.

- Cytokinesis I – the binucleate cell splits into two daughter cells, with each daughter cell receiving one of each pair of homologous chromosomes.

- Prophase II – the meiotic spindle forms between the two centrioles.

- Metaphase II – chromosomes align along the microtubules of the spindle.

- Anaphase II – sister chromosomes separate.

- Telophase II – chromatids arrive at each pole.

- Cytokinesis II – cell division is complete, with four daughter cells resulting, each with half the number of chromosomes of the parent cell.

Figure IV-3-12 graphically illustrates the major events in meiosis.

Figure IV-3-12: *Major events in Meiosis*

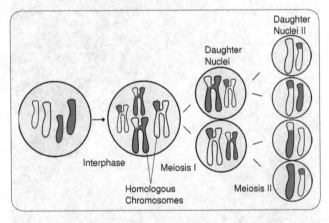

In this lab session, we'll observe the cell cycle by examining prepared slides that are stained to show the stages of mitosis in plant and animal cells. Although the stages of meiosis can also be observed by examining prepared slides, there's little point to doing so because the physical appearance of the various meiotic phases is nearly impossible to discriminate microscopically from the corresponding mitotic phases.

PROCEDURE IV-3-1: OBSERVING MITOSIS

Students are often surprised that mitotic phases can actually be seen and discriminated with a microscope, but the staining protocols for doing so have been in use since the 19th century. Figure IV-3-13 is a sketch that appeared in the 1900 book *The Cell in Development and Inheritance* by Edmund B. Wilson. The original description for this sketch read, "*General view of cells in the growing root-tip of the onion, from a longitudinal section, enlarged 800 diameters. a. non-dividing cells, with chromatin-network and deeply stained nucleoli; b. nuclei preparing for division (spireme-stage); c. dividing cells showing mitotic figures; e. pair of daughter-cells shortly after division.*" Comparing Wilson's sketch with the graphics and images in the preceding section, it's clear that mitotic phases were well understood (albeit under different names) more than 100 years ago.

Figure IV-3-13: *Sketch of onion root tip ls, showing cells in various phases of mitosis*

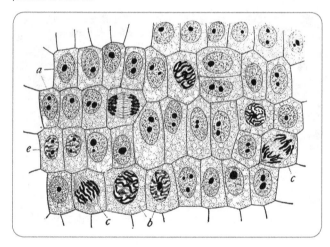

The problem with observing mitosis is that the objects you are viewing are both tiny and three-dimensional. That means you'll need to use high magnification (400X or 1,000X), and that you'll spend a lot of time tweaking the fine focus to accommodate the very shallow depth of focus at high magnification.

As you perform this procedure, record your observations in your lab notebook. Identify the features that should be present in each mitotic phase and verify their presence. Sketch the significant features you observe that are pertinent to each mitotic phase.

In many plants, including Allium (onion), many or most of the cells undergo mitosis. If a specimen is properly sectioned and stained to show mitosis (usually with toluidine blue or Schiff's reagent), the various stages of mitosis are clearly visible in different cells in the section.

1. Obtain a prepared slide of an Allium root tip in longitudinal section, stained to show mitosis.

2. Scan the slide at low magnification to identify an area where mitosis is occurring (or was, at the time the slide was made). That area is one in which cell nuclei have distinctly differing appearances, which are visible even at 40X. Some will appear small, compact, and roughly circular, while others will appear broken or fuzzy.

3. Center that area in the field of view and change to high-dry (400X) magnification. Note the distinctly different appearance of many of the cell nuclei in the field of view. Do a count of the cells visible in the field of view, judging which mitotic phase each of the cells is in.

Here's what you'll actually see. Figure IV-3-14a through Figure IV-3-14e show the same 400X field of view of a longitudinal section of an onion root tip stained with toluidine blue, with several stages of mitosis visible. The visible differences among these images are caused by very minor changes in fine focus, one or two units on our 100-unit fine-focus knob. Even a thin section has depth, and the extremely small depth of focus at 400X means you'll need to tweak your fine focus constantly to bring different parts of the same nucleus into focus.

Figure IV-3-14a: *Onion (Allium sp.) root tip sec mitosis, 400X (a)*

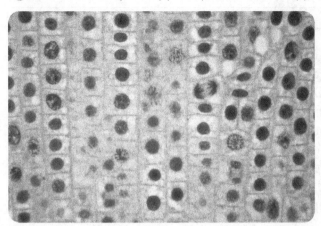

Figure IV-3-14d: *Onion (Allium sp.) root tip sec mitosis, 400X (d)*

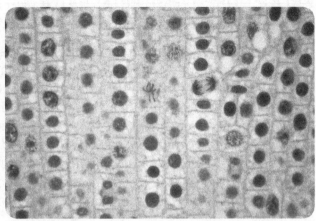

Figure IV-3-14b: *Onion (Allium sp.) root tip sec mitosis, 400X (b)*

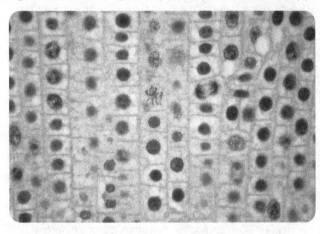

Figure IV-3-14e: *Onion (Allium sp.) root tip sec mitosis, 400X (e)*

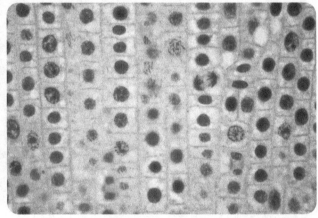

Figure IV-3-14c: *Onion (Allium sp.) root tip sec mitosis, 400X (c)*

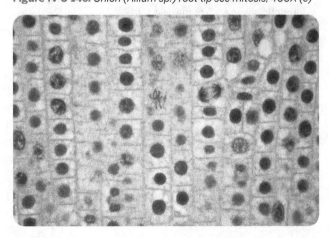

4. Locate a nucleus that is in any of the five stages of mitosis or one that is in cytokinesis. (You needn't observe nuclei "in order"; for example, if the nucleus nearest the center of the field in in telophase, work with telophase first rather than searching out a nucleus in prophase.)

5. Observe that nucleus at 400X, tweaking the fine focus as necessary to bring all features into sharp focus sequentially. Identify as many as possible of the features that should be present in that stage. If possible, count the chromosomes. Make a sketch of that phase.

6. If your microscope has an oil-immersion objective, center the nucleus in question carefully and change to 1,000X magnification to see if any more detail is visible.

Figure IV-3-15: *Roundworm (Ascaris sp.) egg sec mitosis, 400X*

You don't have to dance with the girl you brung. For example, the first nucleus you pick may turn out to be a damaged telophase. Just move along to the next nucleus, whether or not it's telophase. Keep scanning for "good" examples of each phase until you've critically examined and sketched at least one nucleus in each phase.

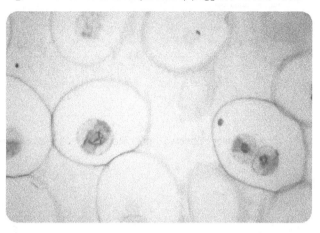

7. Repeat steps 4 through 6 for each of the other stages of mitosis in plants.

8. Obtain a prepared slide of an animal cell section (such as the traditional whitefish blastula or Ascaris), stained to show mitosis. Figure IV-3-15 shows mitosis in a roundworm (*Ascaris* sp.) egg at 400X.

9. Repeat steps 2 through 7 to observe all of the stages of mitosis as well as cytokinesis.

REVIEW QUESTIONS

Q1: Does mitosis occur throughout the onion root tip or in only one specific part of it? How do you know?

Q2: When you counted the onion root tip cells, did you find that one or more phases occurred more frequently in your field of view than others? Did you expect this result? Why or why not?

Q3: Why do some of the onion root tip cells have little or no chromatin visible?

Q4: Were you able to observe more detail at 1,000X than was visible at 400X?

Q5: The diploid number of chromosomes in Allium is 16. Did your observations of a cell in anaphase confirm this number? If not, propose an explanation.

Q6: Although mitosis has been widely studied since the 19th century, the actual number of human chromosomes (46, in 23 pairs) was not definitively determined until 1956. Propose an explanation.

Q7: What structures are present in animal mitosis that are not present in plant mitosis?

Q8: What, if any, difference is there in the duration of mitosis in plants versus animals? How can you tell?

Q9: How does the division of daughter cells differ between animal mitosis and plant mitosis?

Q10: In metaphase, what is the chromatin material that aligns on the equatorial plane for mitosis, meiosis I, and meiosis II?

Q11: In anaphase, what is the chromatin material that aligns on the equatorial plane for mitosis, meiosis I, and meiosis II?

Sampling Plant Populations in a Community

EQUIPMENT AND MATERIALS

You'll need the following items to complete this field session. (The standard kit for this book, available from *www.thehomescientist.com*, includes the items listed in the first group.)

MATERIALS FROM KIT

- Centrifuge tubes (specimen containers)
- Magnifier
- Ruler

MATERIALS YOU PROVIDE

- Gloves
- Assistant
- Camera (optional, with macro feature)
- Hammer or mallet
- Marking pen (for labeling)
- Field guides to plants and trees (for your region)
- Plastic bags (specimen containers)
- Pocket knife
- Scissors
- Stakes (see text)
- String or cord (250 feet or 75 meters)
- Tape measure (50-foot or 15-meter)

BACKGROUND

In this field session we'll define a survey area and do sample counts of the plant populations within that area. From those sample counts, we'll estimate plant populations for the area as a whole.

Such population surveys are an important tool for ecologists, who use them to track the rise and fall in populations of various species in the environment. Like a canary in a coal mine, the rise or decline in population of a species that is particularly sensitive to a particular environmental factor may provide an early warning of significant environmental changes that otherwise may not have been detected until months, years, or even decades later.

Other than collecting household items, no advance preparation is required for any of the procedures in this field session. You can save time in the field by preparing four 10-meter (32' 13/16") lengths and 16 one-meter (39-3/8") lengths of cord ahead of time. Tie loops large enough to fit over the stakes in both ends of each cord, ensuring that the cords are the proper length when stretched from the end of one loop to the end of the other.

Although this field session focuses on surveying plant populations in a community, you can also take advantage of this field trip to gather other specimens, such as fungi, insects, and even animals. (Always know and abide by your local laws and regulations pertaining to contact with wildlife.)

Wildlife is not defenseless, as anyone who has encountered hornets or poison ivy can attest. Always wear gloves, goggles, and protective clothing when you gather specimens, and always know what specimens you are gathering. Capturing a garter snake for your terrarium is one thing; mistaking a water moccasin for a garter snake is quite another.

PROCEDURE V-1-1: CHOOSING AND PREPARING THE SURVEY AREA

In this procedure, we'll select and prepare a defined area for observing plant populations within a community. Choose an area with considerable diversity in plants. For example, a well-maintained lawn is a poor choice because the flora are predominantly one species. In suburban areas, a park or natural area often provides many suitable locations.

Try to choose an area that has a mixture of trees, shrubs, grasses, and ground cover (lichens, mosses, etc.) An area that has parts that are usually shaded and others that are often sunlit usually provides considerably more diversity than an area that is exclusively shaded or sunlit. A sampling area that is on the boundary between a wooded area and a grassy area often provides more diversity than is common in residential areas. Figure V-1-1 shows part of the survey area we chose.

Having chosen the survey area, make sure you have permission to use it. Over the course of a couple hours or more, you will be driving stakes and possibly shooting images and taking plant specimens. Many property owners will freely give permission, particularly if you explain the purpose of your activities and promise to clean up afterward, but some will not.

WARNING

You will be observing and handling wild plants. Wear gloves to prevent toxic or irritating plants from contacting your skin. Do not disturb animal or insect nests. Do not gather specimens that you cannot identify at least generally. Wash your hands thoroughly after handling plant specimens.

Figure V-1-1: *A survey area that exhibits considerable plant diversity*

You can use plant stakes from the garden store, or anything similar that is large enough and strong enough to secure the cord on the corners of the square. For example, tent pegs or large nails are suitable. If nothing else is available, you can use short sections of small fallen branches.

We'll use geometry to ensure that our square is really square. The diagonal of a square is 1.414 times the length of a side, so all we need to do is make sure the 14.14 meter diagonal and the next 10 meter side meet at the same point.

Once you have permission to use the area, proceed as follows:

1. Choose a section of the area that you expect to exhibit considerable species diversity and is 10 meters or more square.

2. Drive stake #1 at one corner of the area you have selected. Measure 10 meters (32' 13/16") along one side of the square you've chosen and drive stake #2 at that point to form the first side of the square. Stretch one of your 10-meter cords between stakes #1 and #2.

3. Loop a second 10-meter cord over stake #2, and stretch it out to about where stake #3 should be placed. Have your assistant hold the free end of the tape measure on stake #1, and pull the tape out to 14.14 meters (14 m, 14 cm or 46' 3/8"). The point where the end of the 10-meter cord and the 14.14 meter mark on the tape measure meet is the next corner of the square. Drive stake #3 at that point, and place the loop on the free end of cord #2 over that stake. (It is not necessary to be exact, but keep as close as you can to a true square that's 10 meters on a side.)

4. Repeat step 3, placing one end of another 10-meter cord on stake #3 and stretching the cord out to about where the final corner of the square should be. Have your assistant hold the free end of the tape measure on stake #2, and pull the tape out to 14.14 meters. The point where the end of the 10-meter cord and the 14.14 meter mark on the tape measure meet is the final corner of the square. Drive stake #4 at that point and place the loop on the free end of cord #3 over the stake. Finally, use cord #4 to connect stakes #4 and #1, completing the square.

PROCEDURE V-1-2: SURVEYING THE PLANT COMMUNITY

After defining the survey area, the next procedure is to locate and identify as many plant species as possible within this defined area. If you have chosen your area well, you should be able to locate at least half a dozen plant species within the survey area, and possibly many more. (We identified more than two dozen species in our own survey area.)

Look for plants in the following categories:

- Woody plants – trees, shrubs, and vines

- Herbaceous plants – grasses, weeds, small flowering plants

- Ground cover – mosses, worts, and lichens (lichens are technically not plants)

Plants representing all three of these categories were present in our own survey area, and will likely be present in your survey area as well. Representatives of two of these categories are readily visible in Figure V-1-1, the trees (woody plants) and the green ground covering (a moss). Figure V-1-2 shows a representative herbaceous plant.

Figure V-1-2: *A representative herbaceous plant*

As you perform your survey, keep in mind that specimens of the same species at different life stages may have different appearances. For example, although they appear very different, the small plant shown in Figure V-1-3 is actually a juvenile example of *Acer cappadocium* 'Aureum', the golden maple tree shown on the left in Figure V-1-1.

How do we know? Well, not being botanists, we couldn't swear to it. (Even botanists can't necessarily identify specimens absolutely by visual means alone.) But there is some strong evidence to suggest that this juvenile plant is in fact *Acer cappadocium* 'Aureum'. First, it is growing within the roots of an adult golden maple. We're reasonably certain of the species (*Acer cappadocium*) and cultivar ('Aureum') of the adult tree because its leaves are five- and seven-lobed (versus the three- or five-lobed leaves of the *Acer rubrum* (red maple) trees that also grow in the immediate vicinity, and because we have watched that tree's leaves change to a brilliant golden color every autumn for the last 20 years (rather than the bright reds and oranges of *Acer rubrum*). The fact that the large upper right leaf of the juvenile plant exhibits seven nodes makes the identification nearly conclusive.

Figure V-1-3: *A juvenile golden maple tree (with moss visible as ground cover)*

As you make your survey, don't be too concerned if you can't identify both genus and species of each plant in the survey area. If you can identify the genus of all specimens, you're doing well.

Species is a fairly slippery concept anyway, with ongoing debate among biologists about just how to define a species. The advent of molecular biology and DNA analysis has had a significant effect on taxonomy. Species that were formerly believed to be closely related based on morphology (physical structure and appearance) are now recognized as being only distantly related, and *vice versa*.

When biologists are certain of the genus but not absolutely certain of the species, they use "sp." (unitalicized) to indicate that uncertainty. For example, we are certain that the tree in question is of the genus *Acer*. We strongly suspect that it is of the species *cappadocium*, but we couldn't swear to that in a court of law. So, if we were writing a formal scientific paper, we would identify this tree as *Acer* sp. rather than *Acer cappadocium*.

Biologists use "spp." to refer to multiple species within one genus. For example, if a biologist is referring to a characteristic shared by multiple (but not all) *Acer* species, she would refer to that group collectively as "*Acer* spp." If she is referring to a characteristic shared by all *Acer* species, she would use simply "*Acer*" without any qualification to indicate that she was referring to all species within the *Acer* genus.

As you do your survey, also log any plant-like lifeforms you observe. For example, we logged the magnificent example of the fungi *Amanita* sp. shown in Figure V-1-4. The cap of this specimen was about 10 cm in diameter. Such fungi can disappear literally overnight, so we also took a sample for later microscopic analysis.

WARNING

Treat all mushrooms as potentially lethal. Many species of *Amanita* are among the most poisonous flora known. Handle them only with gloves, do not inhale the spores, and **never** under any circumstance eat mushrooms that you have picked yourself. Even expert mycologists have died from consuming misidentified mushrooms.

As you encounter each species, use the guidebooks to identify it. Record the identity of each species, along with your notes about its size and other pertinent characteristics in your lab notebook. If you have a camera, record images of each of the species for your records. (You can provide scale within the image simply by including a ruler in the image.) If possible, shoot multiple images of each specimen, including close-up images that show details such as leaf structure. In particular, photographically document any species that you cannot identify with specificity using the guidebooks.

If you have time (and permission), where practical gather specimens of each species, or at least of a representative sample. (No, we don't expect you to carry home a tree...) If it is not practical to obtain a whole specimen, obtain leaves, flowers, seeds, or other parts of the plant. For small lifeforms (such as mosses), cut or dig up a small representative specimen. Store these specimens in tubes or plastic bags, and observe them in more detail when you get home.

Figure V-1-4: *The fruiting body of Amanita sp.*

Use tubes or plastic bags only for temporary storage, when you want to prevent a specimen from drying out. If you want to preserve specimens (such as mosses, lichens, flowers, and so on) long-term, do not store them in sealed containers, which retain water and will allow the specimen to decay. Instead, store such specimens in paper bags, cardboard boxes, or similar containers that will allow water vapor to escape.

PROCEDURE V-1-3: PERFORMING A POPULATION COUNT

Biologists make population counts to record and track the numbers of specific species present in an area, and to determine the changes in those populations over time. The appropriate size for the survey area depends on the size and numbers of the species being surveyed. Our survey area is a square 10 meters on a side or 100 m^2, which is appropriate for medium-size species that are present in moderate numbers. If were were surveying a larger and less common species—oak trees, for example—we might choose a survey area 1,000 meters on a side, or 1 km^2. If we were surveying a small species that was present in large numbers, we might choose a survey area one meter on a side, or even 10 cm on a side.

If—and this is critically important—the selected survey area is representative of the area as a whole, we can determine the population of the species in the survey area and extend that by multiplication to determine the population numbers for the larger area of which the survey area is a part. For example, we might determine the population of *Aconitum* sp. (monkshood) plants in a survey area 100 meters on a side (0.01 km^2) is 22 individuals. From this data, we could estimate the population of monkshood plants in a larger surrounding area of 1,000 meters on a side (1 km^2) as 2,200 individuals.

WARNING

If you do have monkshood plants in your survey area, use extreme caution. Monkshood contains significant amounts of aconitine, one of the deadliest alkaloid poisons known. The lethal dose of aconitine for an adult human is usually stated as between 3 and 6 milligrams (about 1% of the mass of an aspirin tablet), so ingesting even a few grams of the monkshood plant may be lethal. Nor is monkshood the only deadly plant commonly found in residential yards and flower gardens. **Again, use extreme caution in handling plant specimens**.

But such estimates are inherently imprecise, because it's impossible to state with certainty that the survey area is really representative of the surrounding area as a whole. We might, for example, have chosen entirely by coincidence the one 0.01 km^2 area in that larger area that contained any monkshood plants at all.

To reduce the uncertainty inherent in such estimates, biologists sample multiple survey areas (called *quadrats*) within the larger area. For example, rather than counting the population in only one 0.01 km^2 survey area within the larger 1 km^2 area—which is to say only 1% of the larger area—a biologist might randomly choose five 0.01 km^2 survey areas within the larger area. And the "randomly" part is important in reducing uncertainty.

In this procedure, we'll establish four 1m^2 quadrats within the survey area, do population counts of each species for each quadrat, and extend our results to estimate the population of that species in the entire survey area. Although it won't be

random in a strict mathematical sense, we'll establish our four 1 m^2 quadrats simply by measuring specific lengths on the 10-meter lines that define the survey area. (For a distribution closer to actual random, you can stand several meters outside your large survey area with your back to it, and toss an object over your shoulder. Wherever that object lands within the survey area is the center of your first 1 m^2 quadrat; repeat the process to define the remaining three quadrats.

1. Use the measuring tape and marking pen to make four marks on each of two 10-meter cords that are opposite sides of the square. Place marks on each cord at two meters and three meters from each end. That is, each cord should be marked along its length at the following points: 2, 3, 7, and 8 meters.

2. Extend the tape measure between two of the marks directly opposite each other, and drive in stakes at the 2, 3, 7, and 8 meter marks of the measuring tape.

3. Repeat step 2 for each of the other three marks on the two 10-meter cords. When you have finished, you should have 16 stakes driven inside the survey area, forming four squares, each one meter on a side. Wrap cord around each of those four sets of stakes to define the four 1 m^2 quadrats.

4. Referring to your notes, count the number of individuals present in the first quadrat for each of the species you found within the larger survey area. For discrete plants, count the actual number within the quadrat. (If a plant lies exactly on the quadrat border, count it as 0.5.) For grasses, mosses, and other ground cover plants, where it is difficult to discriminate individuals, estimate as closely as possible the percentage of the quadrat surface area covered by that plant.

5. Repeat step 4 for each of the remaining quadrats.

6. Using the data you obtain from the quadrat counts, estimate the population of each species within the larger survey area.

REVIEW QUESTIONS

Q1: Why did we do counts in four quadrats rather than just one? Why did we not use 3, 5, or (say) 17 quadrats?

Q2: How did you calculate the estimated population of each species within the survey area?

Q3: Were all three plant categories (woody plants, herbaceous plants, ground cover) represented in your survey area? Which were the most and least common? Which was the dominant species?

Q4: Did the species counts differ significantly among the four quadrats? If so, what reasons can you propose for the differences?

Q5: If you returned to the survey area and repeated the population count in one month, what changes (if any) would you expect? Six months? One year?

Q6: If a significant change to the survey area occurred, what effect would you expect? (For example, imagine that the tree in our survey area was cut down.)

Observing the Effect of Rhizobia on Plant Growth Lab V-2

EQUIPMENT AND MATERIALS

You'll need the following items to complete this lab session. (The standard kit for this book, available from *www.thehomescientist.com*, includes the items listed in the first group.)

MATERIALS FROM KIT

- Ammonium nitrate[1]
- Fertilizer, nitrogen-free (concentrate A)
- Fertilizer, nitrogen-free (concentrate B)
- Fertilizer, nitrogen-free (concentrate C)
- Graduated cylinder, 10 mL
- Inoculating loop
- Rhizobia inoculum[1]
- Ruler, mm scale
- Seeds, bush lima bean

MATERIALS YOU PROVIDE

- Gloves
- Balance
- Chlorine laundry bleach
- Foam cups, 16 oz./500 mL
- Lamp, fluorescent plant (optional)
- Paper towel
- Pencil
- Plastic wrap (Saran or similar)
- Soft drink bottle, 2-liter
- Vermiculite (or other sterile growth medium)
- Water, distilled

If you purchase materials separately rather than using the kit, make sure to obtain a rhizobia culture that is suitable for lima beans. Rhizobia is available in many variants, each of which is specific to a particular plant or range of plants. For example, rhizobia that is suitable for alfalfa or clover does not work for lima beans, and vice versa. Rhizobia cultures sold in lawn and garden stores may contain a mixture of many rhizobia variants that works with several plants. Read the label.

If you obtain lima bean seeds separately, make sure they are not pretreated with rhizobia or an insecticide. To ensure the seeds you use are not contaminated (including those provided with the kit), surface-sterilize them as described later in the session.

Commercial fertilizers cannot be used in this lab session because they contain nitrogen. If you do not have the kit, you can prepare a suitable nitrogen-free fertilizer by adding the following to 10 liters of water:

- 1.0 g of calcium sulfate

- 0.1 g of iron(III) sulfate (ferric sulfate)

- 2.0 g of magnesium sulfate

- 8.0 g of potassium hydrogen phosphate (K_2HPO_4)

- 2.0 g of potassium dihydrogen phosphate (KH_2PO_4)

For best results, you should also include trace amounts (a few milligrams per liter) of salts of boron, chlorine, copper, manganese, molybdenum, sulfur, and zinc. The three-part fertilizer concentrate included with the kit is made to a different formula that includes all of these trace elements, and is sufficient to produce more than 10 liters of working-strength fertilizer solution.

The best growth medium is sterile vermiculite, available in small bags in any lawn and garden store. You can also use sand (sold in most big-box home centers). If you use sand, wash it first to remove soluble materials by stirring it up in a bucket full of tap water and pouring off the water. Repeat this washing two or three times. After washing, sterilize the sand by spreading it in a thin layer on a pizza pan or similar oven-safe container and heating it for at least 30 minutes at 450 °F. Allow the sand to cool and then transfer it to the planting containers.

BACKGROUND

Biologists study the interactions of species in ecosystems to learn more about each species than they could learn by examining that species in isolation. In many cases, there is little or no direct interaction between species. For example, wolves and grasses neither harm nor benefit each other other than incidentally. (A wolf may damage a grass population by urinating on it; conversely, wolf feces may fertilize another grass population.)

When species do interact directly, one species may benefit to the detriment of the other, a state called *parasitism*. For example, tapeworms may infect a wolf. The tapeworms benefit; the wolf is harmed. In another type of direct interaction, called *commensalism*, one species benefits without benefit or harm to another species. For example, squirrels build nests in oak trees. The squirrels benefit by having a secure haven from predators while the oak trees neither benefit nor are harmed by the interaction.

The final type of direct interaction between species is called *mutualism*, in which each species benefits from the presence of the other. For example, woodpeckers eat insects that infest the bark of trees. The woodpecker benefits from the tree providing a

ready source of food, while the tree benefits by the woodpecker keeping down the population of insects that might otherwise damage the tree.

Note that while one type of interaction between species may be commensalist, another type of interaction between those same species may be mutualist. Returning to the oak trees and squirrels provides such an example. The oak trees provide acorns as a food source for the squirrels, benefiting the squirrels. In turn, the squirrels bury acorns over a wide range. Some of those acorns are forgotten and subsequently germinate, benefiting the oak tree. Overall, then, the relationship between oak trees and squirrels is mutualist.

Also note that when more than two species are involved, the type of interaction depends on the species-pairing point of view. From an insect-tree POV, the relationship is parasitic; the insect benefits to the detriment of the tree. From a bird-insect POV, the relationship is also parasitic; the bird benefits to the detriment of the insect. From a bird-tree POV, the relationship is mutualist; both benefit at no detriment to either species.

In this lab session, we'll examine a mutualist relationship between a plant, the bush lima bean, and a bacterium called rhizobia. Legumes, including the bush lima plant, are commonly infected by rhizobia, which form colonies as nodules in the root structure of the plant. The bacteria benefit by having a nice cozy nodular home amongst the plant's roots. These nodules do no harm to the plant, and in fact benefit it greatly.

To understand why, it's important to understand that plant growth requires six *macronutrients* (arbitrarily defined as those that typically make up 1% or more of the dry weight of the plant) and 10 *micronutrients*, some of which are present only at sub ppm (parts per million) levels. All 10 of the micronutrients—boron (B), calcium (Ca), chlorine (Cl), copper (Cu), iron (Fe), magnesium (Mg), manganese (Mn), molybdenum (Mo), sulfur (S), and zinc (Zn)—must be obtained from the soil. Three of the macronutrients—carbon (C), oxygen (O), and hydrogen (H)—are obtained from carbon dioxide gas and water. Two of the remaining three macronutrients—phosphorus (P) and potassium (K)—must be obtained from the soil.

The final macronutrient, nitrogen (N), may be obtained from the soil or from atmospheric nitrogen with the aid of *nitrogen-fixing bacteria*, such as rhizobia. From a plant's point of view, nitrogen is both abundant and rare. Nitrogen is abundant in the sense that our atmosphere is about 79% nitrogen gas. Unfortunately, plants cannot use nitrogen gas directly but must obtain it in the form of soluble nitrates in the soil, which may be present in inadequate amounts to sustain plant growth. That's where nitrogen-fixing bacteria come in. They convert atmospheric nitrogen gas into soluble nitrates, making the nitrogen available to plants.

In this lab session we'll investigate the growth of the common lima bean, *Phaseolus lunatus*, under controlled conditions. We'll grow two plants in vermiculite, which is a sterile, nutrient-free type of soil that provides only support for the plant. We will provide each plant with nutrients in the form of a fertilizer that contains all of the nutrients named above with the exception of nitrogen. We'll inoculate one of the seeds with rhizobia before planting it, and plant a second seed as a control without rhizobia, and compare the growth of the plants with and without nitrogen-fixing bacteria present.

> This experiment requires successful germination and growth of two lima bean plants, one with and one without rhizobia present. If you start with only two seeds, there is a reasonably high probability that one of the plants will fail to germinate, ruining the experiment. Planting two (or more) seeds each without rhizobia and with rhizobia greatly increases your chance of success.

PROCEDURE V-2-1: GROW LIMA BEANS WITH AND WITHOUT RHIZOBIA

1. If you have not already done so, put on your goggles and gloves.

2. Rhizobia is very common in the environment, so we need to make sure that our lima bean seeds are not contaminated. To surface-sterilize the seeds, prepare enough of a mixture of one part chlorine laundry bleach to 10 parts water to fill a foam cup to depth of 5 cm or so. Immerse the lima bean seeds in the diluted bleach solution and allow them to soak for a couple of minutes, swirling the contents periodically. Empty the bleach from the cup, and rinse the seeds thoroughly in running tap water to remove all bleach residue. Allow the seeds to drain on a paper towel until you need them.

3. Prepare and label two 16-ounce (500 mL) foam cups. Use a pencil to poke several small drainage holes in the bottom of each cup. Cover the holes with a single layer of paper towel.

4. Fill each foam cup with vermiculite to within 2 cm from the top.

5. Press a lima bean seed into the center of the vermiculite in the first cup to a depth of about 2 cm. Cover the seed with vermiculite. Gently pour distilled water onto the surface of the vermiculite until water comes out of the drain holes on the bottom of the cup. Set that cup well aside from your work area.

6. Press a lima been seed into the center of the second cup and use the inoculating loop to transfer a small speck of rhizobia to the well containing the seed. Cover the seed and rhizobia with vermiculite. Again, gently pour distilled water onto the surface of the vermiculite until it drains from the holes in the bottom of the cup.

7. Wash your hands thoroughly before proceeding. If you have an alcohol-based hand sanitizer, also use it to ensure that no rhizobia is transferred to the first cup.

8. Place both cups under a plant grow light or on a windowsill or other area where they'll get sunlight for most or all of the day.

9. Record the details in your lab notebook.

10. Water both cups with distilled water daily. Allow the water to drain completely each time, and then cover the cup with plastic wrap between waterings until the seeds germinate. Observe the two cups at least daily. Note any indications of germination, such as the first appearance of stem or leaves, in your lab notebook. If you have a digital camera, make a photographic record of your observations.

11. After the seeds germinate, transfer 20 mL of fertilizer concentrate A from the kit to an empty, clean 2-liter soft drink bottle and half-fill the bottle with water. Transfer 4.5 mL of fertilizer concentrate B to 500 mL of water, swirl to mix the solution, and add that solution to the 2-liter bottle with swirling to mix the solutions. Transfer 2 mL of fertilizer concentrate C to 500 mL of water, swirl to mix the solution, and add that solution to the 2 L bottle, swirling to mix the solutions. (Do not be concerned if there is some cloudiness or a slight precipitate in the mixed solutions.) Water the young plants every day or two, alternating between distilled water and the fertilizer solution. (If there is cloudiness or a slight precipitate in the fertilizer solution, agitate it to resuspend the precipitate before each use.) Use enough liquid to dampen the vermiculite thoroughly. Do not cover the cups with plastic wrap once the seeds have germinated.

12. After six weeks, measure the heights of both plants and record it in your lab notebook. Carefully remove the plants from the cups, and gently shake off and rinse off all of the vermiculite, making sure to retain all of the root structure and nodules. Allow the plants to dry completely and then weigh them. Record the weight of each plant in your lab notebook.

ENDNOTE

1. Ammonium nitrate is included only in kits shipped to Hawaii. Rhizobia innoculum is included in kits shipped to the other 49 states. Kit users in the continental US and Alaska can use the instructions provided in this chapter to complete this lab session. Kit users in Hawaii ONLY should modify these instructions as follows: rather than using rhizobia, make up the nitrogen-free fertilizer solution as described later in this session. Use half of that solution to grow lima beans without nitrogen. To the other half of the solution, add about a quarter teaspoon of ammonium nitrate per liter of the diluted nitrogen-free fertilizer working solution to make a fertilizer solution that includes nitrogen, and use that solution to fertilize the other half of your lima bean seedlings.

REVIEW QUESTIONS

Q1: Why did we use vermiculite rather than potting soil or ordinary soil from a lawn or garden?

Q2: In the absence of any information about the bacteria, what three hypotheses might you make about the effect of the bacteria on plant growth? Which of these hypotheses is more or less likely? Why?

Q3: What differences did you expect in the two cups? Why? Did your observations confirm or falsify your predictions? Could rhizobia infection of the magic beans account for the extreme size and rapidity of growth of Jack's beanstalk?

Q4: What are the controlled, independent, and dependent variables in this experiment?

Q5: We used height and mass as a proxy in this experiment, on the reasonable assumption that they would fairly represent the actual question, which is more difficult to answer. What is that question, and why is it more difficult to answer?

Q6: Is the interaction between the plant and bacteria commensalist or mutualist? Explain.

Air Pollution Testing

Lab V-3

EQUIPMENT AND MATERIALS

You'll need the following items to complete this lab session. (The standard kit for this book, available from *www.thehomescientist.com*, includes the items listed in the first group.)

MATERIALS FROM KIT

- Microscope slides

MATERIALS YOU PROVIDE

- Gloves

- Saucers or similar containers

- Microscope

- Petroleum jelly (Vaseline or similar)

- Plastic wrap (Saran or similar)

- Spray bottle (optional)

- Watch or timer

BACKGROUND

Air pollution takes many forms. It may result from natural processes or human activity. Gaseous air pollutants such as carbon monoxide, nitrogen oxides, sulfur oxides, and hydrocarbon vapors are produced by automobile exhausts and industrial processes, as well as by wildfires and other natural processes. Solid air pollutants, called particulates, are produced by erosion, volcanoes, fires, and other natural processes and by human activities such as construction, road traffic, and diesel exhaust. In fact, some "particulates" are lifeforms, such as insects, spores, and pollens.

Most living things can tolerate significant levels of many common air pollutants without suffering noticeable harm. In fact, low levels of some materials that are ordinarily considered pollutants may actually be beneficial to some organisms, but chronic exposure to higher levels of these pollutants may harm these and other organisms. Even short-term exposure to high levels of some pollutants may cause acute distress or even death, depending on the organism, the pollutant, and the level of exposure.

In short, whether or not a particular material is considered a pollutant depends on the nature of that material, its concentration level, and the organism in question. For example, some people are severely allergic to ragweed pollen. If you are one of those people, there's no question. Ragweed pollen in the air is a pollutant from your point of view. Conversely, most other organisms, including most people, are unaffected by even heavy concentrations of ragweed pollen. To them, that pollen is not a pollutant, but simply a part of the natural environment.

Similarly, some plants tolerate acid environments much better than other plants. The presence of sulfur oxides and nitrogen oxides in the atmosphere causes acid rain, favoring the growth of acid-loving plants at the expense of other plants that do not tolerate acid conditions as well. From the point of view of the acid-loving plants, the gases that cause acid rain are not pollutants, but instead provide them with a competitive advantage against other plants. As long as the levels of sulfur oxides and nitrogen oxides is relatively low, that is. At elevated levels of those pollutants, the rain is so acidic that it kills even the acid-loving plants.

In this lab session, we'll test the air at various locations for the presence of particulates. We'll use particle traps, the same method that environmental scientists actually use in the field. A particle trap is simply a covered vessel that contains a sticky surface. At each of our sampling locations, we'll remove the cover from a fresh particle trap, expose the sticky surface to the ambient air for a measured time, and then replace the cover. Back in the lab, we'll examine the sticky surface with a microscope and observe the numbers, sizes, and types of trapped particles.

Other than collecting household items, no advance preparation is required for any of the procedures in this lab session. You can save time on the day of the lab by preparing particle traps ahead of time. Actual field and lab time will vary according to the number of sampling areas to be tested and the time required to visit those locations. If you limit yourself to half a dozen nearby sampling areas and run the particle traps simultaneously, you should be able to complete the lab in one two-hour session.

PROCEDURE V-3-1: BUILDING PARTICLE TRAPS

In this procedure, we'll build as many particle traps as we intend to use. (We recommend four to eight traps as a reasonable number to balance time requirements against data diversity.) Because it's important that the traps contain no particles initially, it's best to build the traps indoors in an area where air is not circulating. If possible, turn off the central heating or air conditioning while you build the traps. If you have a spray bottle available, fill it with tap water and mist the air in your work area to clear the air of particulates.

1. Obtain as many shallow vessels (saucers, dessert plates, shallow bowls, etc.) as you intend to build traps. Label each container A, B, C, and so on.

2. Label a clean microscope slide to correspond to one of the containers. Place a tiny drop of petroleum jelly about 25mm from one end of the slide, centered on the width of the slide. Use the edge of a second slide to spread the petroleum jelly to cover a 25mm square area in the center of the first slide.

> It's easy to apply too much petroleum jelly. Use just enough to cover the center third of the slide with an extremely thin layer of petroleum jelly.

3. Immediately place the coated slide, sticky side up, in one of the containers and cover the container to protect the tacky surface of the slide, making sure to allow nothing to contact the surface of the slide.

> Plastic wrap works well to cover shallow bowls and similar containers. If you're using flatter containers such as saucers or dessert plates, substitute sheets of thin cardboard to avoid the danger of the plastic contacting the slide surface.

4. Repeat steps 2 and 3 to build as many particle traps as you intend to use. It's a good idea to build one or two spare traps in case of accidents.

5. Place the particle traps aside until you are ready to use them.

PROCEDURE V-3-2: POSITIONING AND EXPOSING THE PARTICLE TRAPS

Choose as many locations as you have particle traps. Aim for diversity in locations. For example, you might want to expose traps in your kitchen, garage, laundry room, and yard or garden. Consider removing the grill from a ventilation duct and placing a trap there (with the central heat or air running). Other possible locations are a busy street or road, perhaps at an intersection, a park or wooded area, a shopping center, and so on.

If your results are to be comparable, it's important to use the same procedure for each trap. Place the trap in the selected position, note the time, and remove the cover from the trap. Record the number/letter of the trap and the location in your lab notebook. After a standard amount of time (we recommend 30 minutes or one hour) has passed, immediately replace the cover on the trap, making sure that no particles present on the cover are transferred to the trap.

PROCEDURE V-3-3: COUNTING AND IDENTIFYING PARTICLES

When you return to the lab, set up your microscope. Use canned air or a puffer to displace any dust present on the microscope or surrounding work area. Again, try to minimize air currents, and mist the air in your work area to help particulates settle out. Handle the particle trap slides with gloves to avoid transferring particulates from your hands to the slides, and avoid breathing on the slides.

1. Rotate the low magnification objective into position, and place the first slide on the stage with the center part of the tacky area under the objective.

> In doing particle counts, the two major factors that must be kept constant are the exposure time and the sampling area in which particles are counted. We've fixed the first of these variables by exposing each trap for the same period. We'll fix the second by observing a standard sampling area for each slide, defined by the field of view of the low-magnification objective. The fact that we must position each slide on the stage before we can count particle densities is sufficient to randomize the sampling area. Don't reposition the slide to look for more- or less-dense particle accumulations, because doing that makes the sampling area nonrandom.

2. Observe each particle visible in the field of view. Record the count, size distribution, type, and any other data you observe in your lab notebook.

> Most or all of the particles may be difficult or impossible to identify. Others may be identifiable by category. For example, dust particles may appear under magnification like tiny rocks, and it's possible that you'll find tiny insects such as dust mites. Large spores and pollens may be visible at low magnification as tiny dots.

3. After you finish the count at low magnification (but not before), center any unidentifiable small particles in the field of view and observe them at medium or high magnification. You probably won't be able to identify any of them specifically, but you may be able to identify some or all of them by class, such as spores, pollen, and so on.

REVIEW QUESTIONS

Q1: Which of your locations showed the highest and lowest levels of particulate pollution?

Q2: If one of your particle traps showed no trapped particles, can you conclude that no air pollution exists in the area where you exposed the trap? Why or why not?

Q3: What step could we take to increase the reliability of our data for each of the tested locations?

Soil and Water Pollution Testing

Lab V-4

EQUIPMENT AND MATERIALS

You'll need the following items to complete this lab session. (The standard kit for this book, available from *www.thehomescientist.com*, includes the items listed in the first group.)

MATERIALS FROM KIT

- Goggles
- Centrifuge tubes
- Hydrochloric acid
- Pipettes

- Reaction plate, 24-well
- Sodium borate solution, 0.1% w/r to boron
- Turmeric reagent

MATERIALS YOU PROVIDE

- Gloves
- Desk lamp or other strong light source
- Paper towels

- Specimens (see text)
- Water, distilled

BACKGROUND

Soil and water pollution takes many forms. It may be chemical or biological. The source may be natural—as, for example, when heavy metals leach from a crumbling rock face or scat from wild animals contaminates an apparently pristine mountain stream—or the result of human activities such as manufacturing, mining, or even simply doing a load of laundry.

Environmental scientists and technicians frequently test soil and water specimens to determine if they are contaminated by specific pollutants, ranging from specific bacteria, protozoa, or fungi to heavy metals to organic solvents to inorganic ions such as phosphates or nitrates.

These tests are often done in two phases. The first phase, called screening tests, often uses color-test reagents. Screening tests are fast, inexpensive, and can usually be done in the field. Negative results from a screening test are normally accepted as evidence that a particular pollutant is not present in a specimen at levels high enough to be of concern. A positive screening test is accepted as evidence that the pollutant in question is probably (but not certainly) present at levels high enough to be of concern. (For many pollutants, including many heavy metals, any level high enough to be detected by a screening test is by definition high enough to be of concern.) A positive screening test is normally followed by instrumental testing, which is more sensitive and more accurate than a screening test, but is also more expensive, time-consuming, and must normally be done in a formal lab.

For example, our municipal water authority constantly tests our drinking water for a wide range of heavy metals (chromium, lead, and so on) as well as for various biological contaminants, such as *E. coli* bacteria. If a screening test indicates the presence of chromium ions at any level, that specimen is immediately subjected to instrumental testing. Similarly, a positive screening test for coliform bacteria (such as E. coli) is cause for immediate followup tests, including culturing and possibly DNA analysis of the culture to determine which E. coli variant is present.

In this lab session, we'll test soil and water specimens for the presence and concentration of boron, a light element that is widely distributed in soil and water. Most people are familiar with two boron compounds: sodium borate (borax) and boric acid. Borax is widely used as a laundry supplement, and boric acid is used for purposes as diverse as making eye drops, killing small rodents, and treating insect infestations on rosebushes.

At the levels common in soil and water specimens—typically one ppm or less—boron is innocuous. In fact, as we mention in another lab session, boron is an essential micronutrient for plant growth. Nonetheless, at the much higher concentrations sometimes found in soil and water samples, boron hinders the growth of many plants. For that reason, environmental scientists may need to test boron levels in soil and water samples.

The standard screening test for boron, called the curcumin test, has been in use for more than 100 years. This test depends on the fact that boric acid reacts with an organic chemical called curcumin, which is found in the spice turmeric, to form an intensely colored red complex called rosocyanine. The curcumin test is extremely sensitive, able to detect boron at levels of one part per million or less. It's also sufficiently selective to yield reliable results. (Some other chemicals yield a positive curcumin test, but none that are likely to be found in environmental samples.) Finally, the curcumin test is fast and costs only pennies per test.

The curcumin test is specific for boric acid, as opposed to boron in the form of borate salts, which yield negative results with curcumin. For this reason, the first step in the curcumin test is to acidify the specimen with hydrochloric acid, which converts any borate salts present into boric acid.

> Most biological water testing depends on E. coli as a proxy for microbial contamination because E. coli is the longest-lived among the species that commonly contaminate water supplies. If E. coli is absent, it's assumed that all other common microbial pollutants are also absent. If E. coli is present, even one of the nonpathogenic variants, it is assumed that other, pathogenic microbes may also be present. That specimen is cultured to determine the types and numbers of microbes present.

PROCEDURE V-4-1: OBTAINING SOIL AND WATER SPECIMENS

In this procedure, we'll gather soil and water specimens that we'll later test for the presence and concentration of boron. Depending on your local environment, you may later find that all, some, or none of your specimens contain boron at levels that are detectable by the test we'll use.

HANDLE WITH CARE

Always wear gloves when gathering soil or water specimens. In particular, water specimens may contain pathogenic bacteria, protozoa, fungi, or other dangerous organisms.

1. Obtain at least two or three water specimens from different sources, such as a pond, stream, ditch, and so on. Label a centrifuge tube with the date and source of the specimen. Simply fill the tube with the specimen and recap it.

2. Obtain at least two or three soil specimens from different sources. Simply transfer about a tablespoon of soil to a labeled and dated centrifuge tube and recap the tube.

If you run out of centrifuge tubes, substitute clean soda bottles, plastic bags, or similar containers.

3. When you return home, rinse each of the specimen containers thoroughly in running tap water to remove any external contamination and wash your hands thoroughly.

4. Put on goggles and gloves.

5. Uncap each water specimen container and add two drops of hydrochloric acid. Recap the container and invert it several times to mix the solutions.

6. Uncap each soil specimen container, fill it nearly full of tap water, and add two drops of hydrochloric acid. Recap the container and invert it several times to mix the solutions. Allow the solids to settle.

If you have a balance, you can obtain quantitative results by weighing the soil specimens and measuring the volume of water added. For example, you might use 10 g of soil and 40 mL of water and subsequently determine that the boron concentration in the water is 20 ppm. Assuming that all of the boron in the soil has dissolved in the water, you can calculate that a boron concentration of 20 ppm in 40 g of water means that the boron concentration in your 10 g of soil must be 80 ppm.

7. Place the containers aside until you are ready to test the specimens.

PROCEDURE V-4-2: TESTING THE REAGENTS

Before analyzing specimens, careful scientists test their reagents to make sure that they perform as expected. Before proceeding to specimen analysis, we need to verify that our curcumin reagent reacts as expected with a specimen known to contain boron.

1. If you have not already done so, put on your goggles and gloves. Review the hazards of each chemical you will use.

2. Place the 24-well reaction plate on a white paper towel under a desk lamp or other strong light source.

3. Transfer 1 mL of distilled water to each of wells A1, B1, C1, and D1.

4. Transfer five drops of the sodium borate solution to each of wells B1 and D1.

5. Transfer one drop of 6 M hydrochloric acid to each of wells B1 and C1.

At this point, well A1 contains only distilled water, well B1 contains sodium borate and hydrochloric acid, well C1 contains hydrochloric acid, and well D1 contains sodium borate but no hydrochloric acid. By observing which well or wells exhibit a color change, we can determine which of the chemicals or combination of chemicals causes that color change.

1. Add two drops of the turmeric reagent to each of wells A1, B1, C1, and D1.

2. Observe the color changes, if any, in each of the wells, and record your observations in your lab notebook.

PROCEDURE V-4-3: MAKING BORON CONCENTRATION COMPARISON STANDARDS

In this procedure, we'll set up two 3X3 arrays in the reaction plate. We'll later use the center wells of these arrays for testing actual specimens. Each of the eight surrounding wells will have a different concentration of boron ions, so we can easily compare the center specimen well with each of the eight adjacent wells in each array.

1. If you have not already done so, put on your goggles and gloves. Review the hazards of each chemical you will use.

2. Place the 24-well reaction plate on a white paper towel under a desk lamp or other strong light source.

3. Using a graduated pipette, transfer, as accurately as possible, 1.0 mL of distilled water to each of wells A2, A3, B1, B2, B3, C1, C2, and C3.

4. Transfer, as accurately as possible, 1.0 mL of the sodium borate solution to well A1 and another 1.0 mL of the sodium borate solution to well A2. Mix the solution in well A2 thoroughly by stirring it with the tip of the pipette and by drawing up and expelling the solution several times.

At this point, we have 1.0 mL of sodium borate solution in well A1. That solution is 0.1% with respect to boron, which can also be stated as 1,000 parts per million (ppm) boron. Well A2 contains 2.0 mL of solution that is 500 ppm with respect to boron.

1. Draw up 1.0 mL of the solution in well A2 and transfer it to well A3. Mix the solution thoroughly. Well A2 now contains 1.0 mL of 500 ppm boron, and well A3 2.0 mL of 250 ppm boron.

2. Draw up 1.0 mL of the solution in well A3 and transfer it to well B3. Mix the solution thoroughly. Well A3 now contains 1.0 mL of 250 ppm boron, and well B3 2.0 mL of 125 ppm boron.

3. Draw up 1.0 mL of the solution in well B3 and transfer it to well C3. Mix the solution thoroughly. Well B3 now contains 1.0 mL of 125 ppm boron, and well C3 2.0 mL of ~63 ppm boron.

4. Draw up 1.0 mL of the solution in well C3 and transfer it to well C2. Mix the solution thoroughly. Well C3 now contains 1.0 mL of ~63 ppm boron, and well C2 2.0 mL of ~31 ppm boron.

5. Draw up 1.0 mL of the solution in well C2 and transfer it to well C1. Mix the solution thoroughly. Well C2 now contains 1.0 mL of ~31 ppm boron, and well C1 2.0 mL of ~16 ppm boron.

6. Draw up 1.0 mL of the solution in well C1 and transfer it to well B1. Mix the solution thoroughly. Well C1 now contains 1.0 mL of ~16 ppm boron, and well B1 2.0 mL of ~8 ppm boron. To keep the solution level in all wells equal, draw up 1.0 mL of the solution in well B1 and discard it.

At this point, we have eight wells that contain boron solutions, with each well half as concentrated as the preceding well. The concentrations in this first 3X3 array range from 1,000 ppm down to about 8 ppm. The next step is to create a second 3X3 array of wells that contain concentrations ranging from 4 ppm down by halves. To avoid small fractions, we'll shift to using ppb (parts per billion) to specify concentrations for this second array. A concentration of 4 ppm is the same as 4,000 ppb, so that's where we'll start.

1. Transfer, as accurately as possible, 1.0 mL of distilled water to each of wells A4, A5, A6, B4, B5, B6, C4, C5, and C6.

2. Transfer, as accurately as possible, 1.0 mL of the 8 ppm sodium borate solution from well B1 to well A4 to produce a solution that's about 4 ppm (4,000 ppb) with respect to boron. Mix the solution in well A4 thoroughly by stirring it with the tip of the pipette and by drawing up and expelling the solution several times.

3. Repeat the serial dilution procedure to produce concentrations of ~2,000 ppb (A5), ~1,000 ppb (A6), ~500 ppb (B6), ~250 ppb (C6), ~125 ppb (C5), ~63 ppb (C4), and ~31 ppb (B4).

4. Add one drop of turmeric reagent to each of the 18 populated wells in the two arrays. The center wells (B2 and B5), which contain only distilled water, provide a reference to compare the other wells against. Those two wells should have a noticeable yellow color, which is the natural color of the reagent. Depending on lighting conditions, the yellow in those wells may be too pale to see clearly. If that's the case, add one more drop of the turmeric reagent to wells B2 and B5. If that's sufficient to show the contents of those wells as clearly yellow, also add one more drop of the turmeric reagent to each of the other populated wells.

5. Record the concentrations of the wells and your observations of any color changes that occurred (and how quickly or slowly they occurred) in your lab notebook. Include a sketch of the layout and concentrations of the wells.

At this point, you can continue immediately with the following procedure, testing environmental specimens. If you delay performing that procedure for a significant length of time, carefully cover the reaction plate with its lid to prevent evaporation.

PROCEDURE V-4-4: TESTING SPECIMENS FOR BORON

In this procedure, we'll test our environmental specimens for the presence and concentration of boron.

1. If you have not already done so, put on your goggles and gloves.

2. Transfer 1 mL of your first specimen to well B5 and add the number of drops of turmeric reagent that you decided was optimum in the preceding procedure. Note any color change and the intensity of the color relative to the comparison wells, and record your observations in your lab notebook.

> If the color of the specimen is intermediate between the colors of two comparison wells, estimate the concentration of the specimen based on interpolation. For example, if the color of the specimen is about midway between the 4 ppm and 2 ppm comparison wells, you might record the concentration of the specimen as about 3 ppm. If the color is between 2 ppm and 4 ppm but closer to 4 ppm, you might record the concentration of the specimen as about 3.5 ppm.

3. If the specimen showed no color change, or if the color is within the range of the comparison wells surrounding well B5, proceed to the next step. If the color is more intense than any of the comparison wells, repeat the preceding step, starting with 1 mL of the specimen in well B2.

4. Use a pipette to withdraw the liquid from well B5 (and B2, if you used it) and discard it. Use the corner of a paper towel to absorb the last few droplets. Fill the well with distilled water, stir the contents, and empty the well with a pipette. Repeat this rinse twice and then dry the well with the corner of a paper towel. The well is now ready for testing the next specimen.

Incidentally, don't be disappointed if none of your environmental samples contain detectable levels of boron. Some do; some don't. In science, negative results are often as useful as or more useful than positive results.

REVIEW QUESTIONS

Q1: We used acidified tap water to solubilize any boron salts present in the soil specimens. What potential error does this introduce? What change in procedure might we make to eliminate this potential error?

_____

_____

_____

_____

Q2: Did you detect boron in any of your water or soil specimens? If so, at what level?

_____

_____

_____

_____

Q3: Describe an experiment to determine the effect of elevated boron levels on plant growth.

_____

_____

_____

_____

Exploring Mendelian Genetics

Lab VI-1

EQUIPMENT AND MATERIALS

You'll need the following item to complete this lab session. (The standard kit for this book, available from *www.thehomescientist.com*, includes the only item needed for this session.)

MATERIALS FROM KIT

- Phenylthiocarbamide (PTC) test strips

MATERIALS YOU PROVIDE

- None

BACKGROUND

In 1930, the American chemist Arthur Fox had a lab accident while he was purifying an organic compound called *phenylthiocarbamide* that he had just synthesized. Some of the finely powdered phenylthiocarbamide (also called *PTC* or *phenylthiourea*) escaped as a cloud of dust. Fox thought nothing of it until a colleague working nearby complained about the bitter taste of the material. Fox, who had been exposed to much more of the compound than his colleague, had tasted nothing.

Being scientists, they immediately verified their observation by retasting the PTC, confirming that Fox was unable to detect any bitterness while his colleague nearly gagged at the bitterness. Fox soon extended the experiment to determine which of his family, friends, and colleagues at DuPont were able to taste PTC. Some of them were unable to detect any taste, others detected a slight bitterness, and still others found the taste of PTC to be extremely bitter. As it turns out, about 70% of people can taste PTC, but that percentage differs significantly in different populations, ranging from about 58% among Australian Aborigines to about 98% among American Indians.

Until well into the 20th century, chemists commonly tasted new compounds they had synthesized. Also well into the 20th century, accidental poisoning was a common cause of death among chemists. As a college freshman chemistry major in 1971, Robert was told never to taste any compound unless it was known to be safe, but many of his older professors were survivors from a generation that had routinely tasted compounds whose toxicity was unknown.

Geneticists soon determined that the ability to taste PTC is an inherited trait. In fact, the genetic correlation is so strong that, until the advent of DNA analysis, PTC testing was used (along with blood type) in paternity testing. But it wasn't until 2003 that geneticists discovered exactly what determined the ability of an individual to taste or not taste PTC. That ability is determined by one gene, designated TAS2R38.

There are two common *alleles* (forms) of that gene—one for tasting and one for nontasting—along with several rare alleles. Everyone has two copies of each gene, so an individual may have two copies of the tasting gene, two copies of the nontasting gene, or one copy of each gene. *Homozygous* individuals—those who have either two copies of the tasting gene or two copies of the nontasting gene—find PTC extremely bitter or entirely tasteless, respectively. *Heterozygous* individuals—those who have one copy of each allele—can taste PTC, but only as a mild bitterness.

> Actually, there is some debate about whether the ability to taste PTC is a purely Mendelian trait, associated only with the TAS2R38 gene. Humans have about 30 other genes that code for the ability to taste various other bitter compounds. It is possible that the tasting TAS2R38 allele exhibits either *complete dominance*—in which case the partial ability to taste PTC could be coded for by one or more of the other tasting genes—or *incomplete dominance*, in which case the partial ability to taste PTC would manifest in individuals who are heterozygous with respect to the TAS2R38 gene.

PTC is very uncommon in nature—occurring only at levels near or below the detection threshold of even homozygously-sensitive individuals—but the ability to detect PTC is nonetheless a selective advantage because the structure of PTC is similar to the structures of the bitter alkaloids produced as defense mechanisms by many plants. Accordingly individuals who are PTC tasters can detect the bitterness of many alkaloids, many of which are extremely toxic.

You might wonder, then, why the nontasting gene was not eliminated long ago by nontasting individuals unintentionally poisoning themselves by consuming toxic plants. There is still some debate among geneticists as to why this has not occurred. Prevailing opinion is that nontasting individuals are instead able to taste other toxic compounds that tasting individuals are not able to taste. If that is the correct explanation, one would expect individuals who are heterozygous with respect to the TAS2R38 gene to have a competitive advantage because they are able to taste both classes of toxic compounds, albeit not as strongly as homozygous individuals are able to taste one or the other class.

If true, this would tend to preserve the recessive nontasting allele in heterozygous form. From that, it follows that a minority of individuals in a large population should be homozygous with respect to the recessive (nontasting) TAS2R38 gene, which is exactly what we observe.

In this lab session, we'll use test paper strips impregnated with tiny amounts of PTC to test as many individual volunteers as possible for their ability to taste PTC. You can begin with your own family, of course, but your friends, neighbors, church, sports teams, and other social groups that you belong to should provide a ready source of volunteers if you present your project properly to them.

> The kit contains a vial of 100 PTC test strips, which you can cut into halves or thirds if you have more than 100 test subjects. If necessary, you can purchase additional PTC test strips from nearly any lab supplies vendor.

Testing unrelated individuals provides some useful data, but ideally you want to test as many related individuals as possible so that you can follow inheritance of the PTC tasting and nontasting alleles through generations of families. Testing both parents and their children is good; testing parents, children, and all four grandparents is better still. Best of all is testing the full extended family, with aunts and uncles and cousins.

WARNING

Any human genetic testing, including this lab session, potentially has serious ethical implications. Many families have at least one "skeleton in the closet" that they'd prefer to keep hidden from the world at large. You are obligated—morally, ethically, and possibly legally—to maintain the absolute privacy of your test subjects by refusing to disclose the data you obtain to anyone else, including the test subjects themselves.

For example, in one of your family groups of test subjects you might find that both parents are nontasters, as are all of their children except the eldest. You might conclude that that child was adopted or had a different father. Disclosing that conclusion **TO ANYONE, INCLUDING THE TEST SUBJECT**, is a serious ethical violation, and may have direct and indirect consequences you cannot imagine. If you discover such an anomaly, **KEEP IT TO YOURSELF**.

Also consider this: you are not a geneticist, so you may be wrong.

PROCEDURE VI-1-1: TESTING SUBJECTS FOR THE ABILITY TO TASTE PTC

Do not tell your test subjects what to expect, lest you skew your results. If asked, you can assure your test subjects that there is no hazard involved in this experiment, but do not do so unless asked.

1. Working one-on-one if possible, ask your first test subject to place a PTC test strip on his tongue. Once he has done so, ask him to describe what, if anything, he tastes. Some of your subjects will probably report tasting nothing, and others may spontaneously report strong bitterness. (YUCK!) If a subject reports a bitter taste, ask the subject to characterize the bitterness as extremely bitter, strongly bitter, moderately bitter, or slightly bitter.

2. Record the identity and relationships of that test subject in your lab notebook, along with the results of the taste test for that subject.

3. Repeat steps 1 and 2 for your other test subjects.

PROCEDURE VI-1-2: CHARTING INHERITANCE OF PTC-TASTING ABILITY

1. Either with pen and paper or with your computer and graphing software, make up a genealogy chart for each of your family groupings, indicating tasting, nontasting, or partial-tasting for each member of the chart by color-coding or other visual means.

2. Analyze the chart to track the effect of the PTC tasting allele through the generations.

REVIEW QUESTIONS

Q1: Based on your data, can you conclude with reasonable certainty that the ability to taste PTC is a full Mendelian trait? Why or why not?

_____

_____

_____

Q2: In human studies, it is common to anonymize the data, which is to say to remove any information that would allow another party to identify the test subjects. Propose a method for anonymizing your own data, both in raw form and in the final genealogy charts. What overriding factors must you take into account when anonymizing your data?

_____

_____

_____

_____

Observing Specialized Eukaryotic Cells

Lab VII-1

EQUIPMENT AND MATERIALS

You'll need the following items to complete this lab session. (The standard kit for this book, available from *www.thehomescientist.com*, includes the items listed in the first group.)

MATERIALS FROM KIT

- Goggles
- Forceps
- Slides (flat) and coverslips
- Eosin Y
- Gram's iodine

- Hucker's crystal violet
- Methylene blue
- Safranin O
- Sudan III

MATERIALS YOU PROVIDE

- Gloves
- Elodea leaf
- Microscope

- Onion, raw
- Prepared slides (see Note)
- Water, distilled

BACKGROUND

In this session, we use a variety of prepared slides. We have two goals: first, to observe the similarities and differences among different types of cells from the same or similar species. Second, to compare the same types of cells from different species to observe the similarities and differences.

For the first goal, you'll need to purchase prepared slides of various types of human cells. We suggest several of the following: blood, bone, muscle, nerve, ovary, skin, and sperm. Ideally, these should all be human cells, but in a pinch, any mammal cells are acceptable. For the second goal, obtain prepared slides of the same types of cell from other species. For example, compare mammal skin against frog or fish skin, and mammal muscle against bird muscle.

If you don't have any or all the prepared slides required for this session, there are various options. If you have prepared slides that don't exactly correspond to those we used, you can substitute them. (For example, you may have prepared slides of canine cells, which can be substituted for the human cells.) Many home science vendors sell prepared slide sets, which range in quality from mediocre to good, with corresponding price differences. Some of these sets cover a wide range of subjects with little depth, and others cover a much narrower range of subjects with greater depth. You will probably have to buy several different sets to get the breadth and depth you need to cover all of the topics in this book.

Note that images are not a perfect substitute for viewing actual slides. For example, Figures VI-1-1 and VI-1-2 show the cells of an onion epidermis at 100X and 400X, respectively. Because the cells have actual depth, we were forced to choose a compromise focus position when we shot these images. When we were actually viewing the slide, we could tweak the fine focus to bring various cell components into sharper focus, rather than having all of them slightly fuzzy. For most specimens, this problem is obvious at 400X or higher magnification. For many specimens, it's evident at 100X, and for some specimens, it's obvious even at 40X.

In unicellular organisms, such as bacteria and yeast, all life processes must by definition occur within that single cell. Multicellular organisms, such as animals and plants, are made up of many different types of cells, each of which is adapted structurally and biochemically for a specific function or functions.

For example, your epidermal (skin) cells provide a barrier between you and the environment, while your red blood cells are adapted to absorb oxygen as they move through your lungs and then transfer that oxygen to other cells throughout your body. There is a great deal of similarity between cells that have the same purpose in different species. For example, although there are differences, red blood cells from, say, humans, rabbits, and fish are more similar than different, as are (for example) epidermal cells from, say, deer, dogs, and chimpanzees. The similarities reflect the similar purposes of those cells across species, and the differences represent adaptations specific to the species in question.

In this lab session we'll use the skills you learned in session I-3 to prepare several wet mounts of eukaryotic cell specimens. We'll use those slides along with some prepared slides to investigate the structures of various eukaryotic cells and consider how those structures adapt the cells to their purposes. We will also test the effect of various stains on various types of cells.

PROCEDURE VII-1-1: OBSERVING ONION EPIDERMAL AND ELODEA LEAF CELLS

In this procedure, we'll observe the structure and components of cells from an onion epidermis and an Elodea leaf by preparing wet mounts of those cells. We'll also stain the cells with various stains to determine the effect of each stain on different components of the cells.

1. If you have not already done so, put on your goggles, gloves, and protective clothing.

2. Obtain a fresh onion scale (one layer). Snap it in two by bending it in half to leave the epidermal side (the moist side) exposed. At the broken edge of the scale, the epidermis should be visible as a paper-thin, almost transparent layer.

3. Use the forceps gently to pull off a small portion of the epidermis. Do not fold, wrinkle, or crush the epidermis, which may damage the cells. Transfer the piece of epidermis to a slide, add a drop of water, and put a coverslip in place.

4. Adjust the microscope's illuminator and diaphragm for optimum viewing, and observe the epidermis at low, medium, and high magnification. Look for organelles and cell structures such as the nucleus and nuclear envelope, cell wall, cell membrane, cytoplasm, vacuoles, and plastids (such as chloroplasts and chromoplasts). Note that not all of these structures are necessarily present in these cells, and even if present, they may be difficult or impossible to discriminate. Record your observations in your lab notebook, including sketches of the internal cell structure and the arrangement of cells in the epidermal tissue.

5. Repeat step 3 using methylene blue stain. After you have positioned the coverslip, place one drop of methylene blue stain at the edge of the coverslip. Use the corner of a paper towel at the opposite edge of the coverslip to draw water from beneath the coverslip and draw the stain under the coverslip. Allow the stain to work for several minutes to

penetrate the cells. While you wait, you can be preparing additional onion epidermis wet mounts for the following steps.

6. Repeat step 4 using the methylene blue stained slide. Make sure to note which cell structures are stained by the methylene blue.

7. Repeat steps 5 and 6 using eosin Y, Gram's iodine, Hucker's crystal violet, safranin O, and Sudan III stains.

8. Use the forceps to remove one leaf from an Elodea (water weed), place it on a slide, add a drop of water, and put a coverslip in place.

9. Repeat steps 4 through 7 to observe the Elodea leaf cells with and without the various stains.

Figure VII-1-1: *At 100X, the onion epidermis shows its overall structure*

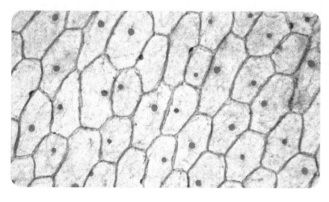

Figure VII-1-2: *At 400X, details of the onion epidermis cells are visible, including distinct cell walls and cell membranes*

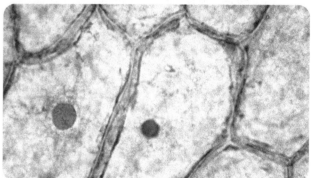

Figure VII-1-3: *At 100X, the structural similarities between the elodea leaf cells (shown here) and the onion epidermis cells are obvious*

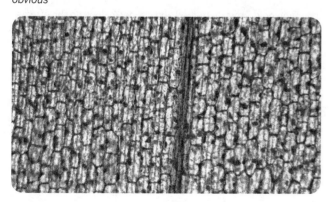

PROCEDURE VII-1-2: COMPARING AND CONTRASTING EUKARYOTIC CELLS

In this procedure, we'll observe the structure and components of cells of different types of cells from the same (or a closely related) organism, as well as similar types of cells from different organisms. Form follows function in cells, both structurally and biochemically. For example, because the cells that make up lung tissue have the same purpose across species, lung cells from, say, a human are closely similar to lung cells from a canine or other mammal. Lung cells from a bird will also be similar to those of a human or a dog, but will also exhibit more differences because the species are not as closely related. Cells of a particular type from two closely related species may be difficult or impossible to discriminate visually. For example, skeletal muscle cells from a lion are very similar to those from a tiger or, for that matter, a house cat.

> Since you'll be working only with prepared slides in this procedure, you needn't put on gloves or protective clothing. However, it's a matter of good practice to wear gloves any time you're working with specimens, if only to avoid getting fingerprints on the slides.

> In particular, note the presence or absence and the appearance of the cell wall, cell membrane, cytoplasm, any vacuoles or plastids present in the cytoplasm, the nucleus, and the nuclear envelope.

1. Place the first of your human (or mammal) cell prepared slides on the stage and examine it at low magnification to locate a good example of the cell in question. Center that area and switch to medium and then high magnification to observe as much as possible about the structures of that cell. Record your observations in your lab notebook, including a sketch or sketches of the cell and its components.

2. Repeat step 1 for each of your other mammal slides, noting the similarities and differences among them.

3. Repeat step 1 for your nonmammal slide(s), comparing them to the similar cell type(s) from mammals.

REVIEW QUESTIONS

Q1: What was the most significant difference between observing the onion epidermis unstained and stained? How did the effects of the different stains vary? How would you prepare a slide to show as much detail as possible in as many different cell components as possible?

_____

_____

_____

_____

_____

Q2: What is the shape of an individual epidermal cell, and what did the structure of the epidermal tissue resemble?

_____

_____

_____

_____

Q3: What similarities and differences did you observe between the onion epidermis cells and the Elodea leaf cells?

_____

_____

_____

_____

Q4: What similarities and differences did you observe between the onion epidermal cells and the mammal epidermal (skin) cells? Reviewing your notes from procedure I-3-1, when you used methylene blue to stain epithelial cells from your cheek, what differences did you notice in staining the onion epidermal cells?

_____

_____

_____

_____

Q5: In procedure VI-1-2, what features did most or all of the cells you observed have in common? What prominent feature in the onion and Elodea cells was absent in the animal cells?

_____

_____

_____

_____

Q6: In procedure VI-1-2, what similarities and differences did you observe while comparing animal cells?

_____

_____

_____

_____

_____

Preparing Culturing Media

Lab VII-2

EQUIPMENT AND MATERIALS

You'll need the following items to complete this lab session. (The standard kit for this book, available from *www.thehomescientist.com*, includes the items listed in the first group.)

MATERIALS FROM KIT

- Goggles
- Agar
- Bottle, polypropylene, 125 mL
- Centrifuge tubes, 15 mL
- Petri dishes
- Spatula
- Stirring rod
- Test tubes
- Test tube rack
- Thermometer

MATERIALS YOU PROVIDE

- Gloves
- Aluminum foil
- Balance (or measuring spoons)
- Bottle, 500 mL soda (clean and empty)
- Broth, chicken (or bouillon cube)
- Cotton balls
- Marking pen
- Measuring cup, 500 mL (microwave-safe)
- Microwave oven
- Pressure cooker (optional; see text)
- Refrigerator
- Sodium chloride (table salt)
- Spray disinfectant (Lysol or similar)
- Sprayer bottle filled with water
- Sucrose (table sugar)
- Tape
- Water (distilled)

BACKGROUND

All living things require a suitable "life support" environment if they are to grow and flourish. Just as people need air, water, food, and protection from temperature extremes, bacteria and other microorganisms grow only in environments that supply their essential requirements.

The process of encouraging growth of particular organisms by providing conditions optimal for those organisms is called *culturing*. For example, growing plants and livestock is called *agriculture*. Fish farming is *pisciculture*, a form of *aquaculture*. In each case, human activity is directed at modifying the environment to suit the needs of the organism.

Culturing microorganisms is no different. We provide everything the microorganism needs to grow and flourish, including nutrients, suitable temperatures and other environmental factors, and a place to grow undisturbed and unthreatened. That place to grow is a gel or a liquid, and is called a *culturing medium*.

Gels are usually based on agar, which when dissolved in hot water and allowed to cool forms a semisolid gel much like dessert gelatin. (Gelatin is seldom used for general bacterial culturing because many bacteria species eat gelatin, turning it into a runny mess. With the minor exceptions of a few marine bacteria, no common microorganisms eat agar.)

Gel media have the advantage of providing a firm, flat surface for bacterial growth. If a mixture of two or more bacteria species is placed on the gel (a process called *inoculation*), those species form separate and distinct *colonies*, the form, color, and other characteristics of which are useful in identifying the particular species, or at least narrowing down the possibilities. Also, because colonies each contain only one species of bacteria, it's easy to obtain a specimen of that single species and use it to inoculate another culturing vessel, thereby producing a *pure culture* of that single species.

Liquid culturing media do not separate growing bacteria into colonies. If multiple bacteria species are introduced into a liquid culturing medium, they grow willy-nilly throughout the medium. But liquid media have the advantage of allowing bacteria to grow much faster than gel media, because bacteria grow throughout the entire volume of the liquid, rather than being limited to just the surface of a gel. For that reason, liquid culturing media are used almost exclusively for producing pure cultures.

The primary characteristic of a culturing medium is the mix of nutrients included. For bacteria, which are what we'll be culturing, a typical nutrient mixture contains some form of carbohydrate (usually a saccharide such as glucose or sucrose, but sometimes glycerol or another carbon source) and nitrogen in the form of proteins or inorganic nitrates. Other components may be added to encourage or discourage the growth of particular classes (or even species) of bacteria.

Such media are called *selective media*, because they favor the growth of one type or species of bacteria at the expense of others. For example, eosin-methylene-blue (EMB) agar contains methylene blue, which is toxic to most Gram-positive bacteria and therefore favors the growth of only Gram-negative bacteria. Conversely, mannitol-sugar agar (MSA) favors the growth of Gram-positive bacteria.

Some media, called *differential media*, provide a visual indication of the types of bacteria growing on the medium, usually by including a dye that changes color when exposed to the waste products of particular types of bacteria. For example, McConkey (MCK) agar is differential for lactose fermentation. It includes the dye neutral red, which is yellow at pH 8.0 or higher and red at pH 6.8 or lower. When it is prepared, McConkey agar is slightly basic, and therefore has a pale yellow color. If McConkey agar is inoculated with lactose-fermenting bacteria, as those bacteria grow and produce acidic waste products, the agar turns red. If the bacteria growing on the McConkey agar are not lactose fermenters, the agar remains a pale yellow color.

Another important distinction in culturing media is whether the identities and quantities of nutrients are known:

Defined medium

A *defined medium* (also called a *simple medium*) contains only known amounts of chemicals whose identities are known exactly. Carbohydrates (the carbon source) are usually supplied as known amounts of glucose, sucrose, or another sugar. The nitrogen source is usually a known amount of a simple ammonia compound or nitrate salt, such as ammonium nitrate or urea. Trace nutrients may be provided in the form of tiny amounts of inorganic salts.

Undefined medium

An *undefined medium* (also called a *complex medium*) contains complex chemicals from plant and/or animal sources. For example, an undefined medium may contain yeast extract or beef broth, both of which contain mixtures of many organic chemicals whose exact identities and amounts are unknown.

The most important advantage of defined media is that they can be tailored to provide optimal growing conditions for specific bacteria or other microorganisms. The most important disadvantage of defined media is that they cannot be used for most microorganisms because those microorganisms require resources that cannot be provided by defined media. (For example, viruses can be cultured only in the presence of the cells that host them.) Even most bacteria cannot be cultured on defined media, simply because we don't know all of their requirements. Conversely, most microorganisms can be cultured on undefined media, if they can be cultured at all, but at the cost of not knowing exactly what we're giving them to eat.

Culturing media are used in vessels that protect the growing cultures from contamination. Liquid media are used in test tubes (culture tubes) or flasks, which are simply filled, stoppered (usually with sterile cotton balls, but sometimes with solid rubber stoppers or screw-on caps), and sterilized. The most common vessels for gel media are the *Petri dish* (also called a *Petri plate*) and the *slant tube*.

Petri dish

A Petri dish is simply a shallow, flat, circular dish with a slightly larger lid. To fill a Petri dish with gel media, you pour the warm, liquid gel into the Petri dish to a depth of a few millimeters, cover the dish, sterilize it in an autoclave or pressure cooker, and allow it to cool until the gel sets.

Petri dishes are available in reusable glass and disposable presterilized plastic versions. The plastic versions are inexpensive individually—typically $0.50 each or less, depending on size—but because they are single-use, their cost adds up quickly if you run many cultures. Conversely, glass models can be sterilized and reused repeatedly, but they cost anything from a couple dollars to $10 or so each. They also tend to suffer breakage.

Slant tube

A slant tube is simply a test tube that functions as a miniature Petri dish. To fill a slant tube, you pour warm, liquid gel into the tube to about one-third full, plug the tube with a sterile cotton ball, sterilize it in an autoclave or pressure cooker, and then place the tube at an angle while

the gel cools and sets. Allowing the gel to set with the tube angled increases the surface area of the gel available for culture growth.

Autoclaving is an important preparation step, because it kills every living thing in the culturing vessel. You might think that boiling the medium during preparation would accomplish that, but it doesn't. The problem is that the boiling point of water—100 °C at standard conditions—is sufficient to kill only bacteria, protozoa, viruses, and fungi. So what's left? Spores, which some microorganisms form when exposed to adverse conditions, such as high temperatures or extreme dryness.

Spores are not killed quickly at 100 °C, a temperature they can survive for anything from several hours to days or more. Fortunately, temperatures just a bit higher are sufficient to kill them quickly, which is where autoclaving comes in. In an autoclave (or pressure cooker), water is boiled under pressure higher than atmospheric pressure, which increases the boiling point. In a standard pressure cooker, water boils at 121 °C, which is sufficient to kill spores after only 15 minutes' exposure, called the *15-15 rule*.

> Most pressure cookers operate at 15 PSI (just over one standard atmosphere) higher than atmospheric pressure, a standard established by the US Department of Agriculture during WWI as being safe for home canning. Some pressure cookers operate at lower pressures, typically 10 to 12 PSI, which produces maximum temperatures of 115 °C to 118 °C.

In fact, this "rule" is not a rule at all, because the actual time needed for sterilization varies with the size of the container being sterilized, its composition, the degree of contamination of the material being sterilized, and other factors. For example, a 250 mL flask takes much longer to sterilize than a test tube because it takes longer for the liquid in the flask to reach the maximum temperature in the pressure cooker. Similarly, a metal tube sterilizes faster than a glass tube because the metal conducts the heat faster, and a polypropylene tube takes longer still because the plastic insulates the liquid inside the vessel. In short, there is no hard and fast rule about how long sterilization requires.

You might think the easy answer would be simply to autoclave or pressure-cook culturing vessels for a long time to ensure they're in fact sterile. The problem with that is that heating the medium too long may destroy nutrients. So, we shoot for a happy medium, so to speak, by autoclaving long enough to kill all or nearly all of the spores present while not cooking the nutrients to death.

Kitchen/lab-size pressure cookers (and autoclaves) were not invented until the late 19th century. So what did biologists do back then to sterilize their culturing media? Well, mostly they just sterilized with boiling water, hoped for the best, and didn't worry too much about a few spores germinating in their "sterilized" media.

But in 1876 an Irish physicist named John Tyndall devised a cunning plan. As usual, he boiled his medium to kill all of the bacteria and most of the spores, but then he allowed the medium to sit undisturbed for a day, during which time many of the spores germinated. He then reboiled the media, killing all the germinated spores. He repeated this for a third day to kill any late-germinating spores. His procedure, called *Tyndallization*, often killed all of the spores in the medium. A few sometimes survived, but even such imperfect sterilization was much better than simple boiling.

Tyndallization is still sometimes used to sterilize materials, such as some seeds, that cannot tolerate the higher temperature in a pressure cooker or autoclave. In fact, it's often not necessary to reach even 100 °C to kill the bacteria of concern on such materials.

We've written this lab session on the assumption that you have (or can borrow) a pressure cooker. Even an inexpensive model from Walmart is fine. Although ordinarily we strictly segregate lab equipment from food-preparation equipment, there is, by definition, no risk to using a pressure cooker both for ordinary cooking and for sterilizing media. The pressure cooker kills anything present inside it, whether you're pressure-cooking a roast or a bunch of culture tubes.

You can also use the pressure cooker to kill live cultures in Petri dishes, slant tubes, and culturing flasks, as long as the containers are glass. When you're finished with cultures, simply autoclave them to kill anything present. The other benefit of this method is that agar gel re-liquefies, making it much easier to clean out the dishes and tubes. Wait until the culturing container has cooled sufficiently to be quite warm to the touch, and then just pour the sterile gel down the drain.

With a pressure cooker, there's no need to follow aseptic procedures while making up the media, filling glass culturing vessels, and so on, because sterilization is the final step. If you don't have a pressure cooker, or if you're using plastic culturing vessels that cannot be autoclaved (as we are in this lab session), aseptic procedure becomes much more important.

Use the following guidelines to keep your culturing media as uncontaminated as possible:

- Turn off the central heat or air conditioning. You don't want the fan circulating air while you're working with culturing media and vessels.

- Immediately before you begin work, sterilize your work surface by misting it with Lysol or a similar disinfectant spray.

- Use a spray mister filled with tap water to mist the air around your work area. The tiny water droplets capture airborne particulates and settle out.

- When you prepare culture medium in a beaker, keep the beaker covered as much as possible with a loosely fitting cap of aluminum foil, plastic wrap, or waxed paper to prevent airborne contaminants from entering the beaker. (Do this whether you are simply boiling the culture media or pressure-sterilizing it. In the latter case, use aluminum foil, because plastic wrap or waxed paper won't survive being pressure-cooked. If you are boiling the medium in a microwave oven, use plastic wrap or waxed paper to avoid damaging the oven and possibly causing a fire.)

- Wear gloves at all times while handling culturing media or vessels. It does no harm to use Lysol or hand sanitizer on the gloves as well to kill any microorganisms present.

If you do not have a pressure cooker to sterilize the culture media, culture plates and tubes should be used as soon as possible after you make them up, and certainly within 24 hours. If you use a pressure cooker, agar plates and tubes remain usable for several days and broth tubes for a week or more. In any event, store unused culture vessels in the refrigerator until you are ready to use them, and then allow them to come to room temperature before inoculating them.

There are also times when you want to preserve a live bacteria culture but not encourage growth. For example, in the following lab sessions we use a purchased mixed culture of three species of bacteria. We would prefer to purchase that culture only once, so we'll preserve the culture by transferring a small amount of it to a tube of sterile *isotonic saline* (also called *normal saline*) solution, in which many species of bacteria can survive for extended periods. This solution is essentially a dilute solution of ordinary table salt. It provides no nutrients, so the bacteria we inoculate into it will not grow (reproduce). Some, however, usually survive by going dormant, essentially putting themselves into hibernation until a food source is again available.

In this lab session, we'll prepare a nutrient broth based on chicken stock (as a nitrogen source) and table sugar (for carbohydrates). We'll use that nutrient broth to prepare broth culturing tubes as well as agar culturing plates and agar culturing slant tubes. We'll also prepare and sterilize isotonic saline tubes that we'll use in the next lab session to preserve our purchased bacteria culture for later use. We'll need about 10 mL of nutrient medium for each broth culture tube, about 5 mL for each agar slant tube, and about 35 mL for each 90 mL Petri dish.

Again, if you do not have a pressure cooker, make up only enough medium for your immediate needs. If you do have a pressure cooker, you can save time by making up additional medium for later use. That's what we'll do in these instructions. Modify them as necessary if you do not have a pressure cooker.

> In science, volumes and masses usually matter a great deal. In this and the following lab sessions, neither are critical.

PROCEDURE VII-2-1: PREPARING NORMAL SALINE AND NUTRIENT MEDIA

1. Label a clean, empty 500 mL soda bottle "Normal Saline" and transfer 500 mL of distilled water to it.

2. Weigh out 4.50 g (a slightly rounded half teaspoon) of sodium chloride (table salt) and transfer it to the bottle. Cap the bottle and invert it several times to dissolve the salt and mix the solution thoroughly.

> Retain this normal saline solution for use in later lab sessions. Although it is not sterile, it may often be used in nonsterile form. When you do require it in sterile form, you can simply sterilize as much as you need for immediate use, as we do next.

3. Transfer about 10 mL of the normal saline solution to each of two labeled 15 mL centrifuge tubes. Cap the tubes loosely and place them in the test tube rack for now.

> Plastic centrifuge tubes are are autoclavable, but never screw the caps down tightly until the tubes have been autoclaved. Autoclaving a tightly capped tube may cause it to burst. Before autoclaving, place the caps on the tubes and turn them just a fraction of a turn—enough to keep the cap from falling off, but not enough to form a gas seal. Once the autoclaving is complete and the tubes have cooled, you can screw the caps down tightly.

4. Calculate the amount of nutrient medium you need for the number of broth culturing tubes, agar slant tubes, and Petri dishes you intend to fill.

For immediate use, we decided to make up enough nutrient broth to fill four broth tubes (4 X 10 mL = 40 mL), and enough nutrient agar to fill two slant tubes (2 X 5 mL = 10 mL), and two Petri dishes (2 X 35 mL = 70 mL). That requires a total of 120 mL of nutrient broth. To save time later, we also decided to make up an extra 125 mL of nutrient agar, which we'll store in the 125 mL polypropylene bottle and autoclave. That medium will remain usable for at least several weeks if refrigerated, and is sufficient to fill four 90 mm Petri dishes, which we'll use in a later lab session. Accordingly, we'll make up 250 mL of nutrient broth.

Use the 125 mL polypropylene bottle from the kit that contains the nitrogen-free fertilizer concentrate A. If you haven't used all of that solution by the time you start this lab session, simply transfer the fertilizer concentrate to another container and label it. Peel off the original label from the 125 mL polypropylene bottle and relabel the bottle with its current contents.

1. Dissolve a bouillon cube (or dilute canned chicken broth) in twice the recommended amount of warm tap water. Use a spoon or paper towel to skim off or blot up any oil or grease floating on the surface of the solution.

2. Transfer 250 mL of the defatted broth to the measuring cup.

3. Add two teaspoons of table sugar to the broth, and stir until it dissolves.

4. Label four 15 mL centrifuge tubes "Nutrient Broth" and fill them to about 10 mL each. Cap each of the tubes loosely and place the tubes in the test tube rack for now.

5. Make sure there is sufficient broth remaining in the measuring cup to make up the desired amount of agar gel medium. In our case, we needed 70 mL for the two Petri dishes, 10 mL for the two slant tubes, and 125 mL for the bottle we're making up for later use, or a total of about 205 mL. We added a bit extra just to be safe. (If there is too little broth remaining in the cup, simply add a small amount of tap water to make up the necessary volume.)

6. Weigh or measure out 2 g of agar powder (~5/8 teaspoons) per 100 mL of broth and stir the agar powder into the cool broth.

WARNING

Be very careful heating any liquid in a microwave oven. The liquid may superheat, which means its temperature reaches higher than the boiling point without the liquid actually boiling. In this unstable condition, the liquid may suddenly start boiling violently and eject itself from the container.

7. Put the measuring cup in the microwave and heat it until it just boils.

8. When the solution is boiling, carefully remove it from the microwave oven and stir until all of the powder is dissolved and the agar solution is clear, homogeneous, and slightly viscous. Allow the agar medium to cool to about 60 °C (hot to the touch).

9. Carefully pour about 5 mL of the agar medium into each of your two slant tubes. Plug them with cotton balls and place the slant tubes in the test tube rack for now.

10. Fill the 125 mL polypropylene bottle with the warm agar medium and cap the bottle loosely.

11. Prepare your pressure cooker, adding the recommended amount of water. Place the pressure cooker rack in the pressure cooker (if your pressure cooker did not come with a rack, use an oven rack that fits it) and transfer the broth tubes, slant tubes, and polypropylene bottle to the pressure cooker.

> If the pressure cooker is large enough, you can leave these tubes in the test tube rack, which is autoclavable. Otherwise, stand the tubes in a beaker or a similar container.

12. Cover the measuring cup loosely with aluminum foil, and transfer it to the pressure cooker.

13. Replace the cover on the pressure cooker, and apply heat until the pressure reaches nominal. Note the time and begin the cycle.

> If your pressure cooker operates at 15 PSI (121 °C), allow 30 to 45 minutes for complete sterilization to occur. If your pressure cooker operates at lower pressure, extend the time appropriately. At 115 °C, we recommend doubling the time.

14. Follow the instructions provided with your pressure cooker for cooling. (Some pressure cookers recommend allowing the unit to cool naturally, but others recommend cooling it under cold tap water.)

15. Once the pressure is released, remove the cover from the pressure cooker. Allow the contents to cool sufficiently that you can handle them with your bare (well, gloved...) hands. They should feel quite warm, but not hot enough to burn you. The goal is to remove them while the agar gel is still flowing as a liquid rather than set up.

16. Tighten the caps on the broth tubes and place them in the rack for now. Place the agar slant tubes at an angle sufficient to allow the agar to almost but not quite reach the cotton plugs. Allow them to remain in this position until they have cooled sufficiently for the agar to set.

17. Carefully remove the measuring cup from the pressure cooker, keeping the contents protected from airborne contamination by the aluminum foil.

18. Carefully remove two of the presterilized Petri dishes from the outer sleeve. Note that the dishes are not individually wrapped, and remain sterile only as long as you keep the covers in place. Be careful not to disturb the covers until you are ready to fill the dishes.

19. Working in a sanitized area as described in the introduction, place the first dish flat on your work surface with the measuring cup of agar medium next to it. Reposition the aluminum foil to expose the pouring lip of the cup. Carefully lift the edge of the Petri dish cover just enough to allow you to pour in gel. Pour sufficient gel to cover the bottom of the dish to a depth of about 0.5 cm, and then replace the Petri dish lid immediately. Repeat to fill the second Petri dish.

20. Allow the Petri dishes to sit undisturbed until the agar gel sets. Placing the Petri dishes in the refrigerator will set the gel within a few minutes. Otherwise, allow at least half an hour for the gel to set completely.

We'll use these Petri dishes, slant tubes, and broth tubes in the next lab session. Store them in the refrigerator until then. Store the Petri dishes inverted (with the gel side up). Store the slant tubes and broth tubes upright, either in the test tube rack or in a beaker or similar vessel. Remove the Petri dishes and tubes from the refrigerator and allow them to warm to room temperature before you use them.

REVIEW QUESTIONS

Q1: Is the nutrient media we made defined or undefined? Why?

Q2: Why do we store unused culturing media in the refrigerator?

Q3: How do unselective and selective media differ? Why might you choose an unselective medium?

_____

_____

_____

_____

_____

Q4: What is the defining characteristic of a differential medium?

_____

_____

_____

_____

_____

Q5: Why might you choose a broth medium rather than a gel medium, and vice versa?

_____

_____

_____

_____

_____

Q6: Why is agar so widely used for making up gel culturing media?

_____

_____

_____

_____

_____

_____

Culturing Bacteria

Lab VII-3

EQUIPMENT AND MATERIALS

You'll need the following items to complete this lab session. (The standard kit for this book, available from *www.thehomescientist.com*, includes the items listed in the first group.)

MATERIALS FROM KIT

- Goggles
- Coverslip
- Inoculating loop
- Microscope slide, flat
- Pipettes

- Stain, Gram's iodine
- Stain, Hucker's crystal violet
- Stain, safranin O
- Stirring rod
- Test tube rack

MATERIALS YOU PROVIDE

- Gloves
- Alcohol (ethanol or isopropanol)
- Alcohol lamp, candle, or butane lighter
- Agar Petri dishes (from preceding lab)
- Agar slant tubes (from preceding lab)
- Bleach, chlorine laundry
- Broth culturing tubes (from preceding lab)
- Clock or watch with second hand

- Large container with lid (for bleach bath)
- Microscope
- Mixed bacteria culture (see introduction)
- Normal saline tubes (from preceding lab)
- Paper towels
- Sanitized work area (see preceding lab)
- Tape (transparent, masking, or similar)

BACKGROUND

In the preceding lab session, we made up culturing media and vessels. In this lab session, we'll use those materials to culture bacteria.

Obtaining bacteria to culture is certainly no problem. We're surrounded by them. Unfortunately, although most environmental bacteria are harmless or actually beneficial to humans, a small percentage of them are pathogenic (disease-causing). The normal human immune system can deal with the small numbers of pathogenic bacteria normally present in the environment, but culturing those bacteria multiplies their numbers exponentially, presenting a potentially serious hazard.

For example, MRSA (methicillin-resistant *Staphylococcus aureus*) bacteria are widely distributed in the environment; there are probably millions of them in and on your body right now. Your immune system can deal with those relatively small numbers, but culturing *S. aureus*, intentionally or unintentionally, multiplies those numbers by orders of magnitude, producing a concentrated colony of *S. aureus*, which is most definitely hazardous.

For that reason, we'll do this lab using a purchased mixed bacteria culture that contains three species of relatively innocuous bacteria. We chose a mixed culture that includes all three shapes of bacteria and examples of both Gram-positive and -negative bacteria:

- *Bacillus subtilis* – Gram-positive rod-shaped bacteria

- *Micrococcus luteus* – Gram-positive sphere-shaped bacteria

- *Rhodospirillum rubrum* – Gram-negative spiral-shaped bacteria

You can purchase this mixed suspension of living bacteria as item #154760 from Carolina Biological Supply (*www.carolina.com*). As we wrote this, the price was $16.90 plus shipping.

In this lab session, we'll inoculate two Petri dishes with the purchased mixed culture. (We'll use two Petri dishes in case one is contaminated by environmental bacteria.) We'll then incubate the Petri dishes at room temperature, observe the growth of our colonies, and thereby determine which of the three bacteria species can be cultured successfully on the nutrient medium we're using.

> To ensure freshness, do NOT order this living culture until you have scheduled this lab session. Carolina Biological allows you to specify an arrival date in the shopping cart. Schedule this arrival date for the day of (or the day before) you intend to start this lab session.

Because we don't want to have to purchase the mixed culture more than once, we'll also inoculate one broth tube and two normal saline tubes with the purchased culture and store the broth tube in the refrigerator. This procedure, called *re-culturing*, allows us to extend the useful life of the original culture.

Because they are made up of living things, cultures pass through life stages. In the *juvenile stage*, relatively few bacteria are present in the culturing medium. They have essentially unlimited food and other resources, and are free to reproduce rapidly. In the *mature stage*, large numbers of bacteria are present, and competition for resources starts to become an issue. In the *senescent stage*, the culture is approaching its death. The bacterial population is much too large to be supported by the available resources, and bacteria must compete for those limited resources. Poisonous waste products are accumulating, and mutations may develop.

By reculturing, we can produce a juvenile culture from a mature culture. By refrigerating the broth tube, we slow the reproduction rate of the bacteria, essentially putting them into hibernation. Because the broth tube contains nutrients, the bacteria will continue to reproduce in the refrigerated broth, but at a greatly reduced rate. (This is why nonsterile food stored in the refrigerator remains edible much longer than nonsterile food stored at room temperature.)

> Under ideal conditions, bacteria can reproduce very rapidly, some in as little as 20 minutes or less per generation. In other words, if you start with one bacterium, after 20 minutes you have two bacteria, after 40 minutes you have four, after 60 minutes you have eight, and so on. That might not seem impressive until you realize that after 10 generations (3 hours and 20 minutes) you have about a thousand bacteria, after 20 generations (6 hours and 40 minutes) you have about a million, and after 30 generations (10 hours) you have about a billion.
>
> Of course, this exponential growth can never last for long, at least in a lab, because the bacteria quickly run out of food and space to grow. But it does illustrate how a juvenile culture can quickly become a mature culture and then a senescent culture.

The saline tubes contain no nutrients. We expect (or at least hope) that the bacteria in the saline medium will remain alive indefinitely but cease reproduction. Some species of bacteria can be maintained in "suspended animation" for months, years, or even decades in phosphate-buffered saline. We'll determine if any or all of our three species can be maintained for at least short periods in the simpler normal saline solution.

And, of course, we'll want to look at our bacteria, so we'll do a smear mount of the original mixed suspension and use Gram staining to differentially stain the various bacteria before observing them under the microscope.

PROCEDURE VII-3-1: STAINING AND OBSERVING THE ORIGINAL CULTURE

1. If you have not already done so, put on your goggles and gloves.

2. In your sanitized work area, lay out your alcohol lamp (or other flame source), the inoculating loop, a microscope slide, the stirring rod, and the original culture tube. Make sure your gloves are sterile before proceeding.

3. Ignite the flame source, and hold the tip of the inoculating loop at the top of the flame until the metal loop glows red. Remove the loop from the flame and allow it to cool. Do not put down the loop; continue holding it in your hand until you complete the transfer of the bacterial culture to the slide.

4. As the loop cools, pass the mouth of the bacterial culture tube quickly through the flame. A second or so suffices. Your goal is not to heat the tube and cap, but to kill any surface bacteria present on the tube or cap.

5. Carefully remove the cap from the culture tube just enough to allow you to insert the cool inoculating loop. (Hold the cap in your hand; placing it on the work surface risks contamination.) Dip the inoculating loop into the culture liquid, remove the loop, and recap the tube immediately.

6. Transfer the drop of liquid culture to the center of the microscope slide and then flame-sterilize the loop before placing it aside.

7. Use the stirring rod or a second microscope slide to make a smear mount by spreading the drop of culture liquid into a thin smear on the slide.

8. Holding the slide by the edges at one end, pass the central section of the slide through the flame several times, culture side up. The goal is not to "cook" the culture liquid, but to dry it out gently and cause the bacteria present to adhere to the slide. When heat-fixing is complete, the central area of the slide should have a dry, slightly hazy appearance. Place the slide aside and extinguish the flame source.

9. Place the heat-fixed slide on a paper towel on a clean, flat surface.

10. Use a clean pipette to place a drop or two of Hucker's crystal violet stain on the smear. Use the tip of the pipette gently to spread the stain until it covers the entire smear. Do not touch the smear with the tip of the pipette.

11. Allow the crystal violet stain to remain in contact with the smear for one minute, and then rinse the slide, smear-side down under a trickle of cold tap water, tilting the slide back and forth to flood the smear with water. Rinse for at most a second or two, and don't allow the water to fall directly on the smear. Drain the slide and place it flat on the paper towel.

12. Use a clean pipette to place a drop or two of Gram's iodine stain on the smear. Again, use the tip of the pipette carefully to spread the stain over the entire smear.

13. Allow the Gram's iodine stain to remain in contact with the smear for one minute.

14. Fill a clean pipette with alcohol (drugstore 70% ethanol or 70% isopropanol is fine). Hold the slide at an angle over the sink and gently flood the smear with the alcohol. Continue until the alcohol runs colorless.

15. Repeat step 11 to rinse all of the alcohol from the slide. (It's important to remove all of the alcohol, because the following step won't work if alcohol is still present.) Drain the slide and place it flat on the paper towel.

16. Use a clean pipette to place a drop or two of safranin O stain on the smear. Use the tip of the pipette carefully to spread the stain over the entire smear.

17. Allow the safranin O stain to remain in contact with the smear for one minute.

18. Repeat step 11 to rinse excess safranin O stain from the slide and then allow the slide to air-dry.

19. You'll probably want to keep this slide for reference, at least until you complete this related group of lab sessions, so place a drop of mounting fluid in the center of the smear. You can use glycerol to make a mount that will last for a week or two, or a permanent mounting fluid if you want to keep the slide indefinitely.

20. Carefully lower a coverslip onto the mounting fluid, beginning with one edge of the coverslip in contact with the slide and then tilting the coverslip downward until it is in full contact with the slide. Make sure to eliminate any air bubbles by pressing gently with the tip of a pipette or forceps to expel the bubble.

21. Position the slide on the microscope stage and use medium magnification to center an area of the smear that shows a large number of bacteria in the field of view. Switch to high-dry magnification, adjust the diaphragm and illuminator brightness to reveal the maximum detail in the bacteria, attempt to identify each of the three types of bacteria that should be present in the smear, and record your observations in your lab notebook.

22. If you have a 100X (oil-immersion) objective, put one drop of immersion oil on the center of the coverslip and carefully rotate the oil immersion objective into position, making sure it comes into contact only with the oil drop. Observe the bacteria at 1,000X, and record your observations in your lab notebook.

PROCEDURE VII-3-2: INOCULATING PLATES AND TUBES

Allow the culturing dishes and tubes to warm to room temperature before proceeding.

1. If you have not already done so, put on your goggles and gloves.

2. In your sanitized work area, lay out your alcohol lamp (or other flame source), the inoculating loop, the original culture tube, both prepared Petri dishes (gel-side down), and the test tube rack with both slant tubes and one of the broth tubes. Ignite the flame source, and make sure your gloves are sterile before proceeding.

3. Hold the tip of the inoculating loop at the top of the flame until the metal loop glows red. Remove the loop from the flame and allow it to cool. Do not put down the loop; continue holding it in your hand until you complete the transfer of the bacterial culture from the original tube.

4. As the loop cools, pass the mouth of the original bacterial culture tube through the flame for a second or two to kill any surface bacteria present on the tube or cap.

5. Carefully remove the cap from the culture tube just enough to allow you to insert the cool inoculating loop. Holding the cap in your hand to prevent contamination, dip the inoculating loop into the culture liquid, remove the loop, and recap the tube immediately.

6. Carefully lift the lid of the first Petri dish just enough to allow access for the inoculating loop. Place the tip of the inoculating loop into gentle contact with the agar gel surface, and move the tip of the loop in a zig-zag pattern across the surface of the agar. The goal is not to gouge a groove in the agar, but simply to move the inoculating loop across its surface to allow bacteria to be transferred from the loop to the agar.

7. Repeat the streaking at a 90° angle to the first streak. If you are using the two-section Petri dishes included in the kit, repeat the streaking in the second section. After you finish streaking, immediately lower the Petri dish lid back into place and place the Petri dish aside for now.

8. Repeat steps 3 through 7 to inoculate the second Petri dish.

9. Repeat steps 3 through 5 to load the inoculating loop.

10. Flame-sterilize the mouth of the broth tube and inoculate the tube with the mixed bacteria culture. Replace the cap and place the inoculated tube in the rack.

11. Repeat steps 9 and 10 to inoculate each of the two slant tubes. Inoculate the slant tubes in the same manner you inoculated the Petri dishes, just touching the loop to the surface of the agar.

12. Repeat steps 9 and 10 to inoculate each of the two normal saline tubes. Simply dip the loaded loop into the tube and swirl gently to transfer the bacterial culture to the saline medium.

> Obviously, you must take care when flame-sterilizing polypropylene tubes to avoid melting the tube and/or lid. Simply rotate the tube quickly in the flame for a second or so to flame-sterilize it.

13. Refrigerate the broth tube and both normal saline tubes. To isolate the inoculated vessels from the food storage area, we recommend standing the tubes in a beaker or similar container inside a sealed wide-mouth jar or plastic food-storage container.

14. Place both Petri dishes (gel side on top) and one slant tube in a dark area to incubate. Allow the second slant tube to incubate where it will be exposed to direct sunlight.

15. Observe the Petri dishes and slant tubes at least daily and record the progress in your lab notebook. (Do not open the containers; observe the growth through the glass.) If you can, shoot images to record the progress. At room temperature some changes should be evident in some or all of the containers after a day or two, with discrete colonies starting to develop after two to three days.

PROCEDURE VII-3-3: PRODUCING PURE CULTURES

You can begin this procedure any time after discrete bacterial colonies appear in your Petri dishes. Allow the remaining three broth tubes to warm to room temperature before proceeding. Label each of the tubes.

1. If you have not already done so, put on your goggles and gloves.

2. Choose one of the Petri dishes that contains discrete colonies of all three bacterial species. Ideally, you want large, widely separated colonies of each of the three.

3. In your sanitized work area, lay out your alcohol lamp (or other flame source), the inoculating loop, the Petri dish you selected (gel-side down), and the test tube rack with the three remaining broth tubes. Ignite the flame source, and make sure your gloves are sterile before proceeding.

4. Hold the tip of the inoculating loop at the top of the flame until the metal loop glows red. Remove the loop from the flame and allow it to cool. Do not put down the loop; continue holding it in your hand until you complete the transfer of the bacterial culture from the Petri dish to the first broth tube.

5. As the loop cools, pass the mouth of the first broth tube through the flame for a second or two to kill any surface bacteria present on the tube. Hold the cap in the same hand as the tube.

> Yes, biologists often need four to six hands. Rather than grow more hands, you might ask someone to assist you in procedures that require more than two hands.

6. When the loop is cool, lift the lid of the Petri dish just enough to insert the tip of the loop. Touch the loop to the first colony type. Immediately withdraw the loop and replace the lid on the Petri dish.

7. Dip the loop into the first broth tube and swirl slightly to transfer the bacteria from the loop to the broth tube. Replace the cap on the tube and place the tube back in the rack.

8. Repeat steps 4 through 7 to inoculate each of the two remaining broth tubes with bacteria from one of the two remaining colony types. When you finish, flame-sterilize the loop before putting it away.

9. Place all three broth tubes upright in a beaker or similar container and put them in a dark area to incubate.

10. Observe the broth tubes every few hours and record the progress in your lab notebook. We'll retain these broth cultures for the next lab session.

As bacteria grow in the broth, the broth gradually changes from transparent to cloudy. (Or not so gradually, depending on species and incubation conditions.) In a senescent broth culture, clumps may actually become visible, but we don't want to let things get to that point. We'd like to have young-mature to mature cultures for the following lab session, so keep a close eye on the broth cultures. Once the broth becomes at least somewhat cloudy in appearance, you can refrigerate the broth tubes to slow bacterial reproduction until you're ready to begin the next lab session. (Use the steps mentioned in the preceding procedure to isolate your cultures from the food in your refrigerator.)

Also retain the original mixed culture tube, the normal saline tubes, and the mixed culture broth tube for later use. You can keep all of these refrigerated to slow bacterial growth and the inevitable senescence of the cultures.

Destroy the remaining cultures by immersing the Petri dishes and slant tubes in a deep vessel filled with a mixture of one part chlorine laundry bleach to four parts tap water. Wear gloves, and open the Petri dishes and tubes only after they are fully submerged. Make sure that there are no air bubbles, and that the bleach solution can reach all parts of the vessels. Allow the vessels to soak in the bleach solution overnight, by which time everything except perhaps a few spores has been killed.

We do not recommend reusing the plastic Petri dishes. Simply rinse them with tap water and diskard them with the household trash. The tubes can be washed with hot (if necessary, boiling) soapy water to clean them. Make sure to remove all traces of the agar gel, which can be persistent.

REVIEW QUESTIONS

Q1: Why did we bother doing the first procedure, rather than simply starting with the second procedure and observing the bacteria after we'd recultured them?

Q2: Based on your observation of the Gram-stained mixture culture, which of the three bacteria species were Gram-negative, and which were Gram-positive?

Q3: In the second procedure, we carefully observed aseptic technique for both the original culture tube and the destination vessels. In the first procedure, we observed aseptic technique for the original culture tube, but transferred the culture medium to a nonsterile microscope slide. Why the difference?

Q4: Which of the three bacteria species were you able to culture successfully on agar and in broth?

Q5: What differences, if any, did you observe in the slant tube incubated in the dark versus the one exposed to direct sunlight? Propose an explanation for any differences you observed and suggest an experiment to verify or falsify your proposed explanation.

Investigating Bacterial Antibiotic Sensitivity

Lab VII-4

EQUIPMENT AND MATERIALS

You'll need the following items to complete this lab session. (The standard kit for this book, available from *www.thehomescientist.com*, includes the items listed in the first group.) In addition to the items listed below, you'll also need the equipment and materials from Lab session VI-2 for making up four nutrient agar Petri dishes and six nutrient broth tubes.

MATERIALS FROM KIT

- Goggles

- Beaker, 250 mL

- Antibiotic capsule, amoxicillin (250 mg cap)

- Antibiotic powder, chlortetracycline (3.3%)

- Antibiotic powder, sulfadimethoxine (88%)

- Antibiotic solution, neomycin (200 mg/mL)

- Chromatography paper

- Forceps

- Petri dishes

- Pipettes

- Ruler

- Spatula

- Test tubes

- Test tube rack

MATERIALS YOU PROVIDE

- Gloves

- Aluminum foil

- Balance (optional)

- Chlorine bleach container

- Hole punch (or scissors)

- Marking pen

- Microwave oven

- Nutrient agar (see Lab VI-2)

- Nutrient broth tubes (see Lab VI-2)

- Pure cultures (from Lab VI-3)

- Refrigerator

- Sanitized work area

- Soda bottles

- Teaspoon (or measuring spoons)

- Water, distilled

BACKGROUND

In bacteria and other organisms that reproduce quickly, genetic mutations are frequent. Most of these mutations are harmful, leading to the death of the individual bacterium. The end of the line, so to speak. Some mutations are neither harmful nor helpful, so that bacterium and its descendants simply continue reproducing. A few mutations are helpful, at least in specific circumstances, by giving that bacterium and its descendants a selective advantage.

WARNING

If you or any member of your family is immuno-suppressed, do not do this session unless your physician tells you that it is safe to do so. The three bacteria species we use in this session are widely used for high-school biology labs, and are generally considered to be nonpathogenic. Regardless, you should exercise extreme caution in dealing with and disposing of them, particularly in the second and third procedures.

For example, in a large population of a particular species of bacteria, the vast majority of those bacteria may be vulnerable to a specific antibiotic. In the presence of that antibiotic, nearly all of the bacteria die off quickly, or at least stop reproducing. But "nearly all" is not "all." A tiny percentage of the original population may have genetic mutations that provide partial or complete resistance to that antibiotic.

So, let's say we start with a population of one billion bacteria, all but one of which is susceptible to the bactericidal antibiotic we introduce into the culture. In short order, 999,999,999 of the bacteria drop dead, leaving only one survivor. From the point of view of the 999,999,999 bacteria, this is an end-of-the-world scenario, but from the point of view of the sole survivor, this is very good news indeed. It now has the resources of the entire "world" available to it. Plenty of food, plenty of room to grow, and no interfering neighbors. Time to have lots of children. Hurray!

That surviving bacterium immediately gets to work. After 20 minutes, there are two of it. After 40 minutes, there are four. And so on. After 30 generations—only 10 hours—there are again one billion bacteria present, and every one of them is resistant to the antibiotic. (This is actually a gross oversimplification of bacterial resistance to antibiotics, but it hits the high points.)

In the last lab session, we produced pure cultures of *Bacillus subtilis*, *Micrococcus luteus*, and *Rhodospirillum rubrum*. In this session, we'll test those three species for susceptibility to four antibiotics: amoxicillin, chortetracycline, neomycin, and sulfadimethoxine. We will then develop a strain of one of the species that is resistant to one of the antibiotics.

This, incidentally, is why it's a very good idea to finish the full course of an antibiotic rather than stop taking it once you feel better. Discontinuing use of the antibiotic prematurely allows the small surviving numbers of bacteria to breed quickly and the infection comes roaring back, this time resistant to the antibiotic.

Antibiotics fall into one of two broad classes (with some blurring, depending on the specific bacterium in question). *Bactericidal* (also called *bacteriocidal* or *bcidal*) antibiotics—including amoxicillin and neomycin—actually kill bacteria via one or more mechanisms. *Bacteristatic* (also called *bacteriostatic* or *bstatic*) antibiotics—including chlortetracycline and sulfadimethoxine—do not kill bacteria directly, but greatly inhibit their reproduction.

In either case, antibiotics do not eradicate the bacteria present. Instead, they simply reduce the population to a small enough number that your own immune system can deal with them.

PROCEDURE VII-4-1: TESTING ANTIBIOTIC SENSITIVITY

In the preceding lab session, we produced pure broth cultures of three species of bacteria. In this procedure, we'll inoculate Petri dishes with those pure broth cultures, flooding the agar surfaces of the dishes to grow a bacterial "lawn" in each dish. After inoculating the dishes, we'll place paper disks infused with various antibiotics on the agar surfaces and incubate the Petri dishes until the bacterial lawn appears. If a particular species of bacteria is sensitive to one or more of the antibiotics, an area free of bacterial growth (called an *inhibition zone*) will appear around the disk or disks. We'll measure the sizes of those zones, which indicate the relative sensitivity of the various bacteria to the various antibiotics.

1. If you have not already done so, put on your goggles and gloves.

2. Sterilize several plastic pipettes. To so so, fill the 250 mL beaker with tap water and fill each of the pipettes with that tap water. Put the beaker in the microwave and bring the water to a boil. (As always, use extreme caution to avoid superheating and sudden violent spontaneous boiling.) Allow the water to boil gently for a minute or two, reducing power as necessary to prevent the contents from boiling over, and then carefully remove the beaker from the microwave, cover it loosely with aluminum foil to prevent airborne contamination, place the beaker in your sanitized work area, and allow it to cool.

3. Loosen the cap on the 125 mL bottle of sterile nutrient agar you made up in Lab VI-2 and place the bottle in the microwave. Heat it carefully, 15 or 30 seconds at a time, swirling the bottle between heatings, until the agar is warm enough to flow freely. Carefully remove the bottle from the microwave oven and place it in your sanitized work area.

4. Carefully remove a sterile Petri dish from its packaging. Keep the base and lid in tight contact to avoid contaminating the dish. Place the Petri dish base (smaller) side up on your sanitized work surface and use the marking pen to label it. Write a tiny "1" at the center of the dish (to indicate the first bacteria type, *Bacillus subtilis*). Designate the divider in the Petri dish as pointing to the noon and 6:00 positions on the circumference. Around the edge of the dish surface write a small "A" at about 2:00, a "C" at 4:00, an "N" at 8:00, and a "S" at 10:00 (for amoxicillin, chlortetracycline, neomycin, and sulfadimethoxine, respectively).

5. Repeat step 4 with a second and third Petri dish, labeled "2" and "3" for *Micrococcus luteus* and *Rhodospirillum rubrum*, respectively. (Retain the fourth Petri dish for use in the following procedure.)

6. Working aseptically, lift the lid of the first Petri dish just enough to transfer sufficient warm liquid agar gel to each half of the dish to fill it to a depth of a few mm. Tilt the dish back and forth to spread the liquid agar, and then place the dish on a clean, flat surface to cool. Fill the other two dishes using the same procedure. When you have filled all three dishes, place them in the refrigerator and allow them to cool for several minutes, until the agar gels sets completely.

In the following steps, which should be performed in your sanitized work area, we'll flood the agar surfaces of the three Petri dishes with a few mL each of the corresponding pure broth cultures. The goal is to produce an even growth of the bacteria across the entire surface of the agar, called a *bacterial lawn*. Having even coverage makes it easier to determine the relative effectiveness of different antibiotics in retarding bacterial growth.

Before proceeding, prepare 1 mg/mL antibiotic solutions, as follows:

- Amoxicillin – dissolve one 250 mg capsule in 250 mL of distilled water in a clean, labeled soda bottle.

- Chlortetracycline – This powder contains 3.3% active chlortetracycline, or 33 mg/g.) Dissolve 3.0 g of the chlortetracycline powder in 100 mL of distilled water in a clean, labeled soda bottle. If you do not have a balance, use one slightly rounded teaspoon. Swirl to dissolve the powder. Not all of the powder will dissolve. It contains corn meal and other fillers. Simply allow the solids to settle for a few minutes before using the solution.

- Neomycin – use a clean graduated pipette to transfer 1.25 mL of the stock 200 mg/mL neomycin solution from the kit to a clean, labeled soda bottle and add distilled water to bring the volume to 250 mL.

- Sulfadimethoxine – This powder contains 89+% active sulfadimethoxine, or about 890 mg/g. Transfer 0.28 g of the sulfadimethoxine powder from the kit to a clean, labeled soda bottle and add distilled water to bring the volume to 250 mL. If you do not have a balance, use one rounded spatula spoon.

> These dilute antibiotic solutions are unstable and should be prepared as soon as possible before use, ideally during the lab session itself. If you must store them for more than an hour before use, keep them refrigerated. We have successfully used all of these solutions after 72 hours of refrigeration, but the antibiotics degrade unpredictably, as do activity levels.
>
> If you incubate your cultures at 37 °C overnight, the refrigerated solutions should be fine. If you incubate at room temperature for several days, mix up fresh solutions for the following procedures.

7. Set up a beaker or similar container filled with one part of chlorine laundry bleach to four parts water. This serves as your disposal vessel during the following steps.

8. Remove the three Petri dishes and the three pure broth cultures from the refrigerator and place them on your sanitized work surface.

9. Remove a sterile pipette from the beaker and expel all of the water from it back into the beaker.

10. Squeeze the pipette bulb to expel as much air as possible, uncap the *Bacillus subtilis* broth tube, insert the tip of the pipette into the tube, and draw up a full pipette of *Bacillus subtilis*. Recap the tube and replace it in the rack.

11. Lift the lid of Petri dish #1 just enough to allow you to insert the tip of the pipette. Expel all of the liquid in the pipette into one section of the Petri dish and tilt the dish back and forth to distribute the *Bacillus subtilis* broth culture across the entire surface of the agar.

12. Tilt the dish to collect the remaining liquid along the edge and draw that liquid back up into the pipette.

13. Expel the liquid into the other section of the Petri dish, tilt the dish back and forth to distribute the liquid across the entire agar surface, and again draw up the excess liquid into the pipette.

14. Replace the lid on the Petri dish. Immerse the tip of the pipette into the chlorine bleach solution in the disposal vessel, expel the liquid in the pipette into the disposal vessel, and draw up a full pipette of the bleach solution. Allow the pipette to remain in the disposal vessel.

15. Repeat steps 9 through 14 to inoculate a bacterial lawn of *Micrococcus luteus* in Petri dish #2 and *Rhodospirillum rubrum* in Petri dish #3.

16. Allow all three dishes to remain undisturbed for several minutes

17. While you're waiting, cut or punch 16 disks (or squares) of chromatography paper 0.5 to 1 cm in diameter (or on a side). We'll use 12 of those in this procedure and the remaining four in the next procedure.

18. Flame-sterilize the forceps, and use it to transfer four of the paper disks to Petri dish #1, positioning them two per section and spaced so as to allow the maximum possible distance between each disk and other disks and the side of the Petri dish. Use the forceps tip to press down gently on each disk to cause it to adhere to the agar surface. Lift the dish lid as little as possible during this procedure, and replace it immediately after you complete this step.

> These disks are not sterile, but we're about to saturate them with relatively concentrated solutions of antibiotics, which should kill or suppress any incidental microorganisms present on the paper.

19. Repeat step 18 for the other two Petri dishes. When you finish, you should have three Petri dishes, each with four paper disks spaced widely on the agar surface.

20. Use a sterile pipette to draw up 0.5 mL or so of the amoxicillin solution. Lift the lid of Petri dish #1 just enough to introduce the tip of the pipette, touch the tip of the pipette to the paper disk that corresponds to the "A" label on the dish, and expel just enough of the amoxicillin solution to dampen the paper disk. Do not allow the pipette to touch anything in the dish other than the surface of the paper disk. Carefully withdraw the pipette and replace the lid on the Petri dish.

21. Repeat step 20 to moisten the "A" disks in Petri dishes #2 and #3.

22. Repeat steps 20 and 21 to moisten the other nine disks in all three Petri dishes with the corresponding solutions of chlortetracycline, neomycin, and sulfadimethoxine.

23. Allow all three Petri dishes to sit undisturbed on a flat surface for one hour or more. The goal is to allow the paper disks to dry out and adhere to the agar surface.

24. Invert the Petri dishes (agar-side on top) and allow them to incubate in a dark area. At body temperature (37 °C), noticeable growth should occur overnight. If you're incubating at room temperature it may require two days or more for growth to become evident.

> If one or more of the paper disks falls off when you invert a Petri dish, don't be too concerned. By that time, the antibiotic solution has infused into the agar.

25. After the bacterial lawn has fully developed in all three of the Petri dishes, use the ruler to measure the diameter of the inhibition zone (the area in which the bacterial lawn is either absent or significantly less dense than the surrounding area), if any, that surrounds each antibiotic disk, and record those values in your lab notebook.

> In some cases, the boundary may be quite distinct, with a completely clear area abutting the unaffected bacterial lawn. In other cases, there may be a more or less gradual transition from clear through slightly cloudy to very cloudy to the unaffected bacterial lawn. In that case, do your best to measure the diameter of the area that includes all of the clear and cloudy portions to the edge of the unaffected lawn.

26. Sterilize the three used Petri dishes as described in the preceding lab session and dispose of them.

PROCEDURE VII-4-2: CULTURING AN ANTIBIOTIC-RESISTANT BACTERIA STRAIN

In the preceding procedure, we tested three species of bacteria for sensitivity to four antibiotics. In this procedure we'll choose one of those 12 combinations and culture a strain of that bacteria that is resistant to that antibiotic.

1. Based on your results in the last procedure, choose a bacteria/antibiotic combination in which the bacteria showed moderate susceptibility to the antibiotic, as indicated by a medium-size inhibition zone, ideally one with a gradual transition from clear to cloudy. For example, if you found that neomycin is moderately active against *Rhodospirillum rubrum*, you might choose that combination.

2. Prepare and sterilize six glass broth tubes, each about 3/4 full of nutrient broth, as described in Lab VI-2. Allow the tubes to cool and label them 1 through 6.

3. Observing aseptic procedure, inoculate each of the six tubes with the pure culture of the selected bacteria.

4. Observing aseptic procedure, use a sterile pipette to transfer two drops of the selected antibiotic solution to tube 1, four drops to tube 2, eight drops (0.25 mL) to tube 3, 0.5 mL to tube 4, and 1.0 mL to tube 5. Retain tube 6 unchanged as a control.

5. Incubate all six tubes in a dark area, observing them regularly until changes become apparent in tube 1. At 37 °C, changes may appear overnight. At room temperature, it may take two or three days for changes to become evident.

6. Once tube 1 shows a visible change, continue incubating the tubes, keeping a close eye on tubes 2 through 5 and comparing them for cloudiness against the control tube. Cloudiness indicates bacterial growth, and our goal is to determine the tube with the highest concentration of antibiotic in which growth of the bacteria is not completely inhibited.

PROCEDURE VII-4-3: RETESTING SENSITIVITY OF THE RESISTANT STRAIN

Having developed a strain of bacteria known to be resistant to one of the antibiotics, our next step is to reculture that strain on a fresh nutrient agar plate and to repeat the antibiotic sensitivity tests with each of the four antibiotics. Obviously, we expect the inhibition zone to be smaller for the selected antibiotic, but we also want to determine what effect, if any, the changes in the resistant bacteria strain have on its sensitivity to the other three antibiotics.

1. Of the broth tubes you cultured in the last procedure, choose the tube with the highest antibiotic concentration that still shows significant growth (cloudiness).

2. Using the contents of that tube, repeat the first procedure to inoculate the entire surface of the fourth agar Petri dish, place four paper disks on the surface of the agar, and infuse those disks with the four antibiotics.

3. Incubate the Petri dish until the bacteria lawn appears. Measure the inhibition zones for each of the antibiotic disks and record your observations in your lab notebook.

Dispose of all cultures in your chlorine bleach container.

REVIEW QUESTIONS

Q1: Why did we label the Petri dishes on their bases rather than on their lids?

_____

_____

_____

_____

Q2: In the second procedure, you cultured any one of the three bacteria species in the presence of different concentrations of any one of the four antibiotics, and then chose the tube with the highest concentration of antibiotic in which growth was evidenced by cloudiness. Why did we go through this process rather than simply adding the highest concentration of the antibiotic to one broth tube and incubating it until growth was evident?

_____

_____

_____

_____

_____

_____

Q3: In the second procedure, why did we not exclude tetracycline from the trial candidates? Tetracycline is bacteristatic, which means that an individual bacterium that is sensitive to it is not killed, but only has reproduction suppressed. Since a broth tube has no immune system to kill those suppressed bacteria, they will be present after reculturing. Why is this not a problem?

_____

_____

_____

_____

Q4: After doing the third procedure, you find (as expected) that the inhibition zone for the selected antibiotic with the resistant culture is smaller than it was with the original pure culture, but you also find that the inhibition zone for another of the antibiotics is smaller than it was with the original pure culture. Propose an explanation.

_____

_____

_____

_____

Q5: After doing the third procedure, you find (as expected) that the inhibition zone for the selected antibiotic with the resistant culture is smaller than it was with the original pure culture, but you also find that the inhibition zone for another of the antibiotics is larger than it was with the original pure culture. Propose an explanation.

_____

_____

_____

_____

Investigating Protista — Lab VIII-1

EQUIPMENT AND MATERIALS

You'll need the following items to complete this lab session. (The standard kit for this book, available from *www.thehomescientist.com*, includes the items listed in the first group.)

MATERIALS FROM KIT

- Goggles
- Coverslips
- Magnifier
- Methylcellulose
- Pipettes

- Slides, flat
- Slide, deep well
- Stain: eosin Y
- Stain: Gram's iodine
- Stain: methylene blue

MATERIALS YOU PROVIDE

- Gloves
- Microscope

- Slides, prepared (see text)
- Specimens, live (see text)

BACKGROUND

In this lab session, we'll examine live specimens and prepared slides of a representative assortment of protista, focusing on the four species listed in Table VIII-1-1. We chose these four species because they illustrate a wide range of protist characteristics, and because they are ubiquitous and easily obtained from a nearby pond or one of the microcosms we built in an earlier lab session. (You can, if you prefer, instead purchase live protist cultures from Carolina Biological Supply or another full-range vendor, either separately or as mixed cultures.) Also, as common representatives of their respective phyla, most or all of these four species are usually included in standard prepared slide sets.

Table VIII-1-1

	Amoebozoa (Amoeba)	Chlorophyta (Spirogyra)	Ciliophora (Paramecium)	Euglenozoa (Euglena)
Major Characteristics				
Autotrophic/Heterotrophic	☐/☑	☑/☐	☐/☑	☑/☑
Unicellular/Multicellular	☑/☐	☐/☑	☑/☐	☑/☐
Freshwater/Saltwater	☑/☑	☑/☐	☑/☑	☑/☑
Reproduction				
Conjugation	☐	☑	☑	☐
Mitosis and cytokinesis	☑	☐	☑	☑
Motility				
Cilia	☐	☐	☑	☐
Euglenoid movement	☐	☐	☐	☑
Flagella	☐	☐	☐	☑
Pseudopodia	☑	☐	☐	☐
Cell structures				
Ectoplasm + endoplasm	☑	☐	☑	☑
Eyespot	☐	☐	☐	☑
Gullet	☐	☐	☑	☑
Nucleus (single)	☑	☑	☐	☑
Nuclei (macro and micro)	☐	☐	☑	☐
Pellicle	☐	☐	☑	☑
Vacuoles, contractile/food	☑/☑	☐/☐	☑/☑	☑/☑

If at all possible, observe both live specimens and preserved slides of all four of these species, as well as any other protist species you have available. Three of these four species are motile. If you observe only prepared slides, you miss seeing the living creatures going about their business. If you observe only live specimens, you miss many of the structural details that become visible when a specimen is mounted and stained.

Although you can isolate these four protists from a microcosm, it's faster and more convenient to purchase live cultures. Carolina Biological Supply (CBS) sells its "Basic Protozoa Set, Living" (#131000) for $19.95 as we write this. It contains individual live cultures of three of the four species we're using in this lab session. It does not include Spirogyra, which you can obtain from nearly any pond (it's "pond scum") or purchase as a living culture (CBS, "Spirogyra, Living Culture", #151321, $8.25). If you purchase either or both of these live culture sets, time them to arrive within a couple days of when you intend to do the lab sessions, and follow all of the care instructions provided with the sets.

Ideally, have at least the following prepared slides available: Amoeba, Euglena, Paramecium, and Spirogyra (vegetative and in conjugation). Also useful for comparison purposes are prepared slides of desmids, diatoms, dinoflagellates, and as many other algae, protozoa, and slime molds as possible.

If you have only live specimens or only prepared slides, don't give up. You can still learn a lot by observing only one or the other.

We'll observe these protists in a particular order. First, Spirogyrae, which are plentiful and don't move. Second, Euglenae, which are also plentiful and do move, but only slowly. Third, Amoebae, which are relatively rare (except in purchased cultures) and like to hide. Fourth, Paramecia, which are more or less plentiful (depending on the particular culture you use), but are so fast that they're difficult to observe, particularly at higher magnifications.

If you don't have prepared slides of any or all of these species, it's easy to make your own. For Spirogyra and Euglena, make simple wet mounts. These species are easily visible without staining, although you can use methylene blue to stain the nuclei and eosin Y to stain the cell contents.

For Paramecium, use a clean pipette to transfer one drop of the culture liquid to each of two flat slides and spread the drop over the central part of the slide. Heat-fix the specimens. Observe one slide without staining. Stain the second slide with methylene blue and eosin Y and then observe the slide.

As the old joke goes, to make an Amoeba stew, first catch an Amoeba. Use a clean pipette to transfer several drops of the culture liquid to a deep well slide and observe the slide at low magnification (~40X) to see if one or more Amoebae are present. If one is, it'll be obvious, because Amoebae are quite large. (Amobae like to hide near the bottom of the culture or in vegetation.)

Once you locate an Amoeba, while still observing through the microscope, squeeze the air out of a pipette, insert the tip into the slide well and suck up the Amoeba with at most a drop or two of liquid. Expel the liquid quickly onto a flat slide, while the Amoeba is still disoriented. (Amoebae like to glom onto things; we've seen them attach themselves to the inside of the plastic pipette, which doesn't seem to offer much in the way of grippable surface.) Repeat to make a second Amoeba slide.

Gently heat-fix the slides, and stain one of them. Observe the Amoeba in unstained and stained versions.

PROCEDURE VIII-1-1: OBSERVING SPIROGYRAE

Spirogyra is immediately recognizable as thin green unbranched filaments made up of a line of individual square or rectangular cells butted end-to-end. Most of the cellular features of Spirogyra are easily observed at low to moderate magnification and without staining. Begin at low magnification, and increase magnification as necessary to observe details of all of the features. As you observe your live and mounted specimens, look for the following features:

- Cell wall – Spirogyra actually has a double cell wall. The outer layer is made up of pectin, which when wet forms a slimy gel. The inner layer, made up of cellulose, remains rigid even when wet. In live specimens, you'll probably see only a single wall. In prepared slides, particularly if they are stained, the outer pectin wall may also be visible.

- Cytoplasmic lining – a thin layer of cytoplasmic lining, also called the plasma membrane, exists on the inner side of the cell wall.

- Vacuole – a large central vacuole is connected to the cytoplasmic lining by cytoplasmic strands. (The strands may be difficult to discriminate in unstained cells at low magnification.)

- Nucleus – the cell nucleus is contained within the vacuole.

- Chloroplasts – band- or ribbon-like chloroplasts are the most prominent feature of the cells. (Depending on the particular cell, the chloroplast(s) may appear scalloped or serrated rather than ribbon-like.) Any particular cell may contain anything from one to several chloroplasts.

- Pyrenoids – these appear as small round dots on the chloroplasts, and are where starches are stored within the cells. Pyrenoids are usually readily visible in stained slides, but may be hard to discriminate in live specimens. If so, you can add a drop of Gram's iodine stain to the slide. As the stain is absorbed by the cells, the starch present in the pyrenoids reacts with the iodine to form an intense blue-black color.

Spirogyra can reproduce asexually or sexually. In asexual reproduction, new filaments are formed by intercalary mitosis. Sexual reproduction in Spirogyra occurs via *conjugation*. In conjugation, one cell acting in the male role extends a tubular protuberance known as a *conjugation tube*. That tube links to a second cell, acting in the female role. Cytoplasm is transferred from the "male" cell to the "female" cell, where the "male" cytoplasm reacts with the "female" cytoplasm to produce a zygospore. (Note that cells are actually neither male nor female, but merely assume one of those roles during conjugative reproduction.)

Conjugation can occur between adjacent cells on a single strand (called *lateral conjugation*) or between two cells on different strands that are in close physical proximity (called *scalariform conjugation*). If you're lucky, you may find conjugation occurring naturally in your live specimen. If not, you can encourage it to occur by adding a drop of sugar water to the slide.

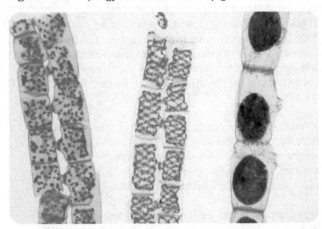

Figure VIII-1-1: *Spirogyra in scalariform conjugation, 100X*

PROCEDURE VIII-1-2: OBSERVING EUGLENAE

From the large to the small. While Spirogyra forms filaments that are often visible to the naked eye, most Euglenoids are tiny creatures. (A few species are much larger, as much as 0.1mm in length, but these are very uncommon.) Euglenae are visible as tiny spots at 40X magnification. You may be able to see some detail at 100X, but observing full detail requires 400X or higher. If you have an oil-immersion objective (1,000X), now is a good time to use it.

> If you don't have an oil-immersion objective but you do have a 15X eyepiece to substitute for the standard 10X eyepiece, you may find you can make out more detail at 600X than you can at 400X.

Euglenae hold an important place in the history of biological taxonomy. Beginning with Linnaeus, biologists classified living things in two kingdoms: *Animalia*, or animal-like creatures, and *Plantae*, or plant-like creatures. Protista were originally classified into one of these two kingdoms, with animal-like protists (such as Amoebae) classified in Animalia and plant-like protists (such as Algae) classified in Plantae.

Euglenae presented these early taxonomists with a real problem, because Euglenae were simultaneously animal-like (were motile and heterotrophic) and plant-like (possessed chlorophyll and were autotrophic). Accordingly, Euglenae were ultimately responsible for taxonomists expanding the original two-kingdom system to a three-kingdom system, with Euglenae and other protists classified in the new kingdom *Protista*.

Unlike Spirogyrae, Euglenae are motile. They use two methods of movement. First and more obvious is their constant alterations in shape, changing back and forth from an extended, worm-like form to a nearly circular form, a form of motion known as *Euglenoid movement*. Second, and less obvious, Euglenae possess flagella, tiny whip-like tails that they use to drive themselves from place to place.

Some of the cellular features of Euglena are visible at medium to high magnification without staining, but most are much more clearly visible in stained specimens. As you observe your live and mounted specimens, look for the following features:

- Pellicle – one of the animal-like characteristics of Euglena is that it possesses no cell wall. Instead, the cell is contained within a *pellicle*, a membranous structure made up of proteins and supported by *microtubules*, which appear as a fur-like layer between the membrane and the interior cell components.

- Nucleus – Euglena contains a single *nucleus* with an embedded *nucleolus*. With proper staining and at very high magnification, DNA strands may be visible as granularity or texturing within the nucleus, and ribosomal RNA as texturing within the nucleolus.

- Chloroplasts – an individual Euglena may contain anything from one or two to several *chloroplasts*, visible as irregularly shaped flat or folded green structures.

- Pyrenoids – *pyrenoids* are non-membrane-bound organelles within the chloroplast where photosynthesis occurs, producing a starchy polysaccharide called *paramylon*. Pyrenoids are present in and characteristic of the genus Euglena, but absent in other Euglenoids.

- Food vacuoles – paramylon produced by the pyrenoids is stored in granules distributed throughout the cytoplasm. During periods of darkness, when photosynthesis cannot occur, the Euglena uses the starch stored in these vacuoles as an energy source.

- Flagellum – the Euglena possesses a single *flagellum*, a thin, whip-like, nearly transparent structure that provides motility. The flagellum is often difficult or impossible to discriminate visually in live specimens with brightfield illumination, even at high magnification, but is often visible in live specimens with darkfield illumination or in stained specimens. You may be able to see flagella in live specimens under brightfield illumination if you use very dim lighting.

- Reservoir – the *reservoir* is a bay-like indentation in the pellicle surface, open to the surrounding environment, where the base of the flagellum is rooted.

- Contractile vacuole – the *contractile vacuole* is adjacent to the reservoir. This vacuole maintains osmotic equilibrium by gathering excess water from inside the cell and periodically expelling it via the reservoir to the exterior environment.

- Paraflagellar body – the *paraflagellar body* is a light-sensitive organelle located within the reservoir at the base of the flagellum.

- Stigma – the *stigma*, also called the *eye spot* or the *red spot*, works in conjunction with the paraflagellar body to allow *phototaxis* (movement toward a light source). As the Euglena moves, the stigma screens some light from the paraflagellar body, allowing the Euglena to determine the direction of the light source and move toward it.

Figure VIII-1-2: *Euglena, 400X*

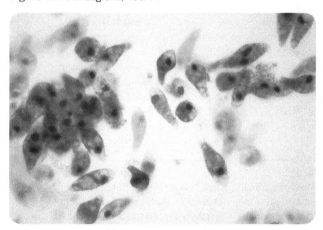

PROCEDURE VIII-1-3: OBSERVING AMOEBAE

From the small back to the very large. The Amoeba is certainly the most recognizable microorganism; even nonscientists can identify it immediately by its appearance. The Amoeba, or at least many of its species, is one of the largest microorganisms. Depending on species and the individual organism, some may be as large as a pinhead, and easily visible to the naked eye. Amoebae from other planets are apparently capable of eating entire buildings (cf. the 1958 horror/SF movie, *The Blob*, which featured one of these gigantic alien amoebae.)

Amoebae live in water, but are seldom found free-floating. Instead, they attach themselves to gravel, branches, leaves, or other solid materials. As heterotrophs, Amoebae must obtain nourishment from their environments. They do so by engulfing prey, including bacteria, other protists, and other tiny organisms.

An Amoeba moves by extending a pseudopod, with which it grasps the surface and then pulls the rest of the organism to the new location, a process called *amoeboid movement* or, more casually, oozing. (Not all microorganisms that use amoeboid movement are amoebae, but all amoebae use amoeboid movement.) Amoebae reproduce by *mitosis* and *cytokinesis*, a process that superficially resembles the binary fission by which prokaryotes (bacteria) reproduce. During reproduction, the nucleus splits in two and the cell "pinches" in until the pinched portion closes on itself and cuts the cell in two, forming two individual cells, which can then repeat the process to form four cells, and so on. If you're patient while observing a live culture at low magnification, you may be fortunate enough to observe cytokinesis.

> So, if Amoebae can simply continue doubling their numbers by cytokinesis, why isn't the surface of the planet covered with really, really old Amoebae? Unlike the Hydra, which does not experience senescence and so is immortal for all practical purposes, Amoebae do undergo senescence. The cytoplasm of vigorous young Amoebae appears almost transparent, while that of elderly Amoebae appears cloudy or chalky in comparison. Although cytokinesis can be thought of as an individual Amoeba cloning itself, that does not mean that those clones are immortal. An Amoeba that has not undergone cytokinesis recently gradually becomes senescent and may then fail to reproduce.

Most people's initial impression of the internal structure of an Amoeba (or lack thereof) is that it is completely chaotic. In fact, Linnaeus originally assigned the binary name for the common Amoeba now known as *Amoeba proteus* as *Chaos chaos*. Because of its size, with proper lighting most of the cellular features of larger Amoebae are visible at low to medium magnification without staining, but most are much more clearly visible in stained specimens. As you observe your live and mounted specimens, look for the following features:

- Cell membrane—like Euglena, one of the animal-like characteristics of Amoeba is that it possesses no cell wall. The amoeba's cytoplasm—ectoplasm and endoplasm— and organelles are contained within a soft cell membrane.

- Cytoplasm – Amoebae contain two types of cytoplasm. *Ectoplasm* is found adjacent to the cell membrane. It is clear and has a jelly-like consistency. *Endoplasm* fills the cell inside the ectoplasm. It is thin and watery and has a granular appearance. Various organelles float within the endoplasm.

- Nucleus – Amoebae contain one *nucleus*. (Some members of the amoeboid genus *Chaos* contain 1,000 or more nuclei.)

- Food vacuoles – Amoebae feed on microorganisms by engulfing them. These prey are stored and digested in organelles called *food vacuoles*.

- Contractile vacuole – the *contractile vacuole* in Amoebae is similar in appearance and function to the same organelle in Euglenae. It maintains osmotic equilibrium by gathering excess water from inside the cell and periodically expelling it to the exterior environment.

Figure VIII-1-3: *Amoeba, 100X*

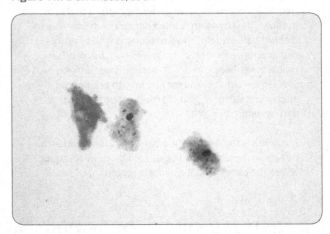

PROCEDURE VIII-1-4: OBSERVING PARAMECIA

The paramecium is the sprinter of this group. Spirogyrae are not motile. Amoebae slowly ooze from place to place. A Euglena, with its single flagellum, might achieve speeds of 100 μm/s or a bit more, taking about 10 seconds to move a single millimeter. But the Paramecium—with its thousands of cilia beating synchronously like thousands of tiny rowers—has been clocked at speeds of 2,000 μm/s, which means it can move completely though a microscope's field of view in a small fraction of a second.

That means observing live Paramecia is challenging, to say the least. You can slow them down significantly by adding a drop of methylcellulose or glycerol to the drop of culture liquid, or by placing a tiny tuft of cotton fibers in the drop of culture to present a physical barrier to their movement. Even then, it's almost impossible to observe fine detail. For that, you'll need a prepared slide.

Paramecia reproduce both sexually and asexually. In sexual reproduction, called *conjugation*, two individuals align longitudinally and exchange nuclear material. In asexual reproduction, called *transverse fission*, one individual splits laterally, with the macronucleus and cell body dividing in two and the micronucleus replicating by mitosis. Fission occurs much more commonly than conjugation, but if your live culture contains numerous individuals you may have the opportunity to observe both forms of reproduction.

As you observe your live and mounted specimens, look for the following features:

- Pellicle – like the Euglena, the Paramecium is contained within a pellicle, but while the Euglena pellicle is extremely flexible, allowing the Euglena to change shape radically, the Paramecium pellicle is relatively rigid, forcing the creature to maintain its slipper-like shape.

- Cilia – the surface of the paramecium is covered with thousands of hair-like *cilia*, which beat synchronously. Like the flagellum of a Euglena, cilia provide motility, but because they are present in much higher numbers, the Paramecium can move much faster than a Euglena.

- Cytoplasm – Paramecia contain two types of cytoplasm. Ectoplasm is found adjacent to the pellicle. Endoplasm fills the cell inside the ectoplasm, with various organelles floating with it.

- Nucleus – The other three protists we've examined each has a single nucleus. Paramecia have two nuclei, each of which has different duties. The *macronucleus* handles routine tasks such as metabolism. The *micronucleus*, much smaller than the macronucleus, is responsible for reproduction.

- Oral grove/mouth pore/gullet – this is a prominent internal structure, beginning near the anterior end of the creature and extending about half its length. Cilia in the *oral groove* sweep bacteria and other unicellular prey through the *mouth pore* and into the *gullet*, also called the *cytopharynx*. When sufficient food organisms have been gathered in the gullet, they are expelled into the endoplasm as a *food vacuole*.

- Food vacuoles – when the cytopharynx is filled with prey, it expels a food vacuole, also called a *phagocytic vacuole*, into the endoplasm. The food vacuole travels the length of the creature, with digestion enzymes acting on the contents of the vacuole. By the time the food vacuole has finished its journey, it has shrunk considerably and contains only waste materials. When the food vacuole reaches the anal pore, it ruptures and expels the waste materials to the outside environment.

- Contractile vacuoles – the Paramecium is contained within a semi-permeable membrane, which passes water freely. The anterior and posterior *contractile vacuoles* maintain osmotic equilibrium by gathering excess water from inside the cell and periodically expelling it to the exterior environment.

Figure VIII-1-4: *Paramecia, 100X*

REVIEW QUESTIONS

Q1: What visual evidence suggests that Spirgoyra and Euglena are autotrophic?

_____

_____

_____

_____

Q2: What additional information might we gain by observing all four of these species in a mixed culture rather than observing individual cultures of each species?

_____

_____

_____

_____

_____

_____

_____

Q3: What differences, if any, did you observe in the motility and motility mechanisms in the different species? Which are the fast and slow movers?

_____

_____

_____

_____

Q4: In the original two-kingdom system (Animalia and Plantae), to which of those kingdoms do you think Algae, Amoebae, and Paramecia were assigned? Why? Using the Internet or other resources, identify a class of organisms that was originally assigned to Plantae, later to Fungi, and now to Protista.

_____

_____

_____

_____

Q5: Using the Internet or other resources, identify a group of organisms that was originally assigned to Plantae, later to Fungi, and now to Protista.

_____

_____

_____

_____

_____

Investigating Fungi

EQUIPMENT AND MATERIALS

You'll need the following items to complete this lab session. (The standard kit for this book, available from *www.thehomescientist.com*, includes the items listed in the first group.) If you are preparing your own live specimens, do so starting a week or so before you intend to do this lab session.

MATERIALS FROM KIT

- Goggles
- Centrifuge tube, 50 mL
- Coverslips
- Inoculating loop

- Magnifier
- Pipettes
- Reaction plate, 24-well
- Scalpel

- Slides, flat
- Slides, deep well
- Stain: eosin Y
- Stain: Gram's iodine

- Stain: methylene blue
- Stain: safranin O
- Yeast

MATERIALS YOU PROVIDE

- Gloves
- Bags, plastic zip
- Bread (without preservatives)
- Butane lighter (or other flame source)
- Microscope
- Microtome (optional)

- Mushroom (fresh)
- Orange slice
- Paper towels
- Slides, prepared (optional; see text)
- Stereo microscope (optional)
- Sugar

BACKGROUND

Fungi are a eukaryotic monophyletic group characterized by some unique features, as well as by other features that they share with plants or animals. Like plants, fungi possess a cell wall and vacuoles, and can reproduce sexually and asexually. Most fungi are *sessile* (fixed rather than motile) and often grow in soil. Like simple plants—mosses and ferns—fungi produce spores and, like mosses and algae, most fungi have haploid nuclei. Some fungi form fruiting bodies that resemble those of plants, and some fungi form *rhizomorphs* that resemble plant roots and perform similar functions. Like animals and some protists, fungi are heterotrophs.

The primary feature unique to fungi is the composition of their cell walls, which contain both *glucans* (a class of polysaccharides) and *chitin*. Although both plants and animals use glucans—for example, the starch used in plant cells for energy storage and the glycogen used in animal cells for the same purpose are both glucans—and some animals use chitin for exoskeletons, beaks, and radulae, fungi are the only organisms that combine glucans and chitin in their cell walls. (Animal cells, of course, are contained by a cell membrane rather than a cell wall.)

In the obsolete two-, three-, and four-kingdom taxonomies, fungi were classified as *Plantae* because they obviously weren't *Animalia* (or, later, *Bacteria* or *Protista*). Over the next 200 years, as the differences between plants and fungi gradually became more difficult to ignore, most taxonomists became convinced that fungi belonged in a kingdom all their own. When Robert and Barbara were in high school in the 1960s, fungi were still officially classified as plants, but that changed in 1969 when Whittaker proposed his five-kingdom system, with *Fungi* as the fourth and final eukaryotic kingdom. (The prokaryotic *Bacteria* and *Archaea* were assigned to the fifth kingdom.) Whittaker's five-kingdom system was rapidly accepted by most biologists, and formed the basis for the later six-kingdom systems.

> The study of fungi is called *mycology*, and a biologist who specializes in fungi is called a *mycologist*. Oddly, many biology books, including recent ones, continue to treat mycology as a subdiscipline of botany, which is the study of plants, although DNA analysis shows indisputably that fungi are more closely related to animals than to plants.

Although it's been more than 40 years since fungi were assigned their own kingdom, a great deal of fundamental disagreement remains about how they should be classified within that kingdom. Even the number of phyla remains in dispute. What is not in dispute is that *Fungi* indeed form a

monophyletic group (a single group of related organisms with a shared common ancestor). *Slime molds* (myxomcetes) and *water molds* (oomycetes), which are structurally similar to and were formerly classified as fungi are now classified in kingdom *Protista*.

The most recent proposal groups true fungi into seven phyla: *Ascomycota* and *Basidiomycota* (which are grouped into the subkingdom *Dikarya*), *Microsporidia*, *Chytridiomycota*, *Blastocladiomycota*, *Neocallimastigomycota*, and *Glomeromycota*.

> In 2001, the polyphyletic phylum *Zygomycota* was broken up, and the former order *Glomales* was promoted to the phylum *Glomeromycota*. About 230 members of *Zygomycota* were reassigned to this new phylum, with the remaining ~750 members assigned to other existing phyla. Despite the fact that it has been deprecated for a decade, many biology books (and biologists) continue to use phylum *Zygomycota* because it provides a convenient morphological grouping of *saprophytic* fungi with *aseptate hyphae*.

In these lab sessions, we'll examine some characteristics of representative fungi from three of these phyla, *Ascomycota*, *Basidiomycota*, and *Glomeromycota* (*Zygomycota*).

The phylum *Ascomycota*, also called *sac fungi*, contains roughly 30,000 species, including some *molds*, *morels*, *truffles*, and *yeasts*. The phylum is named for the sac-shaped reproductive structures called *asci* (singular, *ascus*) present in members of this phylum. Figure IX-1-1 shows a specimen representative of *Ascomycota*, *Morchella deliciosa*, whose common name is *white morel*. Figure IX-1-2 shows the most costly fungus in the world, *Tuber melanosporum*, a.k.a. the black truffle, which commonly sells for $5,000 or more per kilo.

Figure IX-1-1: *Morchella deliciosa (white morel)*

Figure IX-1-2: *Tuber melanosporum, the black truffle*

Although most fungi are know to reproduce both sexually and asexually, the names assigned to classic fungal phyla are based on the structures used for sexual reproduction.

The phylum *Basidiomycota*, also called *club fungi*, contains roughly 25,000 species, including some *mushrooms*, *puffballs*, *shelf fungi*, and many important plant pathogens such as *smuts* and *rusts*. The phylum is named for the microscopic club-shaped reproductive structures called *basidia* (singular, *basidium*) present in members of this phylum. Figure IX-1-3 shows an unidentified species of *Amanita* mushroom growing in a suburban yard.

Figure IX-1-3: *Amanita sp.*

The word *toadstool* is used in casual conversation for toxic (or at least inedible) mushrooms. Biology makes no distinction between mushrooms and toadstools.

The deprecated phylum *Zygomycota* contains roughly 750 species, many of them common bread molds. Members of this group possess sexual reproductive structures called *zygosporangia*, for which the original phylum was named. Figure IX-1-4 shows an unidentified mold growing on pita bread.

Figure IX-1-4: *Bread mold*

Figure IX-1-5: *Rhizocarpon geographicum*

Lichens are composite organisms, made up of a photosynthetic alga or cyanobacterium in symbiosis with a fungus. About two dozen species of algae and cyanobacteria are found in various lichens, but the species present in the vast majority of lichens is either the green alga *Trebouxia* or the cyanobacterium *Nostoc*. Similarly, although exceptions exist, the fungal species present in the vast majority of lichens is one of the *Ascomycete* species.

> Although they are found in similar environments and may appear superficially similar, lichens and mosses are two distinct classes of organisms. Mosses are simple plants.

Lichens encompass about 25,000 "species," many of which are quite colorful. Lichens are informally grouped according to their *growth forms*.

- *Crustose lichens* are flat (crust-like), with the *thallus* (body) growing close to the surface of a substrate such as bark or stone. Figure IX-1-5 shows the crustose lichen *Rhizocarpon geographicum*, vernacular name Map lichen.

- *Filamentous lichens* are hair-like, with thin tendrils or filaments.

- *Foliose lichens*, like those shown in Figure IX-0-6, peel away from the substrate, forming sheets that may superficially resemble leaves.

- *Fruticose lichens* are three-dimensional, with parts of the thallus forming stalks that protrude from the substrate surface.

- *Gelatinous lichens* are gel-like.

- *Leprose lichens* are powdery.

- *Squamulose lichens* are scale-like.

Figure IX-1-6: *A foliose lichen, probably Xanthoria parietina or X. polycarpa*

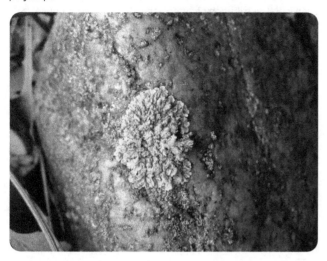

Reproduction in lichens is complicated because technically lichens do not themselves reproduce. Instead, their component algae or bacteria and fungus reproduces—sexually, asexually, or both—according to the characteristics of their particular phyla. In practice, though, lichens can be considered to reproduce asexually because they release fragments or packages that contain both algal (or bacterial) and fungal cells, which cells go on to reproduce using their own methods. The new algae (or bacteria) and fungi may or may not associate symbiotically as new lichen individuals.

Most lichens are extremely hardy because they combine the advantages of cyanobacteria or algae (photoautrophism) and fungi (durable chitinous cell walls). Accordingly, lichens can be found in nearly any environment from the hottest, driest deserts to frigid Antarctic zones. One exception to that hardiness is the extremely high sensitivity of most lichens to air pollutants, probably because lichens have evolved to absorb nutrients efficiently from the atmosphere. For that reason, ecologists consider lichens to be "trigger species," much like a canary in a coalmine. Lichens often suffer mass die-offs at pollution levels low enough that other species are apparently entirely unaffected.

Although you can purchase live specimens of various fungi and molds from Carolina Biological Supply or another vendor, there's really no need to do so. Fungi and mold spores are ubiquitous, and growing them is less a problem than keeping them from growing.

As a representative of *Zygomycota*, we'll use ordinary bread mold, *Rhizopus* sp. To grow a thriving colony, obtain some bread that does not contain preservatives. Bread from a bakery is a good source, as are those pop-and-bake tubes of bread dough sold at supermarkets. (Make sure the label indicates that no preservatives are present.)

> Most bread from the supermarket is loaded with preservatives. When we attempted to grow bread mold on supermarket white bread, we found after two weeks under ideal conditions that no growth had occurred. Those preservatives really do work.

To grow bread mold, simply expose a slice of bread to the air in your kitchen overnight to ensure that it's covered with mold spores and then place it and a wet paper towel in a sealed plastic bag or similar container at room temperature. Keep it in a dark location, or at least out of direct sunlight, and observe the growth over a period of several days. If you have a suitable camera, shoot images every 12 hours or so—say, morning and evening—as the mold develops. Note the variety of textures and colors, each of which represents a different species of mold. The black (or very dark) mold is *Rhizopus nigricans* or *R. stolonifer*. We'll use that in this session, but it's also worth observing and attempting to identify any other mold species growing on your bread.

As a representatives of *Ascomycota*, we'll use the ubiquitous mold *Penicillium* spp. and common baker's or brewer's yeast, *Saccharomyces cerevisiae*. There's a good chance you'll find at least one bluish mold colony on your bread, which is almost certainly a species of *Penicillium*. If not, you can obtain a *Penicillium* specimen from any bleu cheese, such as Camembert or Roquefort. You can culture *Penicillium* obtained from a cheese by inoculating a slice of orange with a small amount of the bluish material from the cheese and incubating it at room temperature in a plastic bag with a wet paper towel. Culturing *S. cerevisiae* is trivially easy. Simply dissolve a teaspoon of table sugar in some tap water in a 50 mL centrifuge tube, transfer a small amount of baker's or brewer's yeast to the tube, fill the tube to the brim with water, and allow the tube to incubate in the dark at room temperature for several days.

As a representative of *Basidiomycota*, we recommend using an edible mushroom from the supermarket, which has the advantage of being known-safe. Using a wild mushroom instead has the advantage of allowing you to obtain the entire mushroom, including the volva and other structural components that may be partially or completely beneath the surface.

Fungi are structurally simple. As you examine your specimens, look for the following (usually) microscopic structural features. Note that not all of these features are present in all species.

Hyphae (singular *hypha*) are the fundamental structures that make up fungi other than yeasts, which are unicellular. Hyphae are (generally) transparent, branching, filament-like structures that comprise cytoplasm and nuclei contained by a chitinous cell wall. Some hyphae, called *septate hyphae*, are divided into cellular sections by walls called *septa* (singular *septum*), with each section containing one nucleus. Other hyphae, called *aseptate hyphae* or *coenocytic hyphae*, have either incomplete or absent septa, with multiple nuclei present within a single cell wall.

A *stolon* is a septate hypha that connects *sporangiophores* (*sporophores* that produce spores within an enclosure) and may also anchor *rhizoids* (also called *holdfasts*), root-like hyphae that are embedded in the soil or other substrate where the fungus is growing. Hyphae used by parasitic fungi to penetrate the cells of the host to obtain nutrition are called *haustoria* (singular *haustorium*).

Asexual reproduction in fungi occurs via different mechanisms and using various specialized structures. Yeasts reproduce by *budding*, shown occurring in Figure IX-1-8, during which mitosis occurs and the cell cytoplasm splits into two uneven portions. The smaller portion detaches from the larger, and

eventually matures as a new individual organism. The similar process called *fragmentation*, resembles budding with a more even initial distribution of cytoplasm. Most fungi reproduce asexually by mitotic production of haploid vegetative cells called *spores*. Spores are produced by specialized hyphae called *conidiophores* and *sporangiophores*, which produce spores in enclosures (*sporangia*, singular *sporangium*) such as sac-shaped *asci* (singular *ascum*) and club- or stool-shaped *basidia* (singular *basidium*).

Mycelia (singular *mycelium*) are intertwined masses of hyphae that make up the vegetative "body" of a fungus. A *monokaryotic mycelium* (also called a *homokaryotic mycelium*) results when one spore germinates; such mycelia cannot reproduce sexually. If two monokaryotic mycelia grow together and merge, forming a *dikaryotic mycelium*, that mycelium can reproduce sexually, forming *fruiting bodies* (e.g., mushrooms), which produce and disperse spores. Depending on the species, mycelia may range from microscopically small to gigantic, weighing thousands of tons.

PROCEDURE IX-1-1: OBSERVING ZYGOMYCOTA

To begin, use the magnifier (or a stereo microscope, if you have one) to examine closely the various molds growing on your bread specimen.

1. Note the color and texture of each species, and record your observations in your lab notebook.

2. Using the Internet or printed reference material, attempt to identify each of the species present, or at least its genus.

3. For each species present, examine and identify any structures visible at low magnification. Are hyphae present that are modified as sporangiophores and sporangia? Is pigmentation evenly distributed throughout each mycelium, or concentrated in specific structures? If you have a suitable camera, record an image, ideally at low magnification through the microscope.

Unless you are very unfortunate, your bread should have at least one dark-colored colony of *Rhizopus* sp. growing on it. We'll use material from that colony in the following steps. If you have time, you can repeat these steps for some or all of the other species present on your bread specimen.

1. Flame sterilize the inoculating loop and use it to transfer a tiny amount of material from that colony to each of five flat slides. Make smear mounts of each slide by adding a drop of water, using the edge of a clean slide to smear the material, and then heat-fixing the specimen.

2. Retain one of the slides as an unstained specimen. Stain each of the other four slides using one of the following stains: eosin Y, Gram's iodine, methylene blue, and safranin O. For each slide, add a small drop of the stain to the specimen area, allow the stain to work a minute or so, and then rinse off the stain with a very gentle stream of tap water.

3. For each slide, position a coverslip, scan at low power to locate a populated area, and observe the specimen at medium and high magnifications. Determine which, if any, of the stains are helpful in revealing additional detail versus the unstained slide. Record your observations in your lab notebook.

Figure IX-1-7 shows *Rhizopus nigricans* (black bread mold) at 100X. Rhizoids are visible at the far left of the image, with sporangiophores supporting the sporangia visible on the far right.

Figure IX-1-7: *Rhizopus nigricans, wm, 100X*

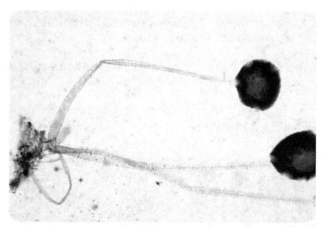

PROCEDURE IX-1-2: OBSERVING ASCOMYCOTA

In this procedure, we'll observe *Ascomycota* (sac fungi), many species of which closely resemble various species of *Zygomycota* with the exception of their spore-forming structures. Many *Ascomycota* species are economically valuable, critical to the production of antibiotics, beer and other alcoholic beverages, cheeses, livestock feed supplements, and many other important products.

The first of the *Ascomycota* we'll observe, ordinary yeast, does not form spores at all, instead reproducing asexually by budding. Other *Ascomycetes* reproduce asexually by forming spores called *conidia* on the surface of *conidiophores*—hyphae adapted for reproduction—in contrast to the enclosed sporangia spore sacs formed by *Rhizopus* and other *Zygomycota*. *Penicillium* spp. and *Aspergillus* spp. are two ubiquitous examples of *Ascomycota*.

1. Carefully open the centrifuge tube that contains your *S. cerevisiae* culture. Note the appearance of the culture and any odor present, and record your observations in your lab notebook.

2. Transfer about 2 mL of the yeast suspension to a well in the reaction plate, add one drop of methylene blue stain, and stir to mix the solutions.

3. Transfer one drop of the unstained live yeast suspension to a flat slide, position a coverslip, scan the specimen at medium magnification to locate budding cells, and then switch to high-dry magnification to observe any visible detail. If you have an oil-immersion objective, also observe the specimen at high magnification. Record your observations in your lab notebook.

4. Repeat step 3 with one drop of the live yeast culture that you stained with methylene blue.

Figure IX-1-8 is a 1,000X differential interference contrast (DIC) microscopy image of *S. cerevisiae* budding.

Figure IX-1-8: *Saccharomyces cerevisiae budding, 1,000X (DIC microscopy)*

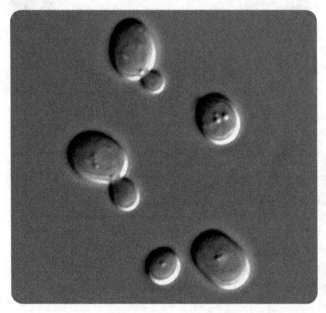

5. Flame-sterilize the inoculating loop and use it to transfer a tiny amount of material from your *Penicillium* culture to a flat slide. Add a drop of water, spread the liquid, position a coverslip, and observe the slide at medium and high-dry magnification.

6. Add a drop of methylene blue at one edge of the coverslip, and use the corner of a paper towel to wick the stain under the coverslip. Allow the stain to work for a minute or so and then add a drop of water at the edge of the coverslip and wick out the excess stain. Observe the slide at high-dry and with the oil-immersion objective, if you have one. Note the appearance of the conidiophores and conidia, and record your observations in your lab notebook.

7. Using the same slide, repeat step 6 with a drop of eosin Y stain.

8. If you have an *Aspergillus* culture, purchased or home-grown, repeat steps 5 through 7 to observe *Aspergillus*. (Like *Penicillium*, *Aspergillus* sp. is ubiquitous, and is likely to be represented among the colonies growing on your bread specimen.)

Figure IX-1-9 is a 100X overview of a *Penicillium* prepared slide. Figure IX-1-10 shows the conidiophores and conidia at 1,000X. Figure IX-1-11 is a 100X overview of an *Aspergillus* prepared slide. Note the similarities between Figures IX-1-9 and IX-1-11.

Figure IX-1-9: *Penicillium sp. wm showing conidiophores, 100X*

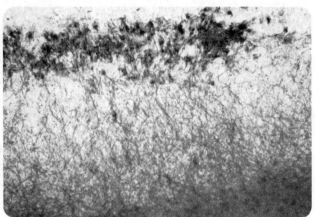

Figure IX-1-10: *Penicillium conidiophores with conidia, 1,000X*

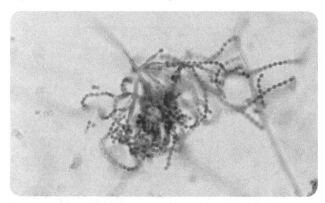

Figure IX-1-11: *Aspergillus sp. wm showing conidiophores and conidia, 100X*

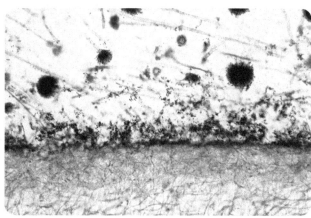

PROCEDURE IX-1-3: OBSERVING BASIDIOMYCOTA

Many *Basidiomycota* (club fungi) species are among the most familiar fungi. *Basidiomycota* include mushrooms, puffballs, and shelf fungi—some of which are edible, but most of which are inedible or even poisonous—as well as various microscopic plant pathogens, including rusts and smuts. Figure IX-1-12 shows a fine example of *Agaricus augustus*, a typical mushroom.

Figure IX-1-12: *Agaricus augustus, a typical Basidiomycete*

Begin by examining your mushroom specimen by naked eye and with the magnifier or a stereo microscope. Identify, name, and describe as many features as possible, including (if present) the *stipe* (stem) and any substructures present including the *annulus* or *volva*, the *pileus* (cap), and the *lamellae* (gills) present on the underside of the pileus. Record your observations in your lab notebook.

The most significant structures of a *Basidiomycete*, and the ones for which the group is named, are the basidia, microscopic spore-producing structures located on the gill. To observe basidia, take the following steps:

1. Cut off the stipe of your mushroom specimen, if present, and place the pileus flat on a hard surface, with the lamellae side (bottom of the cap) facing up.

2. Halve the pileus by using the scalpel or a single-edge razor blade to make a vertical cut through the pileus.

3. Place one of the halves aside, and use the scalpel to cut as thin a vertical section as possible through the remaining half near the edge of the pileus. Ideally, you want this vertical section to be so thin that it is almost transparent.

4. Transfer the section to a flat slide. Position a coverslip on the slide, place the slide on the stage, and observe it at low magnification to locate the basidiophores and basidia. Center that area under the objective and observe it at medium and high-dry magnification. Record your observations in your lab notebook.

Figure IX-1-13 shows a prepared slide of *Agaricus* sp. at 100X. Figure IX-1-14 is a scanning electron microscope (SEM) image showing the stromal layer, basidophores, and basidia of *Agaricus bisporus*.

Figure IX-1-14: *SEM image of Agaricus bisporus, showing stromal layer and basidiospores*

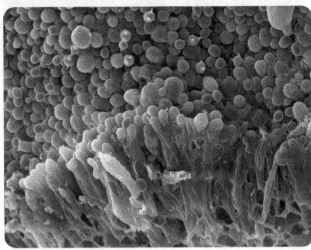

Figure IX-1-13: *Agaricus sp. cs of gills showing basidiospores, 40X*

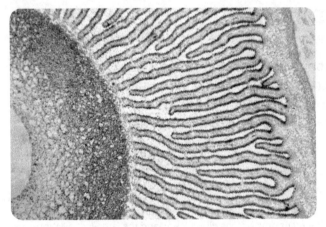

REVIEW QUESTIONS

Q1: Why are fungi not classified as algae? As plants?

_____

_____

_____

_____

_____

_____

Q2: In Lab IX-1-1, which fungal structures were you able to identify? What structure(s) of *Rhizopus* sp. are responsible for its common name, black bread mold?

_____

_____

_____

_____

_____

Q3: In Lab IX-1-2, what odor did you notice when you opened the yeast culture tube? To what is that odor attributable?

_____

_____

_____

_____

_____

Q4: In Lab IX-1-2, which of the stains you tested was most helpful in revealing detail?

_____

_____

_____

_____

_____

Q5: In Lab IX-1-2 while observing the unstained *Penicillium* culture, what did you observe?

_____

_____

_____

_____

_____

Investigating Simple Plants: Mosses and Ferns

EQUIPMENT AND MATERIALS

You'll need the following items to complete this lab session. (The standard kit for this book, available from *www.thehomescientist.com*, includes the items listed in the first group.) If you are preparing your own live specimens, do so starting a week or so before you intend to do this lab session.

MATERIALS FROM KIT

- Goggles
- Coverslips
- Forceps
- Magnifier
- Pipettes

- Scalpel
- Slides, flat
- Stain: eosin Y
- Stain: methylene blue
- Teasing needles

MATERIALS YOU PROVIDE

- Gloves
- Microscope
- Microscope, stereo (optional)

- Slides, prepared (optional; see text)
- Specimen(s), fern (see text)
- Specimen(s), moss (see text)

BACKGROUND

This and the following lab session present an excellent opportunity for a field trip to your nearest botanical gardens. You can observe plant specimens in your own yard, a park, or a nearby woods, but identifying the plant species you find can be problematic. That is, of course, an educational opportunity in itself, but it's easy to get bogged down attempting to identify a few species and thereby miss out on observing the great diversity of plant life available.

> Know and follow all laws with respect to collecting plants. In particular, many federal, state, and municipal parks prohibit collecting specimens.

The advantages to visiting botanical gardens are that all of the species are set out and identified for you, allowing you to concentrate on observing the characteristics of specific known species, and that you will be able to observe many species that aren't found locally in the wild. Also, many botanical gardens have indoor facilities, where you will be able to observe specimens in leaf and in flower regardless of the time of year when you visit. Finally, you'll often find knowledgeable staff members who will be delighted by your interest and happy to help you make the most of your visit. Just let them know that you're interested in studying their collection seriously and you'll probably find they'll go out of their way to answer your questions.

Plan to spend at least half a day studying their collection. You'll probably find that you could easily spend an entire day or more without seeing everything there is to see. Take along your magnifier, camera, and lab notebook, and use them to study the collection in an organized manner. If you see something that you'd really like to have a specimen of, ask the staff. There are no guarantees, but sometimes they'll make an exception for you if they understand that you're doing a serious scientific study of plants.

> More than once, we've come away from such a visit with leaves, flowers, and other plant specimens that we'd not otherwise have had access to. Make sure to come prepared. Carry a supply of plastic zip bags and some plastic bottles or tubes to hold specimens. But always, always ask permission before you collect a specimen.

In the lab sessions, we focus on microscopic examination of plant structures, but that doesn't tell the whole story. The macroscopic structures of plants are also fascinating, and many of those will require getting away from your microscope and into the great outdoors, if only to collect specimens for later study. For example, you should familiarize yourself with the great variety in sizes, shapes, and venation patterns of leaves of the trees and herbaceous plants around you. Doing that makes it much more meaningful when you study the microscopic features of those same types of leaves.

Take every opportunity to study the plant life around you, not just while you're doing this group of lab sessions, but throughout the year. In fact, it's a useful project to choose one or a few plants in your yard or a nearby park and then examine them in detail as the seasons change through the course of year. Some of those changes are obvious, such as the changes in the foliage of a deciduous tree. Others are less obvious, such as the changes in your lawn's grass from spring though summer and autumn to winter.

The earliest multicellular organisms evolved in water. Although they are now classified as protista, multicellular algae can be thought of as proto-plants. Some of these aquatic organisms gradually developed structures and mechanisms to transport, store, and manage water, allowing them to survive in drier environments. These were the ancestors of the first land plants.

The earliest land-dwelling plants were probably *mosses* (phylum *Bryophyta*), whose descendants now flourish in any environment where water is readily available. Like their close relatives, the *liverworts* and *hornworts* (also *Bryophytes*), mosses are structurally simple, lacking the vascular systems of more complex plants (*Tracheophytes*). The absence of a vascular system for water transport limits the physical size of mosses, which seldom grow taller than 10 or 15 cm, and usually much less. Figure X-1-1 shows a common moss, *Polytrichum formosum*, growing on a forest floor.

Figure X-1-1: *Polytrichum moss growing on a forest floor*

When haploid spores germinate, they produce *protonemata* (singular *protonema*), a flat or filamentary mass from which the *gametophore* grows. The gametophore is the body of the plant, and is structurally differentiated into rhizomes ("roots"), stalks, and leaves. The lower part of the plant contains gametophyte structures, the female *archegonia* (singular *archegonium*) and the male *antheridia* (singular *antheridium*), shown in Figure X-1-3.

Figure X-1-3: *Moss (Polytricum commune) antheridia (40X)*

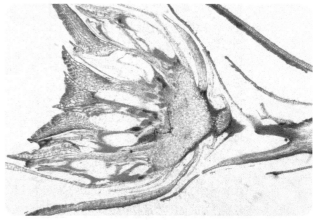

Structurally and reproductively, mosses in many respects more resemble fungi than other plants. Stems are simple herbaceous structures. (Figure X-1-2 shows a cross-section of a moss stem at 400X.) Rather than true roots, mosses employ rhyzoids as *holdfasts* to anchor themselves to their substrates. Rhyzoids do not absorb water or nutrients from the substrate, but are simple physical anchors. Water and nutrients are absorbed directly by leaf-like structures. Mosses lack flowers, and do not produce cones, fruits, or seeds, instead reproducing via spores. In contrast to *Spermatophytes* (seed plants), whose life cycles include a *haploid generation* (ovule and pollen) and a *diploid generation* (the flowering plant itself), mosses are haploid (unpaired chromosomes) for most of the life cycle, with only a short-lived *sporophyte* representing the diploid (paired chromosomes) phase.

The archegonia are small bottle- or pouch-shaped structures at the "leafy" base of the structure, which gather sperm produced by the antheridia. The upper end of the archegonium is constricted like the neck of a bottle to contain captured sperm produced by the antheridia. *Sporophytes* (spore producing structures) consist of long modified stems called *setae* (singular *seta*) atop which sit spore-containing capsules called *calyptrae* (singular *calyptra*). Figure X-1-3 illustrates the life cycle of a typical moss.

Although they are in no sense hybrids, *ferns* (phylum *Pteridophyta*) can be thought of as intermediate between the simple mosses and other *Bryophytes* and the more complex *Spermatophytes* (seed plants). Structurally and in gross appearance (see Figure X-1-5), ferns resemble seed plants (angiosperms and gymnosperms). Unlike mosses, ferns are vascular (have *xylem* and *phloem*), and have true roots, true stems, and true leaves.

Figure X-1-2: *Moss stem, cs (400X)*

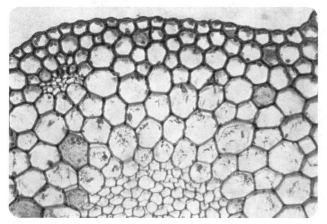

Figure X-1-4: *Life cycle of a typical moss*

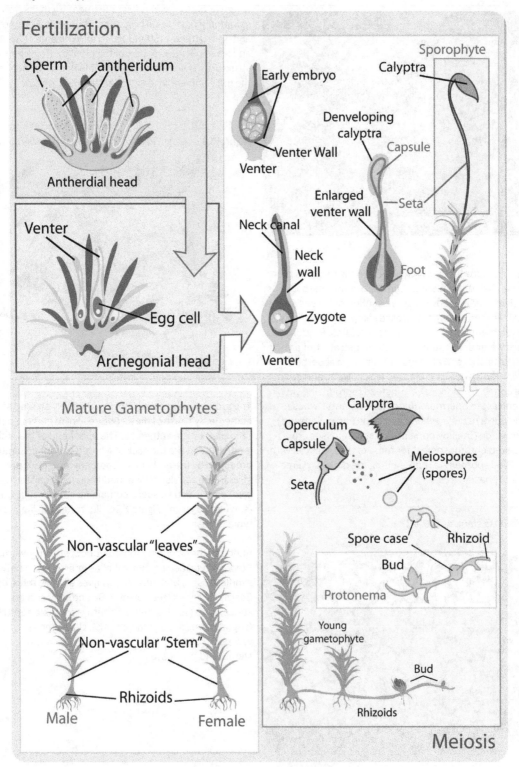

Figure X-1-5: *Cinnamon fern (Osmundastrum cinnamomeum), a typical fern*

Figure X-1-6: *Fern (Dryopteris championii) stem (rhizome), cs (40X)*

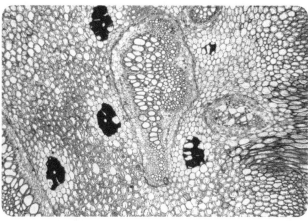

The major components of ferns are similar structurally and functionally to those of seed plants, including:

Roots

While mosses' root-like structures (rhizoids) provide only a physical anchor to the substrate, ferns possess true fibrous roots like those of seed plants, although those roots are actually unicellular rhizoids. Roots absorb water and nutrients from the soil, which the vascular system of the fern transfers to other parts of the plant.

Stems

Again unlike mosses, whose stem-like structures are simple undifferentiated groupings of cells, ferns possess true stems like those of seed plants with differentiated structures comprising specialized cells. (Compare the complex fern stem structure shown in Figure X-1-6 to the simple moss stem structure shown in Figure X-1-2.)

Leaves

Although mosses have leaf-like structures with chloroplasts, they are not true leaves. Ferns, conversely, possess two types of true (vascular) leaves, which are referred to collectively as *megaphylls*. *Trophophylls* are similar in appearance and structure to the leaves on seed plants, and perform the same function: using photosynthesis to produce and store saccharides. *Sporophylls* appear similar to trophophylls, and also engage in photosynthesis, but also produce spores. In that respect, they are functionally similar to the cone scales of gymnosperms and the pistils and stamens of angiosperms.

Reproductively, ferns more resemble mosses than seed plants. Ferns are not flowering plants, so they do not produce seeds. Instead, like mosses, ferns reproduce using spores, although their reproductive life cycle differs from that of mosses. As is true of all vascular plants, including the more complex seed plants, the life cycle of ferns is based upon alternation of generations, with a diploid (paired chromosomes) sporophyte phase and a haploid (unpaired chromosomes) gametophyte phase, as follows:

- Via meiosis, the diploid sporophyte (the fern plant) produces haploid spores

- Via mitosis, the haploid spore produces a haploid gametophyte (the *prothallium* or *prothallus*, shown in Figure X-1-7)

- Via mitosis, the haploid gametophyte produces haploid gametes (eggs and sperm)

- A motile sperm fertilizes a sessile egg (fixed to the prothallus) to produce a diploid zygote

- Via mitosis, the diploid zygote grows to become a diploid sporophyte (the fern plant)

Figure X-1-7: *Fern (Dryopteris-championii) prothallium with young sporophyte (40X)*

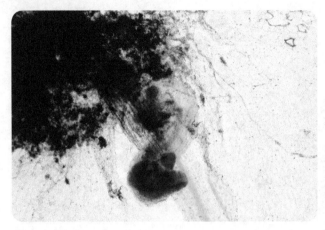

Figure X-1-9: *Fern spores, 100X*

As with mosses, fern sperm are produced by antheridia and eggs by archegonia. A typical prothallus contains multiple antheridia and archegonia. Clusters of sporangia (spore-forming bodies) are called *sori* (singular *sorus*). Sori form on sporophylls, usually as brown or yellow masses on the underside of the leaf. Although it is not present in all fern species, the *indusium* is a membrane that forms a protective cover for the spores contained within the sorus. Figure X-1-8 is a 100X cross-section of a fern sporophyll leaflet showing a sorus with the indusium and spores visible. Figure X-1-9 shows fern spores at 100X.

Figure X-1-8: *Fern sporophyll leaflet cs showing sorus with indusium and spores (100X)*

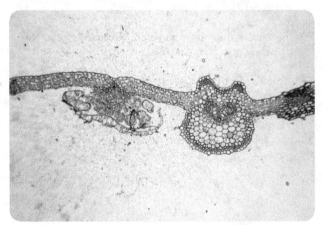

Although you can purchase live specimens of various mosses and ferns from Carolina Biological Supply or another vendor, there's really no need to do so. Mosses are ubiquitous, can easily be gathered at any time of year, and can be dried to preserve them indefinitely. If for some reason you can't gather wild moss, you can purchase sphagnum moss inexpensively at a garden center or DIY store. (Make sure to get "sphagnum moss," which is dried live moss, rather than "peat moss" or "sphagnum peat moss," which is rotted sphagnum gathered from peat bogs.) You can also purchase sphagnum moss from pet stores, where it is sold for use in terrariums. Ferns can easily be gathered in the wild, or you may know someone who keeps them as house plants. You'll find the following prepared slides—which you can purchase or make yourself—useful for this session:

- Moss – stem; protonema; antheridium; and archegonium

- Fern – rhizome; prothallium; sporophyte; sori; and spores

If you prepare your own slides, you can experiment with different stains. We suggest starting with eosin Y and methylene blue, both of which are included in the standard kit.

PROCEDURE X-1-1: OBSERVING MOSS STRUCTURES

To begin, use the magnifier (or a stereo microscope, if you have one) to examine closely your moss specimen or specimens.

1. Note the color and gross morphology of each specimen. Compare any similarities and contrast any differences in your specimens, and record your observations in your lab notebook.

2. Using the Internet or printed reference material, attempt to identify each of the specimens to at least the genus level.

3. For each of your specimens, examine and identify any structures visible at low magnification, including gametophyte structures (rhizoids, stems, and leaves) and sporophyte structures (seta and calyptra). Note the differences between mature male and female gametophytes. If you have a camera with macro capability, record images of the various structures.

4. Using the forceps, carefully remove a leaflet, transfer it to a flat slide, add a drop of water, and observe the specimen at low and medium magnification. and make a wet mount

5. Using the teasing needle and forceps, carefully separate a sporophyte from its gametophyte. Use the scalpel to cut the seta just below the calyptra and transfer the calyptra to a flat slide.

6. Add a drop of water, and use the teasing needle gently to crush the calyptra and release the spores it contains. Position a coverslip and scan the slide at 40X to locate a cluster of spores. Switch to high-dry magnification to observe the spores. Make a sketch or shoot an image of the spores.

7. Using either purchased prepared slides or slides that you have prepared and stained yourself, observe the following structures at suitable magnifications: protonea, rhizoids, stems and leaves, antheridia, and archegonia. Sketch each of the structures or shoot an image of it.

PROCEDURE X-1-2: OBSERVING FERN STRUCTURES

To begin, examine closely your fern specimen or specimens with your naked eye, followed by examination with the magnifier or a stereo microscope.

1. Note the overall appearance and major structural features of each specimen. Compare any similarities and contrast any differences in your specimens, and record your observations in your lab notebook.

2. Using the Internet or printed reference material, attempt to identify each of the specimens to at least the genus level. (The presence or absence of sori, along with their positions, colors, shapes, sizes, and so on are important clues to identifying fern species.)

3. For each of your specimens, examine and identify any structures visible at low magnification, including gametophyte structures (rhizoids, stems, and leaves) and sporophyte structures (seta and calyptra). Note the differences between mature male and female gametophytes. If you have a camera compatible with your microscopes, record images of the various structures.

4. Using the teasing needle and forceps, carefully separate a sporophyte from its gametophyte. Use the scalpel to cut the seta just below the calyptra and transfer the calyptra to a flat slide.

5. Add a drop of water, and use the teasing needle gently to crush the calyptra and release the spores it contains. Position a coverslip and scan the slide on low magnification to locate a cluster of spores. Switch to high-dry magnification to observe the spores. Make a sketch or shoot an image of the spores.

6. Using either purchased prepared slides or slides that you have prepared and stained yourself, observe the following structures at suitable magnifications: protonea, rhizoids, stems and leaves, antheridia, and archegonia. Sketch each of the structures or shoot an image of it.

REVIEW QUESTIONS

Q1: Why might we group mosses and ferns together as "ancient plants"?

_____

_____

_____

_____

_____

Q2: What similarities and differences, if any, did you observe between the moss leaf and the fern leaf?

_____

_____

_____

_____

_____

Q3: What are the female and male reproductive structures in mosses and ferns called?

_____

_____

_____

_____

_____

_____

Q4: How do mosses differ from lichens?

_____

_____

_____

_____

_____

Q5: Mosses seldom grow more than 10 to 15 cm tall. Ferns commonly reach heights of 10 meters or more. What structural feature explains this difference?

_____

_____

_____

_____

_____

Investigating Seed Plants

Lab X-2

EQUIPMENT AND MATERIALS

You'll need the following items to complete this lab session. (The standard kit for this book, available from *www.thehomescientist.com*, includes the items listed in the first group.) Plant the carrot seeds a couple of weeks before you intend to do the rest of this lab session to allow the seedlings to sprout.

MATERIALS FROM KIT

- Goggles
- Coverslips
- Forceps
- Magnifier
- Pipettes
- Scalpel
- Seeds, carrot
- Slides, flat
- Stain: eosin Y
- Stain: methylene blue
- Teasing needles

MATERIALS YOU PROVIDE

- Gloves
- Alcohol (ethanol or isopropanol)
- Carrot
- Flowers (see text)
- Foam cups
- Fruits (see text)
- Knife, paring
- Leaves, monocot & dicot (fresh or dried)
- Leaves, hydrophyte (aquatic plant)
- Microscope
- Microscope, stereo (optional)
- Microtome (optional)
- Paper towels
- Pencil
- Seeds (see text)
- Slides, prepared, dicot structures
- Slides, prepared, monocot structures
- Slides, prepared, gymnosperm structures
- Sand or vermiculite
- Soil (potting or ordinary garden soil)

BACKGROUND

In the preceding lab session, we took a close look at "primitive" plants, mosses and ferns. In this and the following lab session, we'll examine *spermatophytes*, or *seed plants*. Spermatophytes are divided into two major groups: *angiosperms* (flowering plants) and *gymnosperms* (conifers and similar species). Although it's a mistake to think of seed plants as "higher" than mosses and ferns, there's no question that seed plants are structurally more complex than mosses and ferns.

When a seed germinates, the first structure to appear is the primary root, which then branches into a secondary root system. Roots serve to anchor the plant physically to the soil or other substrate and also to absorb water and nutrients. In some species, the root may also serve as a food storage location.

Roots are classified as one of four types: *fibrous roots*, *tap roots*, *adventitious roots*, or *aerial roots*. In a fibrous root system, secondary roots branch from the primary root, eventually reaching similar size. In a tap root system, the primary root becomes the tap root, which is typically much longer and thicker than the secondary roots branching from it. The tap root is longer and heavier because it is a *tuber*, which means that food is stored within it. (Stems in some species also function as tubers.) An adventitious root system is one in which secondary roots develop and grow not only from the primary root, but from other parts of the plant. (For example, if you examine the base of a corn stalk, you'll find that secondary roots are growing from the above-ground portion of the stalk, as well as from the primary root.) Finally, an aerial root system is just what it sounds like: a root system that is partially or completely above ground. As odd as that sounds, aerial roots (which are nearly always also adventitious) are found in many species, ranging from orchids to ivies to mangroves.

> Although we don't explore the various types of roots in this lab session, you should use your textbook or Internet resources to familiarize yourself with appearance and characteristics of the various root types.

Aerial roots aside, you can think of stems as basically above-ground roots. Their purpose is to support the leaves and to connect the roots to the leaves. The leaves, of course, are where most of the work is done, using water and nutrients supplied by the roots and stems and combining those with atmospheric carbon dioxide via photosynthesis to produce the saccharides that feed the plant. Leaves vary widely in appearance, from the familiar leaves of trees and bushes to the needles of pine trees to the blades of grass that make up your lawn. Regardless of their appearance, they serve the same function of providing food for the plant (and, ultimately, for the animal kingdom as well).

Transfers of water and nutrients from the root system to the leaves and transfers of saccharides from the leaves to the stems and roots take place via the *vascular system* of the plant, which is analogous to the arteries and veins that make up the circulatory system in animals. Plants have two types of transport tissue, both of which are found throughout the plant. The primary purpose of *xylem* tissue is to transport water within the plant, although it does also transport some nutrients. The most familiar form of xylem tissue is wood. The primary purpose of *phloem* tissue is to transport soluble nutrients via a process called *translocation*, which of course also involves transporting water. The most familiar form of phloem is the inner bark of trees, which lies between the protective outer bark layer and the wood that makes up the trunk (stem).

Although phylogenetics tells us that the concept is not valid for classification, most biologists find it convenient organizationally to group angiosperms into *monocot* and *dicot* species. Those terms refer to the number of cotyledons, or embryonic leaves, contained in the seed, with monocots having only one and dicots two. Although this difference may sound trivial, its implications are profound for the structures of mature plants, which reflect their origins as monocotyledonous or dicotyledonous.

In this session, we'll look at the primary structures of seed plants: roots, stems, leaves, and reproductive structures.

PROCEDURE X-2-1: OBSERVING GERMINATION OF A SEED PLANT

In this procedure, we'll plant carrot seeds and observe their germination. Once the seedlings have sprouted, we'll examine the structures of the seedlings.

1. Use a pencil or similar object to poke a few small drainage holes in the bottoms of two foam cups. Cover the holes with pieces of paper towel to keep soil in the cups.

2. Label one foam cup "A" and fill it to near the brim with clean sand or vermiculite. Label a second foam cup "B" and fill it to near the brim with potting soil or ordinary garden soil.

3. Place a few carrot seeds just under the surface of the soil in each cup.

4. Water the soil with tap water until it is thoroughly dampened.

5. Place the cups in an area where they will be exposed to daylight but not excessive direct sunlight, and where they will remain near room temperature.

6. Observe the cups at least daily, and record your observations in your lab notebook. In particular, note how long it takes for the first visible evidence of germination to occur and the initial appearance of the seedlings.

7. Water the soil gently every day or two, dampening it each time but not using enough water to turn the soil into mud. Continue observing the cups over a period of two to three weeks, noting any differences between the two cups.

8. After two to three weeks, gently remove one of the seedlings from each cup, making sure not to damage the visible part of the seedling or its root structure. Wash the seedlings gently with a trickle of tap water to remove any soil that is adhering.

9. Use the magnifier (or a stereo microscope, if you have one) to observe the structures of the seedling. Note the size and appearance of the roots, root hairs, stems, and leaves, and note any differences between the seedlings from the two cups.

PROCEDURE X-2-2: OBSERVING ROOT STRUCTURES

In this procedure, we'll examine root structures, comparing and contrasting the similarities and differences between monocots and dicots. To begin, we'll examine the root structure of *Daucus carota sativus*, the garden carrot. The tuberous root structure of the carrot is the part that is ordinarily consumed, and what most people think of as the "carrot" itself.

1. Obtain an ordinary carrot from a grocer. Choose a large one, if possible, to make the structures more easily visible.

2. Using the scalpel, carefully cut the carrot across its diameter to produce a piece about 10 cm long.

3. Using the scalpel, carefully cut from one end of that piece the thinnest possible cross-section, ideally thin enough to be almost transparent.

4. Transfer the cross-section to a flat slide. Add a drop or two of methylene blue stain (sufficient to cover the section) and a similar amount of eosin Y stain.

> If you have a microtome, use it to cut the cross-section. It's much easier to obtain very thin and consistent sections using a microtome than if you cut the section by hand.

5. Allow the stains to work for a minute or so, and then rinse off excess stain by flooding the specimen with alcohol from a pipette.

6. Observe the stained cross-section with a magnifier or stereo microscope, noting the *stele* (the core, which contains the xylem and phloem components of the vascular system), the *cortex* (surrounding material where food is stored), and the *epidermis* (skin covering the cortex), and any visible substructures within them. Note your observations in your lab notebook.

7. Use the scalpel carefully to cut a longitudinal (lengthwise) section of the carrot. Repeat steps 4 through 6 to observe the stained longitudinal section.

We'll next examine prepared slides of monocot and dicot root structures, and compare and contrast the analogous structures in each type. (Figure X-2-1 shows an example of a soybean cross-section.) As you observe the slides, use appropriate magnifications. For large specimens, all features may not be visible in one field of view even at low magnification, so scan around the slide as necessary to view all points of interest. Use medium and high magnification, as appropriate, to view details. As you work, record your observations in your lab notebook. Make sketches or shoot images to document what you find.

> Using prepared slides that contain monocot and dicot structures side-by-side on the same slide makes it easier and faster to compare and contrast the structures. Such slides are available commercially, but most prepared slide sets include monocot and dicot structures on different slides. That's actually a very good reason to make your own slides.
>
> If you have a microtome (or are very skilled at cutting freehand sections), you can prepare your own monocot and dicot sections and stain them using methylene blue and eosin Y. Onions, garlic, and many flowers (such as amaryllis, daffodils, irises, lilies, and tulips) are examples of monocots that are readily available locally. Most garden vegetables—such as beans, peas, pumpkins, squash, and tomatoes—are dicots, as are many common flowers, including daisies, geraniums, marigolds, roses, and snapdragons.

1. Place a prepared slide of a dicot root cross-section on the microscope stage and scan it at low magnification to locate areas of interest. Identify the stele, cortex, epidermal layer, and root hairs. Note the details of any structural features present in each.

> For a large dicot specimen, the cross-section may be too large to view in one field. For a small specimen, you may need to start at medium magnification rather than low.

2. Examine the epidermis and root hair at medium and high magnification, noting the distribution, size, and shape of the cells and the cell wall thickness(es).

Figure X-2-1: *Soybean (Glycine max) root tubercle cs, 40X*

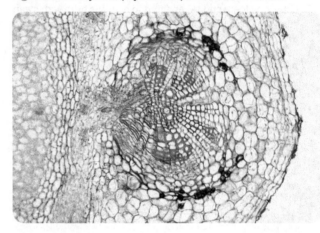

3. Observe the cortex at medium or high magnification and note the differences, if any, between the size, shape, distribution, and wall thickness of cortex cells versus epidermal cells and root hairs.

4. Center the stele in the field of view and observe it at low magnification to note areas of interest. Identify the xylem (thick-walled) and phloem (thin-walled) cells and observe them at medium or high magnification to note their size, structure, and arrangement.

> The cross-shape arrangement of the larger, thick-walled xylem cells and the position of the smaller, thin-walled phloem cells between the arms of that cross are characteristic of dicot roots.

5. Examine any other dicot root cross-sections for which you have slides available, and note the similarities between dicot roots of different species.

6. Place a prepared slide of a monocot root cross-section on the microscope stage, and repeat steps 8 through 12 to observe and note its characteristics.

Figures X-2-2 and X-2-3 show root cross-section examples.

Monocot stems are also characterized by the position and arrangement of their xylem and phloem cells. In monocots, both xylem and phloem cells are scattered in groups or clusters throughout the stele.

Figure X-2-2: *Corn (Zea mays) root cs, 100X*

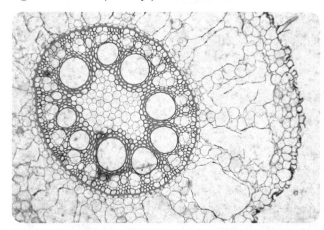

Figure X-2-3: *Potato (Solanum tuberosum) section, 40X*

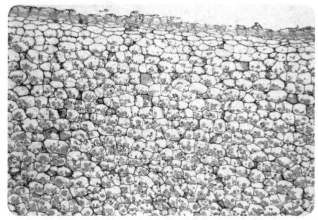

PROCEDURE X-2-3: OBSERVING STEM STRUCTURES

As is true of root structures, stem structures have characteristic differences between monocot and dicot species. In this procedure, we'll examine prepared slides of monocot and dicot stem structures, and compare and contrast the analogous structures in each type. Again, as you observe the slides, use appropriate magnifications, scanning at lower magnification and then switching, as appropriate, to higher magnification to view details. As you work, record your observations in your lab notebook. Make sketches or shoot images to document what you find.

1. Place a prepared slide of a dicot stem cross-section on the microscope stage, and scan it at low magnification to locate areas of interest. Identify the pith (central section), vascular bundles, cortex, and epidermal layer. Note the details of any structural features present in each.

2. Examine the epidermal layer and cortex, and compare them to the analogous structures in the dicot root specimen(s). Note that the cortex is a relatively thin layer separating the pith from the epidermal layer. Note also the arrangement of vascular bundles in a ring toward the outer edge of the pith. This arrangement is characteristic of a dicot stem.

3. Center a vascular bundle in the field of view, and change to higher magnification to observe the details. Note that the vascular bundle comprises a group of larger, thick-walled xylem cells surrounding the smaller, thin-walled phloem cells.

4. Place a prepared slide of a monocot stem cross-section on the microscope stage, and repeat steps 1 through 3 to observe and note its characteristics. Figure X-2-4 shows monocot and dicot stem cross-sections side-by-side for comparison.

Figure X-2-4: *Monocot (left) and dicot stems compared, 40X*

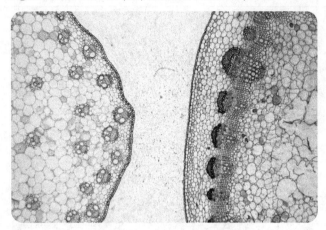

Figure X-2-6: *Corn (Zea mays) stem cs detail, 400X*

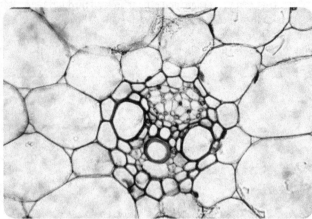

5. Examine as many prepared slides or specimens as you have available of dicot and monocot stems in cross-section and longitudinal section. Compare and contrast the similarities and differences with your other dicot and monocot specimens.

Figures X-2-5 through X-2-9 show the monocots *Zea mays* (corn) and *Triticum aestivum* (wheat) in cross- and longitudinal sections. Figures X-2-10 through X-2-13 show the dicots *Cucurbita* sp. (squash or pumpkin) and *Hibiscus syriacus* (hibiscus) stems in various cross-sections and longitudinal sections.

Figure X-2-7: *Corn (Zea mays) stem ls, 40X*

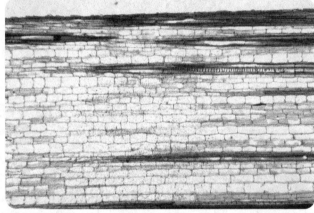

Figure X-2-5: *Corn (Zea mays) stem cs, 40X*

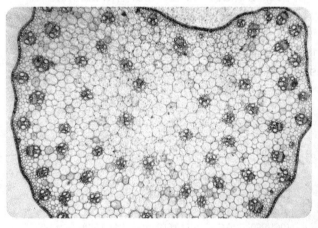

Figure X-2-8: *Wheat (Triticum aestivum) stem cs, 40X*

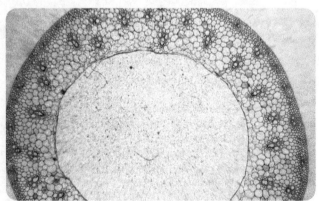

Figure X-2-9: *Wheat (Triticum aestivum) stem cs detail, 100X*

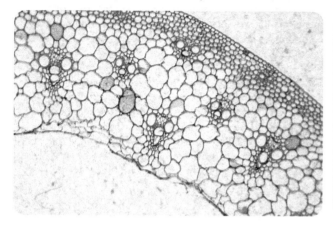

Figure X-2-12: *Hibiscus (Hibiscus syriacus) stem cs, 40X*

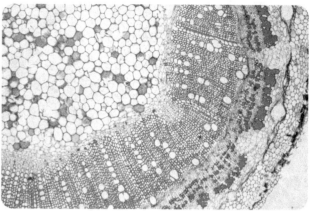

Figure X-2-10: *Squash (Cucurbita sp.) stem cs, 40X*

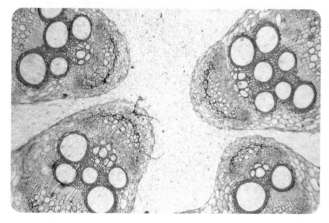

Figure X-2-13: *Hibiscus (Hibiscus syriacus) stem ls, 40X*

Figure X-2-11: *Squash (Cucurbita sp.) stem ls, 40X*

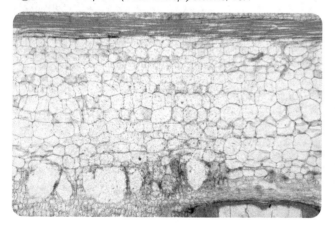

1. Many angiosperms have herbaceous (soft) stems, which die off each year. Other angiosperms like trees and bushes have woody (hard) stems, which remain alive from year to year, growing by adding annual layers, called growth rings. If you have prepared slides or specimens available, compare and contrast the similarities and differences in the structures of one or more angiosperm woody stems with those you've examined of angiosperms with herbaceous stems. If possible, examine cross-sections of woody stems of different ages and note the growth rings. Figures X-2-14 through X-2-16 are cross-sections of basswood stems showing annual growth rings.

Figure X-2-14: *Basswood (Tilia sp.) one-year stem cs, 40X*

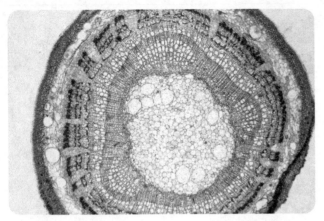

Although the majority of seed plants that surround us are angiosperms (flower-bearing seed plants), gymnosperms (cone-bearing seed plants) are also familiar to all of us, most commonly as conifers such as pines, cedars, cypresses, and related species.

1. If you have prepared slides or specimens available, compare and contrast the similarities and differences in the woody stem structures of one or more angiosperms against those of one or more gymnosperms. If possible, examine cross-sections of woody stems of different ages and note the growth rings. Figures X-2-17 through X-2-21 are cross-sections of basswood stems showing annual growth rings.

Figure X-2-15: *Basswood (Tilia sp.) two-year stem cs, 40X*

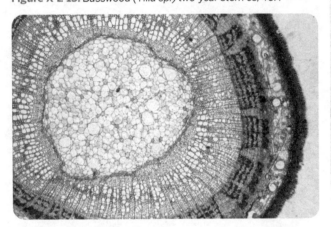

Figure X-2-17: *Cedar (Cedrus sp.) stem cs, 40X*

Figure X-2-18: *Cedar (Cedrus sp.) stem cs detail, 100X*

Figure X-2-16: *Basswood (Tilia sp.) three-year stem cs, 40X*

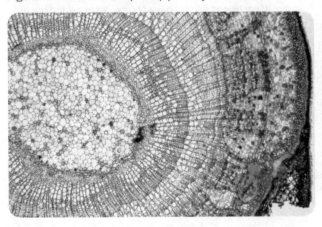

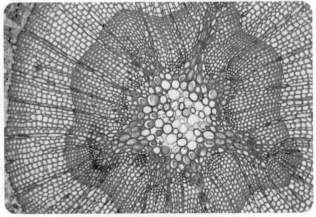

Figure X-2-19: *Young pine (Pinus sp.) stem cs, 40X*

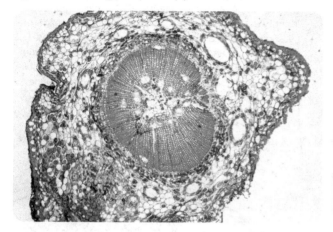

Figure X-2-21: *Older pine stem showing annual growth ring, 40X*

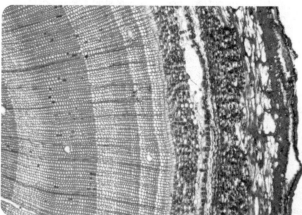

Figure X-2-20: *Young pine (Pinus sp.) stem cs detail, 100X*

PROCEDURE X-2-4: OBSERVING LEAF STRUCTURES

Like root and stem structures, leaf structures exhibit characteristic differences between monocot and dicot species, most obviously shape and venation pattern. In this procedure, we'll examine the macroscopic and microscopic characteristics of monocot and dicot leaf structures, and compare and contrast the analogous structures in each type. As you work, record your observations in your lab notebook. Make sketches or shoot images to document what you find.

1. Obtain as large a variety as possible of fresh and/or preserved leaves from monocot and dicot species.

2. Using your naked eye and the magnifier or a stereo microscope, examine the gross structural features of each specimen. Examine both the top and bottom surfaces. Note the size, shape, and arrangement of the leaf, the

margin (edge) pattern, the vascular system arrangement, and the size, pattern, and density of stomata. Typical monocot leaves are narrow and have parallel venation (see Figure X-2-22). Typical dicot leaves are broader and have branching venation (see Figure X-2-23).

> Use your textbook or another reference source to locate a chart of leaf morphology. You can find an excellent leaf morphology graphic at *http://en.wikipedia.org/wiki/File:Leaf_morphology.svg*.

Figure X-2-22: *Parallel vascular structure of a monocot (lawn grass) leaf wm, 40X*

Figure X-2-23: *Branched vascular structure of a dicot (silver maple) leaf wm, 40X*

3. Examine prepared slides of cross-sections of monocot (Figure X-2-24) and dicot (Figure X-2-25) leaves. For each, identify the following structural features: cuticle, epidermal cells, mesophyll cells, air spaces, vascular bundles and their sheathes, and stomata and their guard cells. For your dicot specimen, locate and identify spongy mesophyll cells and palisade mesophyll cells. Contrast their appearance with the monocot mesophyll cells.

Figure X-2-24: *Typical monocot leaf cs, 100X*

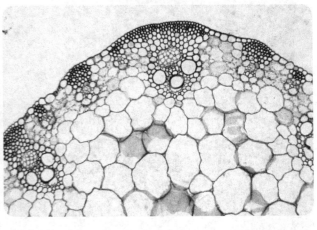

Figure X-2-25: *Typical dicot leaf cs showing obvious spongy and palisade mesophyll cells, 100X*

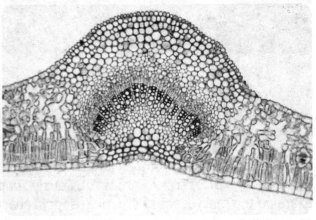

4. Examine as many other monocot and dicot leaf cross-sections as you have access to, comparing and contrasting their structures, and attempt to identify the specimens as monocot or dicot based on their appearances in cross-section. The correct answer isn't always obvious. For

example, Figure X-2-26 shows a camellia (dicot) leaf cross-section, in which the spongy and palisade mesophyll cells aren't as obvious as in the preceding image.

Figure X-2-26: *Camellia (Camellia sp.) leaf cs, 100X*

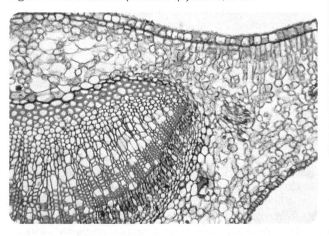

5. Examine as many whole-mount leaf epidermis slides as you have available. Use low magnification to find the significant features and then increase magnification to view details of the venation, stomata, chloroplasts, and so on. Compare and contrast the similarities and differences between your specimens. Examine two specimens of the same type of leaf, one that has recently been exposed to bright sunlight and the other kept in the dark. Note any differences in the appearances of the stomata. Figures X-2-27 through X-2-30 are whole-mount epidermis images of various monocot and dicot species.

It's easy to make your own whole-mount epidermis slides from leaves you obtain locally. Simply cut two small pieces of the leaf and make a wet mount with one piece obverse-side up and the other reverse-side up. If the leaves curl so much that a simple wet mount with a coverslip is insufficient to keep them flat, substitute a second microscope slide for the coverslip.

Many leaves are thin enough that they can be observed by transmitted light if you set the illuminator to its brightest setting. Others are opaque and will require reflected light. Use a high-intensity desk lamp or other top illumination to view those leaves. (We use an LED Mighty Bright book-reading light, which can be clamped to the microscope stage and has a flexible stalk that allows the light to be directed as necessary.)

Figure X-2-27: *Wheat (Triticum aestivum) leaf epidermis wm showing stomata, 100X*

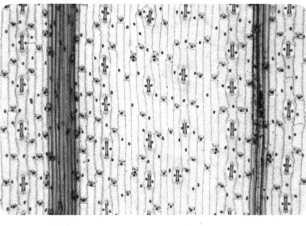

Figure X-2-28: *Iris (Iris sp.) leaf epidermis wm showing stomata, 400X*

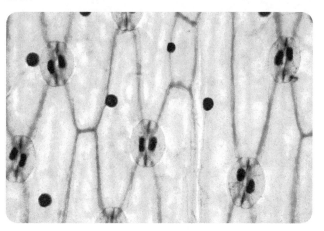

Figure X-2-29: *Stonecrop (Sedum sp.) leaf epidermis wm showing stomata, 400X*

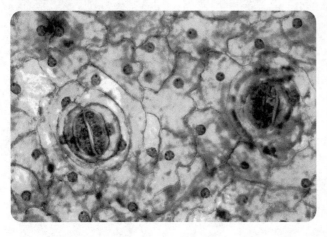

Figure X-2-30: *Broad bean (Vicia faba) leaf epidermis wm showing stomata, 400X*

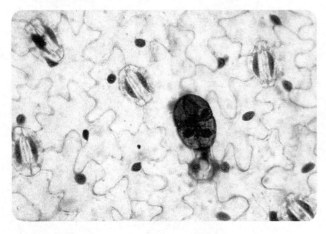

Aquatic plants (also called *hydrophytic plants* or *hydrophytes*) are those that have adapted to living in aquatic environments, whether freshwater or saltwater. True aquatic plants are vascular plants (which excludes algae), and may be ferns or angiosperms, both monocots and dicots.

Hydrophyte structures—roots, stems, and leaves—often differ significantly from those of land plants. Hydrophytic roots, for example, are often small and feathery compared to those of land plants, because they don't have to anchor or support the plant and because water can diffuse directly into the leaves rather than being supplied by the root system. Similarly, stems are often smaller and finer than those of land plants, because most of the mass of the plant is supported by the surrounding water. In floating aquatic plants, roots or stems (or both) contain air channels, which both aid buoyancy and provide a channel for exchanging gases between the surfaced and submerged portions of the plant.

But the greatest differences between aquatic plants and land plants are often visible in the leaves, which are highly adapted in most hydrophytes to suit them for their environments, submerged in or floating on liquid water. The cuticle in most hydrophyte leaves is very thin because a thick cuticle is needed only to conserve water, which is not an issue for hydrophytes. Similarly, the stomata in most hydrophytic leaves are numerous, present on both surfaces of the leaves, and open all or nearly all the time, in contrast to land plants, which close their stomata to conserve water when photosynthesis is not occurring. Finally, hydrophytic leaves are often relatively large and flat and contain air pockets and air-trapping hairs, all of which aid in increasing buoyancy to support the plant.

You can obtain leaves from an aquatic plant from any nearby pond or aquarium. Common species include various water lilies, duckweed, water cabbage, lotuses, and various aquatic grasses. Many species of aquatic plants are fast-growing, invasive, and considered weeds, so you'll probably have no trouble getting permission to harvest specimens.

1. To begin, use the magnifier or a stereo microscope to observe the macroscopic features of your aquatic plant leaf specimen(s). Observe both surfaces of the leaf, and pay particular attention to the number, size, and distribution of stomata on the surfaces.

2. If you have a prepared slide available of a *hydrophytic leaf* cross-section, examine it and compare and contrast its structural features against those of your land plant specimens. Compare and contrast the structural similarities and differences against your section slides of land plant leaves. Figure X-2-31 shows a hydrophytic leaf in cross-section at 100X.

Figure X-2-31: *Hydrophytic leaf (Nymphaea tetragona georgi) cs, 40X*

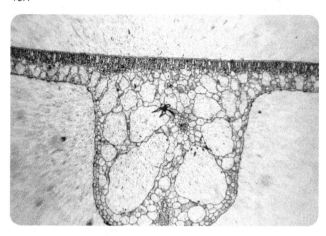

Like other seed plants, gymnosperms have leaves with full vascular systems and complex internal structures. The difference from angiosperms is that the leaves of most gymnosperms are highly modified. For example, leaves of the most familiar gymnosperms, the pine family, are needle-shaped.

1. Obtain a gymnosperm leaf specimen and examine it with the magnifier, with a stereo microscope, or with a standard microscope at low magnification using reflected light. Identify the major structural features, and compare and contrast them with your angiosperm leaf specimens.

2. Using prepared slides or cross and longitudinal sections you prepare yourself, compare and contrast your gymnosperm leaf sections against both your angiosperm leaf sections and your gymnosperm root and/or stem specimens. Figure X-2-32 shows a cross-section of a pine leaf (needle) at 100X.

Figure X-2-32: *Pine (Pinus sp.) leaf (needle) cs, 100X*

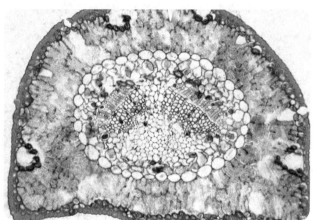

PROCEDURE X-2-5: OBSERVING REPRODUCTIVE STRUCTURES

In this procedure, we'll dissect and examine the reproductive structures of angiosperms: flowers, fruits, and seeds. Flowers are the primary reproductive structures. Flowers are hugely diverse in shape, color, size, and other characteristics, but all produce the fruits and seeds from which new plants originate.

It's useful to have a wide selection of live flower, fruit, and seed specimens available. Here are some good choices for each:

Flowers
    Any live flower can be used for dissection, but it's easier to work with flowers whose internal structures are relatively large. (Flowers like carnations and daisies, despite their large petals, have small internal structures.) Gladioli are a good choice, as are azaleas, daffodils, lilies, roses, or tulips. You can find suitable specimens at a florist, the supermarket, or perhaps in your yard.

Fruits
    If possible, have a range of fruit types available for dissection. Good candidates include a berry (grape, tomato, or banana) and a pepo (melon, squash, or pumpkin), a drupe (apricot, cherry, nectarine, olive, peach, or plum) and a polydrupe (strawberry or blackberry), a hesperidium (lemon, lime, or orange), and a pome (apple or pear).

Seeds

The easiest seeds to dissect are those that are large enough to be easily manipulable and lack a hard shell. You will probably obtain some seeds suitable for dissection when you dissect the fruit specimens. If you want additional specimens you can easily obtain them at the supermarket. Good candidates include beans, peas, popcorn, whole pumpkin or sunflower seeds, peanuts in the shell, pecans, and pistachios.

1. Begin by examining your flower specimen with your naked eye and the magnifier. Using your textbook or another reference source, identify as many as possible of the following structural features: anther, filament, ovary, ovule, petals, pistil, pollen, receptacle, sepal, stamen, stigma, and style. (Not all of these features will be visible in all flowers.) As you work, record your observations in your lab notebook. Make sketches or shoot images to document what you find.

2. Determine where possible if each flower: is monocot or dicot; is complete or incomplete; is composite or simple; is male, female, or both; possesses a superior or inferior ovary; and possesses a single ovule or multiple ovules.

3. Using the scalpel and forceps, carefully section the flower by cutting vertically through the ovary area from the top of the flower downwards. Observe the structures revealed by this longitudinal section, and identify as many as possible of the structures and characteristics mentioned in steps 1 and 2.

4. Repeat steps 1 through 3 with any other flower specimens you have available.

5. View prepared slides of flower structures, particularly the carpel and stamen and their substructures. Figure X-2-33 shows a *Lilium* anther (part of the stamen) in cross-section at 40X. Also view at least one slide of pollen, like the *Lilium* pollen shown in Figure X-2-34.

> If you don't have a prepared slide of pollen available, it's easy to make your own. Simply touch the stamen tip of a live flower with your forceps, transfer the pollen that adheres to it to a flat slide, add a drop of water, and position a coverslip. Pollen from different species varies greatly in appearance and size, from less than 5 μm to more than 250 μm. (At about 120 μm, the *Lilium* pollen shown in Figure X-2-34 is actually rather large; pollen from other species may appear almost dust-like even at 400X.)

Figure X-2-33: *Lily (Lilium sp.) anther cs, 40X*

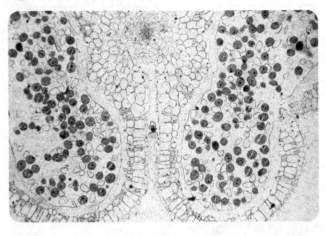

Figure X-2-34: *Lily (Lilium sp.) pollen wm, 400X*

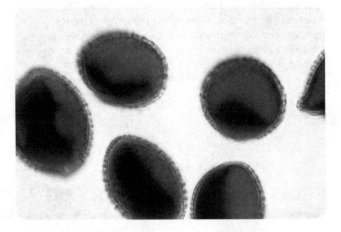

The function of flowers is to produce fruits, which are ripe ovaries that contain a seed or seeds. Fruits are classified into two major types: *Simple fruits* develop from one ovary, which contains one or more chambers that contain ovules and (rarely) other flower parts. *Compound fruits* develop from multiple ovaries, from one or multiple flowers. Each of those ovaries contains one or more chambers that contain ovules and (usually) other flower parts. Compound fruits are further subdivided into *aggregate fruits*—in which one flower has several ovaries, each of which develops into a small fruit that is joined tightly with the fruits of other ovaries to produce a larger fruit cluster—and *multiple fruits*, in which the ovaries of multiple flowers develop into small fruits that fuse into a larger fruit.

In berries and drupes, the seeds are surrounded by the *pericarp*, which forms the edible portion of the fruit. The pericarp is typically divided into three distinct layers.

Exocarp

The *exocarp* (sometimes called the *epicarp* or, in citrus fruits, the *flavedo*) is the outer layer, the skin or peel of the fruit.

Mesocarp

The *mesocarp* (sometimes called the *sarcocarp* or, in citrus fruits, the *albedo* or *pith*) is the middle layer, which often makes up the bulk of the fruit and is usually the edible part.

Endocarp

The *endocarp* is the inner layer that surrounds the hollow ovary and seeds, bounded on the exterior by the mesocarp and on the interior by the *endopericarpal layer*, which contains the seeds.

In *accessory fruits*—such as apples, pears, strawberries, and figs—some or all of the flesh derives from structures other than the ovaries, such as the *receptacle*, *calyx*, or *hypanthium*. In the fruits of grasses—such as barley, rice, and wheat—the pericarp and seed wall are fused into one layer.

1. Begin by examining a whole fruit specimen with your naked eye and the magnifier, if necessary. Locate the *pedicel* (stem) and carefully use the paring knife and/or scalpel, as appropriate, to cut a vertical section through the entire fruit, transecting the pedicel.

2. Using your textbook or another reference source, identify as many as possible of the following structural features, not all of which are present in all fruits: endocarp, endopericarpal layer, exocarp, mesocarp, ovary (including outer and inner layers, if present), ovary wall, pedicel, receptacle, and seeds. Also note the presence or absences of any remaining flower parts.

3. Repeat steps 1 and 2 with any other fruit specimens you have available.

A typical seed consists of the plant *embryo* surrounded by *endosperm* (food for the embryo) with those components in turn surrounded by the *seed coat*, which contains and protects the embryo and endosperm. The embryo is a multicellular diploid structure that represents the plant in its earliest stage of development, following initial cell division but preceding germination. Dissecting seeds gives us the opportunity to view these structures *in situ*.

WARNING

Use extreme care when dissecting seeds, particularly those that are small and/or have a hard seed coat. Work on a nonslip surface, and always use forceps to hold the seed while you are dissecting it.

Some seeds, particularly nuts and pits, have extremely resistant seed coats. For safety, you may prefer to open these seeds as carefully as possible using a nutcracker. The cutting blade on needle-nose pliers also works well for this purpose.

Many seeds are easier to dissect if you soak them first in warm water to soften the seed coat.

1. Begin by examining a whole seed specimen with your naked eye and the magnifier. Identify the point where the seed was attached to the ovary, called the *hilum*, and the point where the pollen tube entered the ovule, called the *micropyle*.

2. Working under a stereo microscope, if available, or with the magnifier, grip a seed with the forceps and carefully use the scalpel or a single-edge razor blade to section the seed longitudinally. (You may find it helpful to recruit an assistant to hold the seed with forceps while you use the magnifier with one hand and use the scalpel with the other.) For some seeds, this section will suffice to reveal all of the structures of the seed. If not, section additional seeds into cross and lateral sections.

Some seeds, particularly nuts, may require more than one or two simple sections to reveal all of their structures. You may need to "disassemble" such seeds piece-by-piece to view the details of their structures.

3. Identify the endosperm, *cotyledon* or cotyledons (embryonic pseudoleaves), and embryo. Within the embryo, locate and identify the *radicle* (embryonic root), *hypocotyl* (embryonic stem), and *epitcotyl* (embryonic true leaves).

4. Repeat steps 5 and 6 with any other seed specimens you have available, comparing and contrasting their structures with the other seeds you have dissected.

REVIEW QUESTIONS

Q1: Your two seedlings were grown in different environments. The sand or vermiculite contains no nutrients, while the potting or garden soil contains nutrients. Did you expect to see any differences between the seedlings from the two environments? Why or why not? Did you observe any differences?

Q2: What structures were you able to identify in your carrot seedlings?

Q3: Based on your observation of the seedlings, is the carrot a monocot or dicot? Why?

Q4: The arrangement of xylem and phloem cells in monocots and dicots differs, and is characteristic of each type. Did you observe any similar characteristic differences between monocots and dicots in their epidermal or cortex cells?

Q5: What adaptation to root epidermal cells increases their surface area? What is the benefit of this increase in surface area?

_____

_____

_____

Q6: The cross-sections of both monocot and dicot roots and monocot and dicot stems show that each of the four has an epidermal layer that bounds the structure, and yet the functions of those layers differs between roots and stems. Knowing what you do about plants, which specific characteristic do you think differs between roots and stems? What structural difference(s) between roots and stems supports your speculation?

_____

_____

_____

Q7: What primary difference did you observe between dicot and monocot stems?

_____

_____

_____

Q8: When you examined whole mounts of similar leaves kept in sunlight and darkness, what major difference did you observe? What does that difference indicate?

_____

_____

_____

Q9: When you examined the whole mounts and cross-sections of hydrophytic plant leaves and compared them to those of land plants, what significant differences, if any, did you observe?

_____

_____

_____

Investigating Porifera and Cnidaria

Lab XI-1

EQUIPMENT AND MATERIALS

You'll need the following items to complete this lab session. (The standard kit for this book, available from *www.thehomescientist.com*, includes the items listed in the first group.)

MATERIALS FROM KIT

- Magnifier

- Pipettes

- Slide, deep cavity

- Teasing needles

MATERIALS YOU PROVIDE

- Culture: Daphnia or brine shrimp (optional)

- Culture: Hydra (optional)

- Microscope

- Microscope, stereo (optional)

- Slides, prepared, Hydra (budding, sections)

- Slide(s), prepared, sponge sections

- Specimen: Grantia (preserved)

- Vinegar, distilled white

BACKGROUND

In this lab session, we'll investigate *Porifera* (poor-ih-FUR-uh) and *Cnidaria* (nye-DAIR-ee-uh). Members of these two phyla are simple invertebrate animals.

Porifera, the sponges, are the simplest animals. They have no true tissues, no digestive tract, and no circulatory or nervous system. Instead, sponges depend on a constant flow of water through their bodies, from which they obtain oxygen and food (bacteria and other microorganisms) and into which they discharge their wastes. Water is taken in through pores called *ostia* (singular *ostium*) that cover the body of the sponge and exhausted through a larger opening at one end of the sponge called the *osculum*.

Structurally, sponges are quite simple, consisting of two layers of cells that contain a jelly-like, nonliving substance called *mesohyl* (formerly called *mesenchyme*). The outer and inner cell layers are called the *ectoderm* and the *endoderm*, respectively. Rigid filamentary structures called *spicules* provide structural support to the sponge. In most species, spicules are composed of stone-like calcium carbonate or glass-like silica. In a few species—the only natural sponges used as bath or kitchen sponges—the spicules are composed of softer fibers of a material called *spongin*.

The central cavity of a sponge, through which water flows from the ostia to the osculum, is called the *spongocoel*. In some species, the spongocoel is a single, large central cavity; in others, the spongocoel is complex, with numerous internal branches and channels. The spongocoel is lined with specialized cells called *choanocytes* (also called *collar cells*), whose central flagella create currents that drive water through the sponge. As the flagella draw water across the choanocytes, it passes through the coanocytes' surrounding collar of *microvilli* (singular *microvillum*), which trap and digest microorganisms present in the water.

Sponges reproduce asexually by fragmentation or by producing *gemmules*. A few sponge species reproduce asexually by budding, a process similar to the asexual budding reproduction used by some protista and cnidarians.

Fragmentation

In reproduction by fragmentation, fragments of the original sponge torn loose by wave action or predators may float freely until they attach themselves to a substrate and begin growing by producing new cells. Eventually, the "new" sponge may grow to the size of the original sponge or more.

Nineteenth century biologists discovered reproduction by fragmentation when they passed sponges through cheesecloth, breaking them up into small clusters of cells, and found that those clusters eventually grew into larger sponges. A sponge fragment can reproduce only if both *collencytes* (which produce mesohyl) and *archeocytes* (which produce all other cell types) are present in the fragment. Without the former, the sponge fragment cannot produce mesohyl; without the latter, it cannot produce any of the other cell types present in a complete sponge.

Gemmules

Gemmules can be thought of as "spores" that are produced by some sponges. A gemmule is a cluster of archeocytes surrounded by nutrients and enclosed by a protective membrane of spongin. Like spores, gemmules are extremely resistant to environmental extremes, capable of surviving conditions that would kill a live sponge.

Many species, primarily freshwater sponges, produce gemmules in huge numbers when the sponge is under extreme stress or dying. Other species produce gemmules routinely before the onset of winter. In either case, the gemmules bide their time until conditions are suitable and then germinate, producing young sponges.

Although sponges have no reproductive organs, they can reproduce sexually. In most species that reproduce sexually, individual sponges function as both male and female. Choanocytes produce cysts that contain sperm and archeocytes produce eggs. During reproduction, the cysts burst and the sperm are expelled through the osculum into the surrounding water. If another sponge of the same species encounters those sperm, it ingests the sperm, which subsequently fertilizes eggs present in the second sponge. Depending on the species, those fertilized eggs may be expelled into the surrounding water, but most species retain them until they hatch. The resulting larvae disperse and settle to the bottom, where they attach themselves to suitable substrates and grow into adult sponges.

Cnidaria—which include hydras, corals, sea anemones, jellyfish, and similar species—are one step up in complexity from sponges. Like sponges, cnidarians possess neither digestive tracts nor circulatory systems. Also like sponges, the bodies of cnidarians are made up of two layers of cells separated by a jelly-like material, which in cnidarians is called *mesoglea* rather than mesohyl.

Unlike sponges, cnidarians possess simple nervous systems that allow them to respond to external stimuli, and a few possess rudimentary sensory organs. Also unlike sponges, cnidarian cell layers are bound by intercellular connections and basement membranes. Cnidarians are unique in possessing *cnidocytes* (also called *cnidoblasts* or *nematocytes*), venomous cells that function like tiny poisoned harpoons for stunning and capturing prey and defense against predators. Some sessile cnidarians also use cnidocytes for attaching themselves to substrates. Cnidocytes are present in large numbers on the tentacles (called *cnidae*, singular *cnida*) of cnidarians.

Each cnidocyte cell contains an organelle called a *nematocyst* (also called a *cnidocyst*), which contains a barbed penetrator connected to a coiled hollow filament inside a capsule. The capsule is covered by a lid called an *operculum*, which has a trigger structure called a *cnidocil*. When the cnidocil contacts a prey organism, it causes the operculum to spring open, much like a missile silo cover. In about a microsecond, the spring-like filament ejects the penetrator from the capsule and into the prey. The barbed penetrator impales the prey and injects a neurotoxin, which quickly stuns and eventually kills the prey. The barbs on the penetrator prevent the prey from escaping while it is being subdued. After the prey is subdued, the cnidae (tentacles) draw the prey toward the mouth of the cnidarian, which engulfs the prey.

Cnidarians reproduce asexually and sexually, as detailed in Procedure XI-1-2.

PROCEDURE XI-1-1: OBSERVING PORIFERA

As you do this procedure, record your observations in your lab notebook. Sketch or shoot images of the significant features you observe.

1. Use your naked eye and the magnifier to examine your *Grantia* specimen closely. Locate and identify the osculum. Examine the osculum closely with the magnifier, noting the spicules that surround the osculum and project through the outer surface of the sponge. On the body of the sponge, locate an ostium and examine it closely with the magnifier.

2. Examine a prepared slide of a section of *Grantia* or another sponge at low or medium magnification. Identify the following features: endoderm and ectoderm; mesohyl; spongocoel; choanocytes and flagella; and spicules. Figure XI-1-1 shows a sponge section at 100X.

Figure XI-1-1: *Sponge section, 100X*

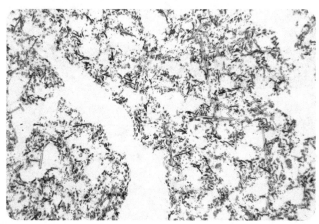

3. Center a choanocyte in the field of view and change to high magnification. Examine the flagellum and microvilli. Observe several spicules and note their general appearance and the degree of variation in their structures. Figure XI-1-2 shows a Grantia spicule at 400X.

Figure XI-1-2: *Sponge (Grantia sp.) spicule detail, 400X*

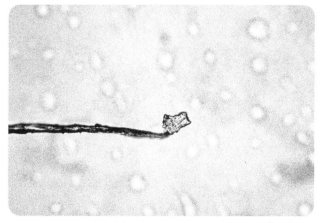

4. If you have a prepared slide of a section of another sponge species, repeat steps 2 and 3 with that slide. Compare and contrast the appearance and structure of the spongocoel and choanocytes between the two species.

PROCEDURE XI-1-2: OBSERVING CNIDARIA

As you do this procedure, record your observations in your lab notebook. Sketch or shoot images of the significant features you observe.

1. Examine a prepared slide of a whole mount hydra at low magnification. At the posterior end (on the left in Figure XI-1-3), identify the basal disk, which the hydra uses to anchor itself to a substrate. At the anterior end, identify the mouth and cnidae (tentacles), shown in Figure XI-1-4.

> Hydras bridge the boundary between microorganisms and macroorganisms. Small specimens may be invisible without magnification, but larger specimens are easily visible to the naked eye. You may have to move the slide around to view the entire animal, even at low magnification.

Figure XI-1-4: *Hydra (Hydra sp.) wm mouth and cnidae, 40X*

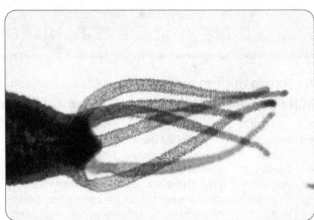

Figure XI-1-3: *Hydra (Hydra sp.) wm body, 40X*

Figure XI-1-5: *Hydra (Hydra sp.) wm cnidae detail, 100X*

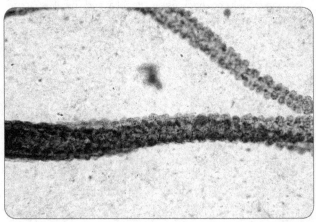

2. Change to medium magnification and examine the details of the cnidae (tentacles). Identify the nematocysts (bumps) on the cnidae. Figure X1-1-5 shows cnidae at 100X.

3. Examine prepared slides of a hydra in cross and longitudinal sections at low magnification. In both sections, locate and identify the outer epidermis, the inner gastrodermis, the jelly-like layer of mesoglea that separates the two dermal layers, and the gastrovascular cavity. Figure XI-1-6 shows a hydra cross-section at 100X.

Figure XI-1-6: *Hydra cs, 40X*

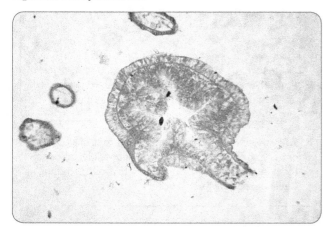

Figure XI-1-7: *Hydra budding, 40X*

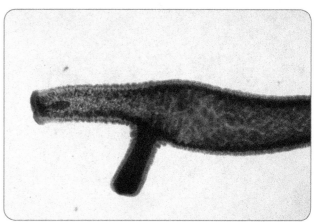

4. If you have a longitudinal section slide, use low magnification to locate and identify the basal disk, mouth, and cnidae. Center a sectioned portion of a cnida in the field of view, change to medium magnification, and locate a sectioned nematocyst. Change to high magnification and examine the details of the nematocyst.

   Under benign conditions (plenty of food, no stress, and so on) most *Hydra* species reproduce asexually by budding. A small swelling in the body wall, called a bud, develops into a miniature hydra, which, when mature, detaches from the parent hydra and begins life as a new individual.

5. Examine a prepared slide of a hydra budding at low magnification, shown in Figure XI-1-7. If you have a longitudinal section slide available that shows a bud section, examine it and note the internal structure of the bud.

   Under harsher conditions, many hydras reproduce sexually. Sexual reproduction begins much like asexual reproduction, with a small bud developing in the body wall of the parent hydra. Rather than developing as a new individual, the bud develops into an ovary or testis. When mature, a testis releases sperm into the surrounding water. If one of those sperm encounters an ovary on another hydra, it penetrates the ovary and fertilizes the egg within. When the female parent dies, the eggs are released and settle to the bottom, where they remain dormant until conditions improve. At that point, the egg hatches releasing a juvenile hydra, called a nymph. In most *Hydra* species, individuals are either male or female, but in some species individuals are hermaphroditic (possess both male and female sex organs).

6. On your whole-mount or longitudinal section slide, examine the hydra to determine if an ovary, testis, or both are present. The ovary, if present, is a bump on the posterior (basal disk end) half of the body. The testis, if present, is a conical structure on the anterior (mouth end) half. If you have section slides available, examine the ovary and testis structures. Figure XI-1-8 shows a hydra ovary in cross-section at 400X; Figure XI-1-9 shows a testis in cross-section at 400X.

Figure XI-1-8: *Hydra ovary cs, 400X*

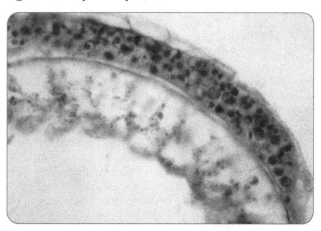

Figure XI-1-9: *Hydra testis cs, 400X*

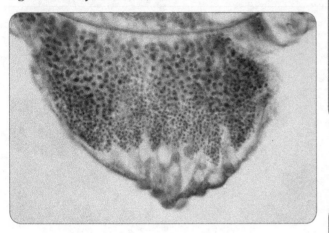

Depending on the hydra culture you use, some of the hydras may be free-floating or they may all have attached themselves to substrates. If the latter, you can dislodge a hydra by drawing up the culture liquid into the pipette and expelling the liquid forcefully toward the hydra. When it loses its grasp on the substrate, draw it up with the pipette and transfer it to the cavity slide.

8. While observing the hydra, add a few drops of a *Daphnia* or brine shrimp culture to the well. If possible, introduce the prey far enough from the hydra to avoid disturbing it. Observe any activity that occurs.

When disturbed by feeding or other external stimuli, hydras need time to recover from the disturbance and return to their normal behavior. Allow at least several minutes for this to occur.

The following part of this procedure is optional. It requires live hydras, either purchased or obtained from your pond-water microcosm. You'll also need a live *Daphnia* or brine shrimp culture to feed the hydras, which are very picky eaters.

Live hydras are relatively difficult to culture successfully and to observe. Hydras are shy and easily disturbed. Even healthy hydras may become unresponsive if the type of water, temperature, and so on is not exactly what they prefer. If they feel threatened, they may immediately draw in upon themselves, retracting their cnidae and forming what looks like a small hemispheric blob.

If you do this part of the exercise, we recommend you do so using purchased cultures. Follow the instructions supplied with the cultures to ensure that the hydras are healthy and active at the time you have scheduled the procedure.

9. While observing the hydra, use a pipette to cause a gentle current in the liquid to impinge upon the hydra. Note how the hydra responds, if at all.

10. While observing the hydra, gently touch the tip of your teasing needle to the base of the hydra. Note how the hydra responds, if at all.

11. While observing the hydra, gently touch the tip of your teasing needle to a cnida of the hydra. Note how the hydra responds, if at all.

12. While observing the hydra, place the tip of your teasing needle close to but not in contact with the hydra. Observe until the hydra contacts the tip of the needle and note how the hydra responds to that contact.

7. Using a pipette, transfer some of a hydra culture that has gone unfed for a day or so to a deep cavity slide, filling the well about one-quarter full. Observe the slide with the magnifier or under a stereo microscope to verify that at least one hydra is present in the liquid. Observe its appearance and behavior.

13. While observing the hydra, slowly add distilled white vinegar dropwise to the well until an obvious response occurs. Note how the hydra responds.

14. When you complete your investigations, dispose of the hydras by flushing them. Do not return them to the original culture container.

REVIEW QUESTIONS

Q1: Because they are sessile, sponges were originally believed to be plants. What evidence did you observe in this lab session that establishes that sponges are in fact animals?

Q2: What purpose do the spicules of a sponge serve?

Q3: Compare and contrast the symmetries and structures of a sponge and a hydra.

Q4: Compare and contrast the characteristics of flagellated or ciliated protists and sponges.

Q5: What, if any, activities was the hydra engaged in when you initially observed it?

_____

_____

_____

_____

Q6: How did the hydra respond when you added the _Daphnia_ or brine shrimp to the well? How does the hydra feed?

_____

_____

_____

_____

Q7: What response, if any, did you observe when you disturbed the hydra with a water current or the tip of your teasing needle?

_____

_____

_____

_____

Q8: What response, if any, did you observe when you added vinegar to the well?

_____

_____

_____

_____

Investigating Platyhelminthes, Nematoda, and Annelida

Lab XI-2

EQUIPMENT AND MATERIALS

You'll need the following items to complete this lab session. (The standard kit for this book, available from *www.thehomescientist.com*, includes the items listed in the first group.)

MATERIALS FROM KIT

- Magnifier

MATERIALS YOU PROVIDE

- Microscope
- Slides, prepared, Platyhelminthes
- Slides, prepared, Nematoda
- Slides, prepared, Annelida

BACKGROUND

The word *worm* was applied by early taxonomists to any *invertebrate* (spineless) animal that was not an *arthropod* (insect, spider, crustacean, etc.). In common usage, worm is sometimes used to refer to any legless creature with a tubular body, including caterpillars, grubs, maggots, centipedes and millipedes, legless amphibians, and even snakes, none of which are true worms in the original sense of that word. Although taxonomists no longer consider worms a valid grouping, not least because the various species formerly classified as worms do not make up a monophyletic group, the word is still used by biologists as a term of convenience. Worms range in size from microscopic to many meters in length, and can be found in diverse terrestrial and aquatic environments. Some species are free-living predators, while others are parasites that occupy the bodies of various animal hosts.

Like other invertebrates, worms are classified according to the number of structural cell layers they contain and the type of body cavity, if any. The *Porifera* and *Cnidaria* we examined in the preceding lab session contain only two structural cell layers, the outer ectoderm and the inner endoderm. As more complex animals, the worms we'll examine in this lab session contain an additional structural cell layer called the mesoderm, which is located between the ectoderm and endoderm.

In worms, a fluid-filled body cavity may or may not be present in the mesoderm. Worms without a body cavity are called *acoelomates*. Those with a body cavity that is partially lined with mesoderm are called *pseudocoelomates*. Those with a body cavity that is fully lined with mesoderm are called *coelomates*.

In this lab session, we'll examine some characteristics of representatives of three major worm phyla: *Platyhelminthes*, or *flatworms*; *Nematoda*, or *round worms*; and *Annelida*, or *segmented worms*. These three phyla provide examples of acoelomates, pseudocoelomates, and coelomates, respectively.

PROCEDURE XI-2-1: OBSERVING PLATYHELMINTHES (FLATWORMS)

As you perform this procedure, record your observations in your lab notebook. Sketch or shoot images of any significant features you observe.

1. Examine a whole-mount prepared slide of a representative Planarian, such as *Dugesia* sp., shown in Figure XI-2-1, with the magnifier and at low magnification with your microscope. Identify any external features visible, such as the *pharynx*, *ocelli* (eye spots), *auricles* (small flaps near the anterior end, used for detecting water currents), and so on.

2. Examine a cross-section of the Planarian at low magnification. Locate and identify the exterior cell layer, the *ectoderm*, which surrounds and contains the middle cell layer, the *mesoderm*. The mesoderm is the cellular connective tissue that makes up the bulk of the body, and contains all of the organs. Examine the structures visible within the mesoderm, including the collagen fibers that provide structural support for the body and the muscles that attach to those fibers. Identify the interior cell layer, called the *endoderm*, a single layer of cells that surrounds the *gastrovascular cavity*. Figures XI-2-2 through XI-2-4 show Planarian cross-sections of the anterior, pharynx, and posterior regions, respectively.

Figure XI-2-1: *Planaria (Dugesia sp.) anterior and pharynx regions wm, 40X*

Figure XI-2-2: *Planaria anterior region cs, 100X*

Figure XI-2-3: *Planaria pharynx region cs, 100X*

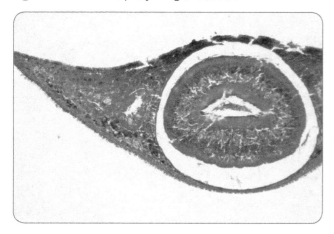

Figure XI-2-4: *Planaria posterior region cs, 100X*

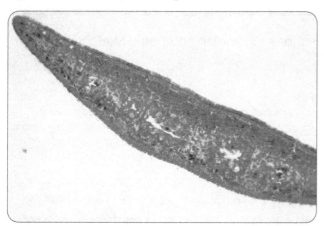

Although *Platyhelminthes* in the paraphyletic sub-phylum *Turbellaria* (like the *Planaria*) are generally free-living and predatory, many flatworm species are parasites that colonize animal hosts and cause serious diseases in humans and animals. These parasites are members of the monophyletic *Neodermata* sub-phylum, which contains three classes: *Cestoda* (*tapeworms*, such as *Taenia* sp.), *Trematoda* (*flukes*, such as *Schistosoma* sp.), and *Monogenea* (fish parasites).

Cestoda are fascinating structurally. Their bodies are made up of a series of segmented plate-like structures called *proglottids*, which may overlap each other or simply be butted edge-to-edge. The *Cestode* continues to produce proglottids throughout its life cycle. As proglottids mature, they are expelled from the posterior of the *Cestode*, mixed with feces. Interestingly, because proglottids contain both male and female reproductive structures, an individual proglottid can reproduce independently. That leads some biologists to consider individual *Cestodes* to be proglottid colonies rather than single organisms.

3. If you have prepared slides available, examine whole-mounts and cross-sections of a representative *Cestode*, such as *Taenia* sp. or another tapeworm. Compare and contrast the structures with those of the *Turbellaria*. Identify the proglottids and examine proglottids in all three states: immature, mature, and gravid (filled with fertile eggs). Figure XI-2-5 is a 40X wm of gravid proglottids from *Taenia pisiformis*, a tapeworm commonly found in dogs, cats, and other carnivores.

Figure XI-2-5: *Tapeworm (Taenia pisiformis) proglottids wm, 40X*

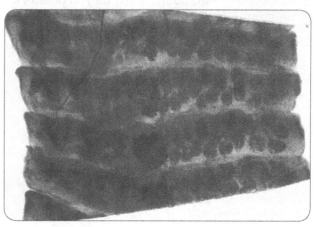

4. If you have prepared slides available, examine whole-mounts and cross-sections of a representative *Trematoda*, such as *Schistosoma* sp., shown in Figure XI-2-6, or another fluke. Compare and contrast the structures with those of *Turbellaria* and *Cestoda*.

Figure XI-2-6: *Blood fluke (Schistosoma mansoni) male wm, 40X*

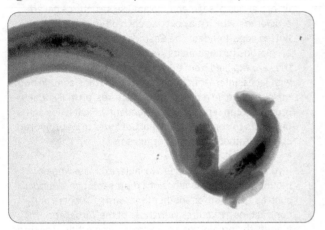

PROCEDURE XI-2-2: OBSERVING NEMATODA (ROUNDWORMS)

As you perform this procedure, record your observations in your lab notebook. Sketch or shoot images of any significant features you observe.

Using low magnification, examine a prepared slide of a cross-section of a representative *Nematode*, such as *Ascaris lumbricoides*, shown in Figures XI-2-7 and XI-2-8. Working from the outer layers toward the inner layers, identify the following structural features:

Cuticle

The *cuticle* is the outermost layer. Depending on the species, the cuticle may be quite thick and have a complex structure, or it may be relatively thinner and simpler. In parasitic species, the impervious cuticle protects the worm against attack by the digestive system or other active defenses of the host.

Ectoderm

The *ectoderm* (or *epidermis*) lies just inside the cuticle, and is a thin single layer of cells.

Mesoderm

The *mesoderm* layer lies just inside of the ectoderm, and has a filamentary or fiber-like appearance.

Pseudocoelom

The *pseudocoelom* occupies most of the body, from the inner edge of the mesoderm to the outer edge of the *endoderm*, and is cluttered with other structures scattered throughout it like raisins in a pudding.

Endoderm

The endoderm is a single layer of cells, bounded on the outside by the pseudocoelom, and surrounding the *gastric cavity*, which is the digestive tract of the worm and runs its length from the mouth to the anus.

Figure XI-2-7: *Roundworm (Ascaris lumbricoides) cs, 40X*

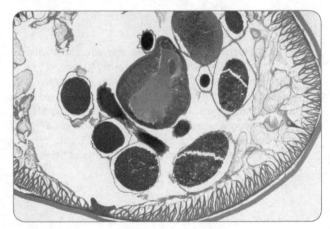

Figure XI-2-8: *Roundworm (Ascaris lumbricoides) cs, 40X*

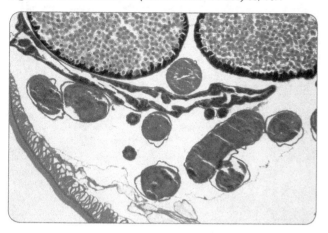

PROCEDURE XI-2-3: OBSERVING ANNELIDA (SEGMENTED WORMS)

As you perform this procedure, record your observations in your lab notebook. Sketch or shoot images of any significant features you observe.

Using low magnification, examine a prepared slide of a cross-section of a representative *Annelida*, such as *Lumbricus terrestris*, (the common earthworm) shown in Figure XI-2-9. Working from the outer layers toward the inner layers, identify the following structural features:

Chaetae

Chaetae (singular *chaeta*), also called *chetae* or *setae*, are thin, stiff bristle-like structures on the exterior surface of Annelids, which provide a grip that allows the worm to move across surfaces, through soil, and so on. The chaetae in Figure XI-2-9 surround the worm but are barely visible as tiny bumps at 40X. Depending on your specimen, you may have to boost magnification to see any detail in the chaetae.

Cuticle

The *cuticle* is the thin, outermost layer, which encloses and protects the inner structures of the worm.

Ectoderm

The *ectoderm* lies just inside the cuticle, and is a thin single layer of cells.

Muscle layers

Lying just inside the ectoderm are two layers of muscles that form the outer boundary of the mesoderm. The thinner outer layer is the lateral muscles, which run around the diameter of the worm; the thicker inner layer is the longitudinal muscles, which run the length of the worm.

Mesoderm, coelum, nephridia, endoderm, and digestive cavity

The *mesoderm* layer is bounded on the outer side by the muscle layers, and on the inner side by the muscles that surround the central circular, oval, or horseshoe-shaped *digestive cavity*. The body cavity between the longitudinal muscles and the digestive cavity is called the *coelum*. (The true coelum body cavity present in *Annelida* differs from the pseudocoelum cavity present in *Nematoda* in that the true coelum is fully lined with mesoderm.) The two filamentary structures, one per side, inside the coelum are called *nephridia* (singular *nephridium*).

Figure XI-2-9: *Earthworm (Lumbricus terrestris) cs, 40X*

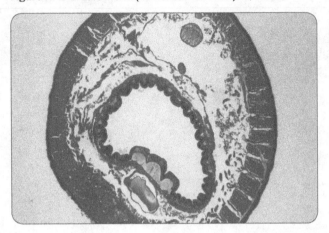

REVIEW QUESTIONS

Q1: If you are given a cross-section slide of an unidentified worm to classify, what single internal feature would allow you to identify unambiguously the worm as Platyhelminthes, Nematoda, or Annelida?

_____

_____

_____

_____

Q2: What characteristic do members of the three phyla we examined in this lab session have in common with each other that they do not share with members of Porifera or Cnidaria?

_____

_____

_____

_____

Q3: Given your knowledge of Annelida structures, propose an explanation for how an earthworm moves forward.

_____

_____

_____

_____

_____

Investigating Arthropods

Lab XI-3

EQUIPMENT AND MATERIALS

You'll need the following items to complete this lab session. (The standard kit for this book, available from *www.thehomescientist.com*, includes the items listed in the first group.)

MATERIALS FROM KIT

- Centrifuge tubes (as collection containers)
- Coverslips
- Forceps
- Magnifier
- Needles, teasing
- Scalpel

- Slides, flat
- Slides, deep cavity
- Stain: eosin Y
- Stain: methylene blue
- Thermometer

MATERIALS YOU PROVIDE

- Ethanol, 70%
- Food for mealworm beetles (see text)
- Jars (see text)
- Microscope
- Refrigerator

- Slides, prepared, Arthropoda (see text)
- Specimens, assorted arthropods (see text)
- Specimens, mealworm (see text)
- Stereo microscope (optional)

BACKGROUND

The phylum *Arthropoda*, whose members are called *arthropods*, comprises *invertebrate animals* including insects, arachnids (spiders, scorpions, mites, ticks, and related creatures), crustaceans (crabs, crayfish, lobsters, shrimp, and related creatures), and related subphyla and classes. With more than 1,000,000 described species and possibly several million more species as yet unclassified, *Arthropoda* is by far the largest phylum in Kingdom *Animalia*, making up more than three quarters of all classified *Animalia* species.

Arthropods are characterized by their chitinous exoskeletons, jointed appendages, and bodies made up of jointed segments, each with a pair of appendages. Each segment is covered by a cuticle made up of chitin, with numerous setae present on the surface. More than any other lifeform, arthropods look as if they were built from an Erector Set. In fact, this jointed structure is characteristic of arthropods, and is the origin of the phylum name: arthro (jointed) + pod (foot).

Internal organs are usually also jointed segments contained within the *hemocoel* (body cavity) present in the body segments, through which blood circulates. Their simple nervous systems comprise a central pair of ventral nerve cords that run through and join the segments, extending to form paired ganglia in each segment. Their simple brains are a mass of ganglia from the fused segments that make up their heads. The excretory and respiratory systems of arthropods differ significantly from species to species, depending on the environments for which they are adapted.

In this lab session, we'll begin with a survey, looking at as many examples of whole arthropods as possible and comparing/contrasting their external structures. In the second procedure, we'll observe insect *metamorphosis*, the process by which arthropods develop from eggs to mature adults. Some arthropods undergo three-stage *hemimetabolism*, also called *incomplete metamorphosis*, during which they develop from eggs through the immature (*nymph*) stage to the adult *imago* stage. Other arthropods undergo four-stage *homometabolism*, also called *complete metamorphosis*, during which they develop from the egg/embryo stage through the infantile (*larval*) stage to the immature (*pupal*) stage and finally to the adult imago stage.

For the first procedure, you'll need specimens of as many arthropods as possible. You can collect live specimens of a huge variety of arthropod species from around your house and yard, from a local park, and so on. Unless you live in an area with severe winters, you can collect widely diverse specimens year-round—ants, flies, bees, spiders, caterpillars, butterflies, mosquitoes, fleas, grasshoppers, and so on. Visit the supermarket for fresh crabs or shrimp.

Use appropriate collection vessels for each species. Test tubes or centrifuge tubes are suitable for smaller specimens and foam cups with lids suffice for larger ones. (**Remember that live arthropods are by no means defenseless, and take appropriate precautions while collecting and handling them. Some of them, such as the brown recluse and black widow spiders and many scorpion species, have bites or stings that are always dangerous and potentially lethal.**) To kill your specimens humanely and preserve them for later use, simply add 70% ethanol to the collection vessel that contains the specimen.

If you live in an area with severe winters and will do this lab session during winter, you have a few alternatives. First, you can collect your specimens during warmer weather. The ethanol that kills them will also preserve them indefinitely. Second, you can use commercial prepared slides and specimens. Third, if you have a local botanical gardens with indoor facilities, if you ask politely and explain your purpose, the staff may be willing to allow you to collect some specimens.

For the second procedure, we'll use mealworm beetles in egg, larval, pupal, and adult stages. These can be purchased at most pet stores, where they are sold as food for reptiles, fish, and birds. You can also purchase mealworms in any bait store, at many stores that carry bird feeders, or on the Internet. If you start with mealworm beetles in all four stages, you can complete this procedure over the course of two to three weeks. If you wish, you can start with only mealworm beetle eggs or larvae (usually sold as just "mealworms") and watch the process over the course of three or four weeks to perhaps three months, depending on temperature and other variables.

If your goal is to fix (preserve) specimens rather than simply kill them, take the size of the specimen into account. Small specimens will be thoroughly infused with the ethanol within a day or two. Larger specimens may require a week or more, and may require several changes of the ethanol.

The ethanol fixative may decolorize specimens. To add color contrast, you can stain fixed whole-mount specimens with methylene blue, eosin Y, or both. For small specimens (such as ants, fruit flies, and so on), simply place the fixed specimen in the cavity of a deep well slide and add a drop or two of methylene blue or eosin Y, or both. Allow the stain to work for a minute or so and then withdraw as much of the stain as possible with a pipette. Rinse and decolorize the specimen with two or three changes of ethanol. Do not over-decolorize.

If you are processing several specimens, you may find it saves time to make up half a dozen miniature "staining jars" using deep-cavity slides. Put several drops of eosin Y in the first well, several drops of methylene blue in the second, and several drops each of eosin Y and methylene blue in the third. Fill the wells in the other three slides halfway with ethanol. Use a pipette or forceps, depending on the size of the specimen, to transfer the specimen to and from the wells.

After staining, put the specimen in the first ethanol well, allow it to remain for a few seconds to remove most of the excess stain, and then transfer it to the second ethanol well. After 30 seconds, drain the specimen and transfer it to the third ethanol well, and thence to a slide for observation.

The mealworm beetle life cycle begins with eggs, which typically require 4 to 14 days to hatch as larvae. The larval stage, during which the larvae feed continuously and molt several times as they outgrow their exoskeletons, may last from two weeks to seven or eight weeks, after which the larvae emerge as pupae. After three days to three or four weeks, the pupae emerge as adult mealworm beetles.

To raise mealworm beetles, you need only some containers and sources of food and water. Nearly any container can be used, but 500 mL (pint) or 1 L (quart) tall wide-mouth glass or plastic jars work well and are readily available. Cover the containers with something that will allow adequate ventilation while preventing the mealworms from escaping. Mealworms thrive on nearly any grain. Plain (no sugar) breakfast cereals such as crushed corn flakes, oatmeal, and so on are convenient food sources. The mealworms also require a source of water, such as small chunks of potato or orange placed in the container and removed as soon as they start becoming moldy. We've even used a small piece of kitchen sponge dampened frequently.

PROCEDURE XI-3-1: OBSERVING AND COMPARING ARTHROPOD STRUCTURES

As you perform this procedure, record your observations in your lab notebook. Sketch or shoot images of any significant features you observe.

Examine as many whole arthropod specimens as you have available (or prepared slides of the same) using your naked eye, the magnifier, and your microscope, as appropriate for the size of the specimen. Pay particular attention to the structures of the body, head, tail, and appendages (legs, wings, antennae). Figures XI-3-1 through XI-3-22 are whole mount images of small arthropod specimens of the sort that are readily available locally for collection.

Figure XI-3-1: *Water flea (Daphnia magna) wm, 40X*

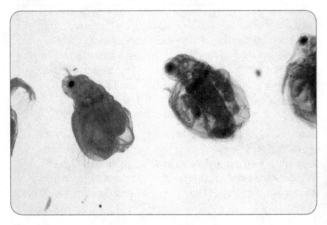

Figure XI-3-4: *Mosquito (Culex sp.) larva anterior wm, 40X*

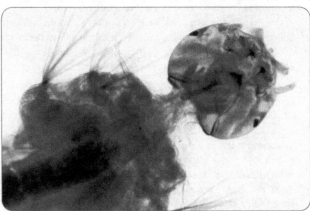

Figure XI-3-2: *Itsy-Bitsy Spider (Aranea ventricosa) sometimes found in waterspouts wm, 100X*

Figure XI-3-5: *Mosquito (Culex sp.) larva body wm, 40X*

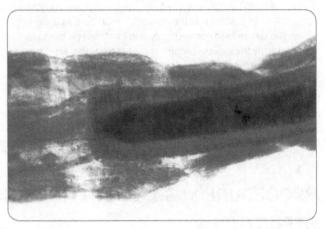

Figure XI-3-3: *Spider (Aranea sp.) leg detail wm, 40X*

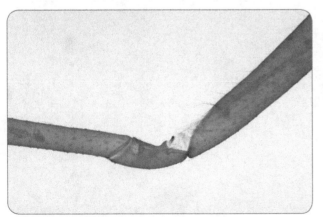

Figure XI-3-6: *Mosquito (Culex sp.) larva posterior wm, 40X*

Figure XI-3-7: *Mosquito (Culex sp.) adult male head and mouthparts wm, 40X*

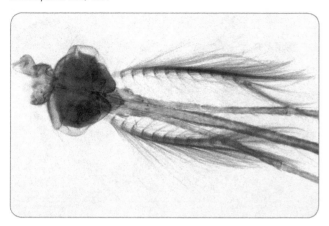

Figure XI-3-8: *Mosquito (Culex sp.) adult male head and mouthparts wm, 40X*

Figure XI-3-9: *Housefly (Musca domestica) head detail wm, 40X*

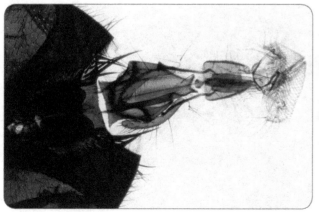

Figure XI-3-10: *Housefly (Musca domestica) head and leg detail wm, 40X*

Figure XI-3-11: *Housefly (Musca domestica) body detail wm, 40X*

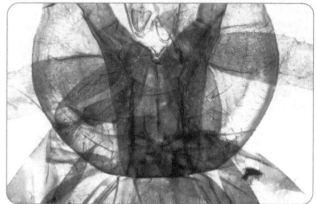

Figure XI-3-12: *Housefly (Musca domestica) wing and leg detail wm, 40X*

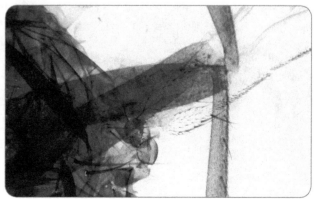

Figure XI-3-13: *Honeybee (Apis mellifica) first leg wm, 40X*

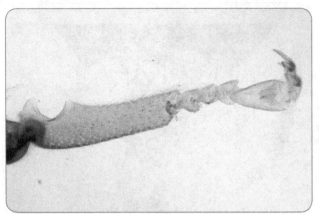

Figure XI-3-14: *Honeybee (Apis mellifica) third leg wm, 40X*

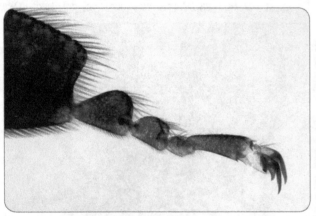

Figure XI-3-15: *Honeybee (Apis mellifica) stinger wm, 40X*

Figure XI-3-16: *Ant (Monomorium phalaonis) wm, 40X*

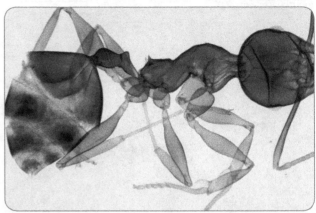

Figure XI-3-17: *Butterfly (Pieris rapae L.) head and mouthparts wm, 40X*

Figure XI-3-18: *Flea (Pulex irritans) wm, 40X*

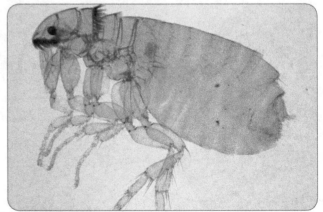

Figure XI-3-19: *Fruit fly (Drosophila melanogaster) anterior wm, 40X*

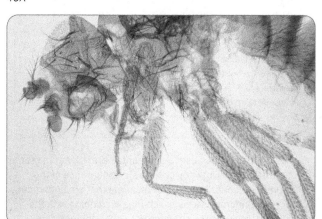

Figure XI-3-21: *Fruit fly (Drosophila melanogaster) legs wm, 40X*

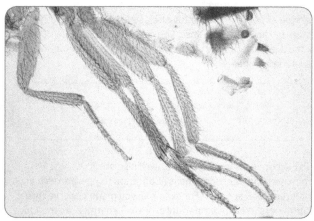

Figure XI-3-20: *Fruit fly (Drosophila melanogaster) wings wm, 40X*

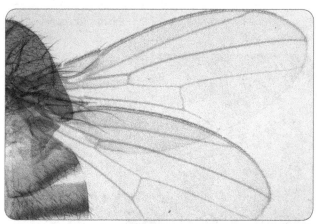

Figure XI-3-22: *Fruit fly (Drosophila melanogaster) posterior wm, 40X*

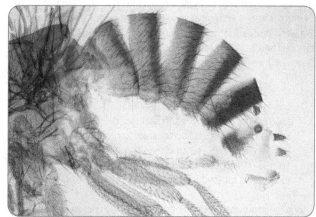

PROCEDURE XI-3-2: OBSERVING INSECT METAMORPHOSIS

In this procedure, we'll observe metamorphosis in the mealworm beetle and determine the effect of temperature, if any, on the rate of metamorphosis by observing specimen groups kept at different temperatures. The specific temperatures are not important as long as there is a significant difference, say 15 to 18 °C or more, with the lower temperature above freezing. For most people, the easiest way to accomplish this is to keep one group at room temperature (typically 20 to 25 °C) and a second in a standard refrigerator (typically ~2 °C).

If you have the means to do so, also observe a third group kept at an elevated temperature (we suggest 37 to 40 °C). If you do this procedure during the cold weather months, that can be as simple as keeping the container close to a heating vent, heat strip, or radiator. Otherwise, you can weight the container and keep it partially immersed in a warm water bath maintained with an aquarium heater.

> Mealworm beetles are primarily nocturnal, so keep their environment darkened as much as possible.

To introduce a second variable, we'll also observe metamorphosis of specimens at different temperatures using different food sources. For example, you might observe three containers at each temperature, one with oatmeal as the food source, a second with crushed corn flakes, and a third with crushed wheat bran.

As you perform this procedure, examine specimens regularly with your naked eye, magnifier (or stereo microscope) and microscope, and record your observations in your lab notebook. Sketch or shoot images of any significant features you observe. We will assume that you are beginning with mealworms in the larval stage. You can modify these instructions according to the time you have available and the life stage you begin with.

1. Prepare containers for each of the food/temperature combinations you intend to use. Place a layer of the chosen food on the bottom of each container and distribute a few small chunks of potato or orange to provide a water source for the worms. (Keep an eye on these chunks over the course of the procedure, and replace them when they start to become moldy.)

2. Using forceps or a small spatula, gently transfer the same number of mealworm larvae to each of the containers. (Don't squish the larvae.) Judge the number according to the size of the container. You want as many larvae as possible without crowding the container. Record the number of larvae transferred to each container.

> As an interesting extension to the experiment, consider using two similar containers, each with the same food and kept at the same temperature. To one of these containers, transfer just enough larvae to populate the container well without crowding. To the second container, transfer several times as many larvae, crowding the larvae. Observe the metamorphosis of the larvae in both containers and note any differences based on competition for resources.

3. For the first week, observe the containers daily at the same time each day. Note any changes in appearance of the larvae and count the number of larvae that are undergoing or have undergone a visible change. Beginning with the second week, you can change your observations and counts to every other day. Continue these observations until the first larvae begin to emerge as pupae, shown in Figures XI-3-23 and XI-3-24, and then return to the daily schedule.

Figure XI-3-23: *Mealworm beetle (Tenebrio molitor) pupa with shed larval skin*

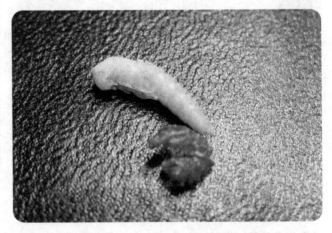

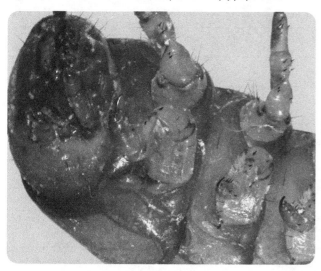

Figure XI-3-24: *Mealworm beetle (Tenebrio sp.) pupa detail*

4. As the larvae emerge as pupae, transfer the pupae to additional containers, and continue to observe both the original and new containers. Note any differences between newly emerged pupae and those that are several days or more old.

5. As the pupae begin to emerge as adult beetles, transfer the adult beetles to new containers and continue to observe the containers of all three stages.

There is, of course, no natural ending point for this procedure. Eggs become larvae, which become pupae, which become adult beetles, which lay eggs, which become larvae, and so on. That's why it's called a life cycle.

When you have observed the mealworm beetle life cycle long enough to have seen all stages, conclude the procedure. Use your detailed notes to determine what, if any, effect temperature and/or food has on the timing of the life cycle.

REVIEW QUESTIONS

Q1: What unique features and similarities did you observe in your arthropod specimens?

Q2: What similarities might mislead a student to mistake mealworm beetle larvae for Annelida? What external anatomical features allow the larvae to be distinguished from Annelida?

Q3: How can you distinguish newly emerged pupae from older pupae?

_____

_____

_____

_____

Q4: What external anatomical features that characterize them as insects do the adult mealworm beetles share with the other insects you observed in the preceding procedure?

_____

_____

_____

_____

Q5: Are mealworm beetles hemimetabolic or homometabolic? Why?

_____

_____

_____

_____

Q6: What characteristic of mealworm beetles suggests that temperature may affect the timing of their life stages? Did you observe such an effect?

_____

_____

_____

_____

Q7: How would you expect climate to affect the mealworm beetle life stages in the wild?

_____

_____

_____

_____

Q8: Did you observe any variations in the life cycle with different foods?

_____

_____

_____

_____

Investigating
Vertebrate Tissues

Lab XI-4

EQUIPMENT AND MATERIALS

You'll need the following items to complete this lab session.

MATERIALS FROM KIT

- None

MATERIALS YOU PROVIDE

- Microscope
- Slides, prepared, Vertebrata (see text)

BACKGROUND

Vertebrates are members of the subphylum *Vertebrata* of the phylum *Chordata*. Vertebrates are the organisms that most people think of as animals, including the classes fish, birds, reptiles, amphibians, and mammals.

The defining characteristic of vertebrates is their possession of a *vertebral column* (or spine) consisting of a segmented series of hard, bony *vertebrae* (singular *vertebra*) separated by *intervertebral discs*, which provide mobility and allow the spine to move and flex. (In other chordates, the spine is a stiff rod of uniform composition that is incapable of flexing.) Vertebrates are also the only members of *Animalia* that possess true brains.

There are many differences among vertebrates. For example, some are *oviparous* (giving birth to eggs that subsequently hatch into live young) while others are *viviparous* (giving birth to live young). Some vertebrates are *ectotherms* (cold-blooded) while others are *homotherms* (warm-blooded). But the similarities among vertebrates—particularly within classes, but even across classes—are much greater than the differences.

In studying vertebrate structures, it's convenient to treat them as a hierarchy: a collection of cells forms a tissue; a collection of tissues forms an organ; and a collection of organs forms an organ system. For example, a collection of cardiac muscle cells makes up cardiac tissue, and a collection of cardiac tissues make up the heart, which in turn is part of the circulatory system.

Form (macro- but particularly micromophology) often follows function. Despite sometimes gross morphological differences across species and even classes, the analogous organs perform very similar functions and so tend to resemble each other, more or less. For example, it is difficult or impossible for someone who is not a vertebrate anatomist to distinguish between, say, cross-sections of a fish liver and a mammal liver, and even that anatomist might be unable to distinguish readily between cross-sections of a pig liver and a chimpanzee liver.

All vertebrate organs are made up of one or more of the four basic tissue types:

Epithelial tissues

Epithelial tissues cover the surfaces and cavities of body structures, and also form many endocrine and exocrine glands. For example, your skin, the inside of your mouth, and your stomach, liver, and pancreas are made up, in whole or in part, of epithelial tissues. Epithelial tissues perform four general functions. First, they protect underlying tissues from dehydration and physical damage. Second, they provide a selectively permeable barrier that prevents or allows the passage of particular molecules, such as water or nutrients. Third, they secrete specific fluids, such as digestive fluids. Fourth, they provide sensory inputs to the nervous system.

Connective tissues

Connective tissues are found throughout vertebrate bodies, serving in many roles. Bone and bone marrow, cartilage, tendons, adipose tissue, and blood are just some examples of structures made up partially or entirely of connective tissue.

Muscle tissues

Muscle tissues are made up of muscle cells, which contain *contractile filaments* that can contract and expand to change the sizes and shapes of muscle cells. Muscle tissues are classified as *cardiac muscles*, *smooth muscles*, or *skeletal muscles*. The first two types are not under voluntary control, and perform such functions as circulating blood, moving food through the digestive tract, and dilating or contracting pupils in response to changing light levels. Skeletal muscles are under voluntary control and are used to move the body, from tiny motions such as shifting your eyes to gross movements such as kicking a soccer ball. (Some muscular activity—such as focusing your eyes, breathing, or urinating—take place under partial voluntary control and partial involuntary control.)

Nervous tissues

Nervous tissues are the primary component of the brain, spinal cord, and nerves, which together regulate body functions, including thought. Nervous tissues are made up of *neurons*, which transmit electrical impulses, and *neuroglia cells*, which support the neurons. Although all cells have the ability to react to external stimuli, the cells that make up nervous tissues are adapted to transmit impulses over long distances to whichever organs are responsible for dealing with the specific stimulus.

As you work on each procedure, use your textbook or Internet resources to determine which features, structures, organelles, and so forth to look for. For example, when you are examining slides of epithelial tissues, compare and contrast simple, stratified, and pseudostratified epithelial tissues, and do the same for squamous, columnar, and cuboidal epithelial cells. Identify the basement membrane and any glands present in the epithelial tissue, and note the presence or absence of blood vessels. Record your observations in your lab notebook. Sketch or shoot images of any significant features you observe.

Although nervous impulses are electrical, they propagate at a tiny fraction of the speed of electrical impulses in copper wire or similar conductors. Electrical impulses in insulated wire typically propagate at about 0.7 light speed or roughly 210,000,000 m/s, millions to billions of times faster than the electrical impulses in nerves, which propagate at between 0.1 m/s and 100 m/s.

That's why it takes a noticeable fraction of a second for the reaction when someone taps your knee to test your reflexes. The major reason for that delay is that your nervous system isn't a bunch of individual wires. Instead, it's zillions of nervous cells that are connected together and function more like a bucket brigade than a cable. One cell delivers the message to the adjoining cell, which turns around and delivers the message to the next cell in line, with a lag involved at each step along the way.

To minimize the number of prepared slides required, we've included many images in this lab session. As always, remember that images are not perfect substitutes for viewing actual slides, so we recommend that you obtain as many prepared slides as possible and use the images only to fill in for slides you don't have.

The standard kit for this book, available from *www.thehomescientist.com/biology.html*, includes a disk with high-resolution JPEG files of the photomicrographic images used in this book. Those images can be viewed on any PC (Windows, Mac OS X, or Linux) or tablet.

PROCEDURE XI-4-1: OBSERVING EPITHELIAL TISSUES

Epithelial cells and tissues are found throughout vertebrate bodies. Epithelial structures are classified by cell type and tissue layer type:

- *Squamous epithelial cells* are flat and resemble fried eggs.

- *Cuboidal epithelial cells* resemble dice in whole mounts or squares in sectional mounts.

- *Columnar epithelial cells* in longitudinal sections resemble tall, narrow columns, but in cross-section they may be confused for squamous or cuboidal cells.

These epithelial cell types may be organized in the following ways:

- *Simple epithelium* has only one layer, which may comprise any of the types of epithelial cells, and is found in tissue that performs absorption or filtration functions. The cell layer in simple epithelium is in direct contact with the *basement membrane*.

- *Stratified epithelium* has two or more layers above the basement membrane, may comprise any combination of epithelial cell types, and is found in any type of tissue that must be resistant to physical abrasion or chemical attack.

- *Pseudostratified epithelium* has only one columnar cell layer, but the locations of the nuclei of those cells may make it appear in sectional mounts that the epithelium is stratified.

- *Ciliated epithelium* is a type of pseudostratified epithelium in which the outward-facing membrane possesses tiny hair-like structures called *cilia*. Cilia may function as extra-cellular sensors, and may also provide propulsive functions such as moving ova from the ovary down the Fallopian tubes to the uterus or distributing mucus produced by *goblet cells* in the epithelial tissue.

- *Glandular epithelium*—found in *endocrine glands*, *exocrine glands*, and other organs—is made up of highly adapted epithelial cells that do not function as a protective layer. Instead, these cells produce hormones, milk, mucus, digestive fluids, sweat, and other substances.

Examine prepared slides showing the various types of epithelial cells and epithelial structures. As you examine each slide, identify the epithelial cell types present and their arrangement. Figures XI-4-1 through XI-4-21 show examples of epithelial cell types and numerous epithelial structures.

Figure XI-4-1: *Flat simple epithelium wm, 40X*

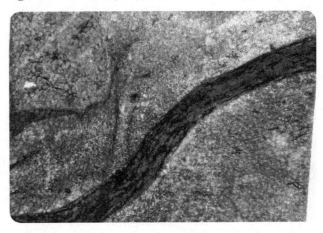

Figure XI-4-2: *Flat stratified epithelium sec, 100X*

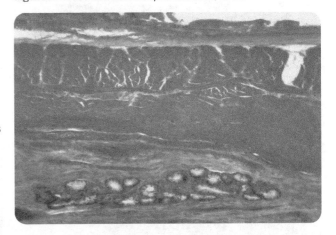

Figure XI-4-3: *Ciliated epithelium sec, 100X*

Figure XI-4-4: *Human (Homo sapiens) skin sec through hair follicle, 40X*

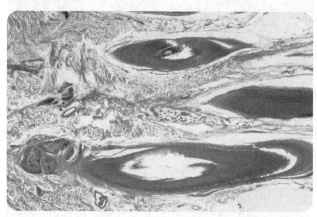

Figure XI-4-5: *Human (Homo sapiens) skin sec through sweat gland, 40X*

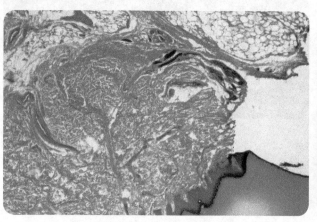

Figure XI-4-6: *Mammal tongue cs, 40X*

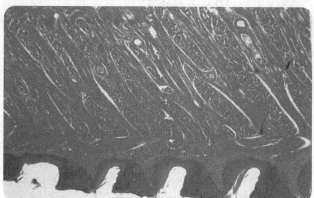

Figure XI-4-7: *Pika (Ochotona sp.) taste buds sec, 40X*

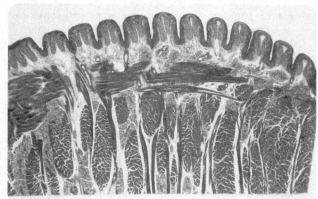

Figure XI-4-8: *Mammal esophagus cs, 40X*

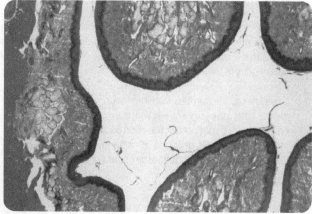

Figure XI-4-9: *Dog (Canis lupus familiaris) stomach wall sec, 40X*

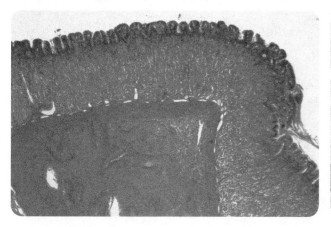

Figure XI-4-12: *Mammal kidney ls, 40X*

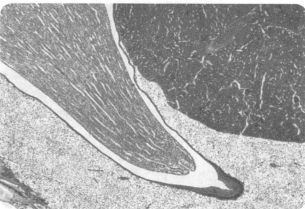

Figure XI-4-10: *Dog (Canis lupus familiaris) small intestine cs, 40X*

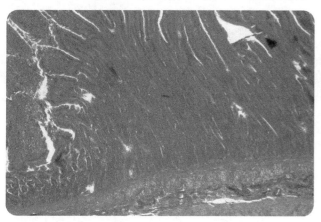

Figure XI-4-13: *Mouse (Mus muscula) kidney ls, 40X*

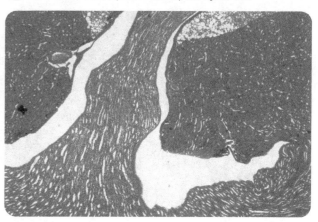

Figure XI-4-11: *Dog (Canis lupus familiaris) pancreas cs, 40X*

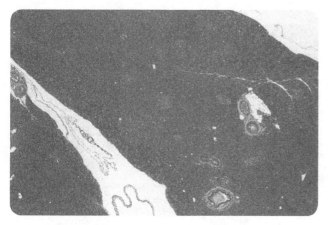

Figure XI-4-14: *Kidney (dye-injected) cs, 40X*

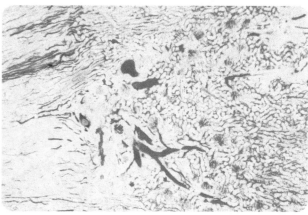

Figure XI-4-15: *Liver sec, 40X*

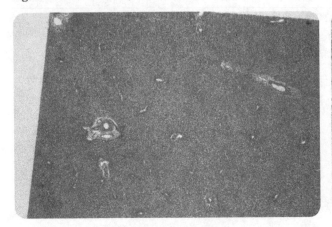

Figure XI-4-18: *Mammal testis sec, 100X*

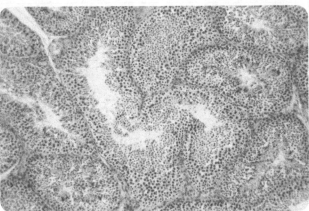

Figure XI-4-16: *Mammal ovary sec, 40X*

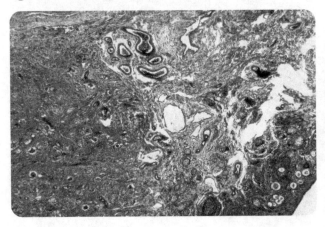

Figure XI-4-19: *Mammal uterus sec, 40X*

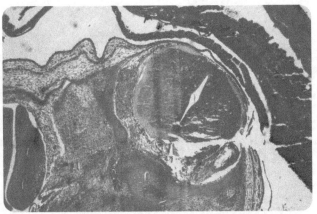

Figure XI-4-17: *Mammal testis and epididymis sec, 40X*

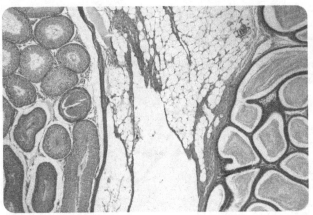

Figure XI-4-20: *Frog (Rana catesbeiana) blastula (young embryo) sec, 100X*

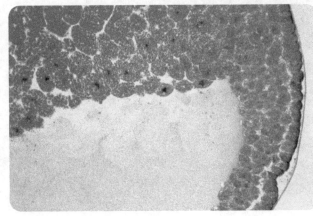

Figure XI-4-21: *Frog (Rana catesbeiana) late gastrula (older embryo) sec, 40X*

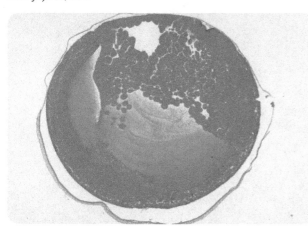

Various vertebrate classes have adapted by forming specialized exterior surface epithelial structures such as the scales of fish, the feathers of birds, and the hair of mammals. Examine prepared slides of as many of these structures as you have available, and consider how the morphology of the various structures is adapted to suit particular functions. Figures XI-4-22 through XI-4-26 show several of these specialized structures.

Figure XI-4-22: *Goldfish (Carassius auratus) scale wm, 40X*

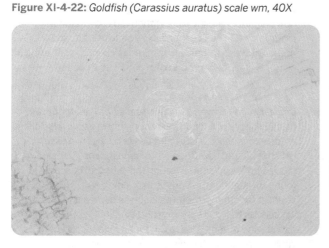

Figure XI-4-23: *Sparrow (Passer rutilans) contour (pennaceous) feather wm, 40X*

Figure XI-4-24: *Sparrow (Passer rutilans) down feather wm, 40X*

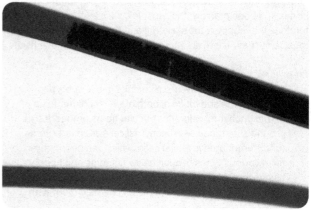

Figure XI-4-25: *Sheep (Capra sp.) wool wm, 100X*

Figure XI-4-26: *Sheep (Capra sp.) wool cs, 400X*

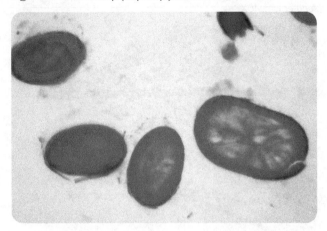

PROCEDURE XI-4-2: OBSERVING CONNECTIVE TISSUES

Connective tissue (CT) is the most diverse of the four tissue types, and is found throughout vertebrate bodies. CT performs numerous functions, including providing a structural framework for the body, storing food energy, protecting organs against physical damage, and distributing oxygen to and removing carbon dioxide from other tissues and organs.

CT is made up of specialized cells and an extracellular matrix produced by those cells. That matrix is made up of a transparent, homogeneous liquid known as *ground substance* and fibers of any of three types: *collagenous fibers*, *elastic fibers*, and *reticular fibers*. These fibers strengthen the CT in the same way that rebar strengthens reinforced concrete. Because they can be stretched and then return to their original shape and size, elastic fibers provide flexibility to CT.

Connective tissue makes up numerous body structures in whole or in part, including adipose tissue; blood and lymphatic tissue; bones and bone marrow; the capsules, cartilage, ligaments, and tendons that support the skeleton; and the fibrous framework of muscle tissues. There are three categories of connective tissue.

Embryonic connective tissue
Embryonic connective tissue occurs in two subtypes, *mesenchyme* and *mucoid*, both of which are made up of mesenchymal cells and reticular fibers embedded in ground substance. Mesenchymal cells are pluripotent (undifferentiated juvenile) cells, which can be triggered to form various types of adult cells, including adult connective tissue cells. Mucoid CT, found in the umbilical cords of mammals, is very similar to mesenchyme CT, but with a much sparser distribution of cells and fibers. Because reticular fibers are invisible without special staining, embryonic CT (ECT) is normally visible in standard prepared slides as only an apparently random scattering of cells in a clear background of ground substance.

Ordinary connective tissue
Ordinary connective tissue (also called *proper connective tissue*) subtypes include *loose connective tissue*, also called *areolar connective tissue*, and *dense connective tissue*, which is further subdivided into *regular dense connective tissue* and *irregular dense connective tissue*. Loose CT is found throughout the body, surrounding and providing support for other structures, such as nerves and blood vessels, as well as anchoring and binding together other tissues and organs. Dense CT, found in tendons and similar structures, has a much higher density of collagen connective fibers.

Special connective tissue
Special connective tissue subtypes include adipose (fatty) tissue, blood, bone, cartilage, hematopoietic tissue (bone marrow), and lymphatic tissue, each of which has a characteristic extra-cellular matrix.

Although adipose tissue is classified as a special CT, it is found in conjunction with loose CT, which supports the fatty adipose cells (visible in Figures XI-4-28, XI-4-31, and many other figures in this session). At first glance, an adipose cell may appear to be just a cell membrane with no nucleus or other contents. In fact, the center of the adipose cell is occupied by lipids (fats), which push the nucleus to the edge of the cell, where it is visible in suitably stained sections as a tiny dark spot.

Examine prepared slides showing loose and dense connective tissue. Figures XI-4-27 through XI-4-30 show examples of these types of connective tissue.

Figure XI-4-27: *Loose connective tissue wm, 40X*

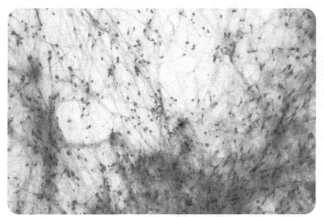

Figure XI-4-28: *Artery and vein cs showing loose CT and adipose CT, 40X*

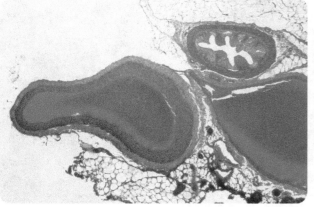

Figure XI-4-29: *Dense connective tissue wm, 100X*

Figure XI-4-30: *Rabbit tendon (dense CT) sec, 100X*

Figure XI-4-31: *Lymph node sec, 40X*

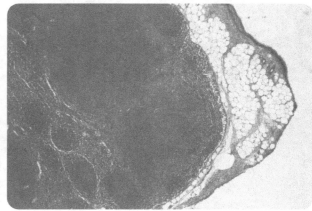

Although it may seem odd to classify blood as a special connective tissue, that is in fact what it is. Like other special CTs, blood consists of individual specialized cells rather sparsely distributed in an extra-cellular matrix. For blood, that matrix is called *plasma*.

Erythrocytes or *red blood cells*—which carry oxygen to the body's cells and carbon dioxide away—are by far the most common cell type found in blood. *Leukocytes* or *white blood cells*—which are part of the immune system—are much less numerous, as are tiny *platelets*, which are anuclear fragments of bone marrow cells that assist in blood clotting.

Examine a prepared slide of a human blood smear. Identify erythrocytes (numerous, small, and of regular shape), leukocytes (relatively rare, larger, and of less regular shape), and platelets (relatively rare, tiny, and irregular). Examine other prepared slides of blood smears from other vertebrates. Figures XI-4-32 through XI-4-35 show examples of blood from several species.

> The red color of the cells in these figures results from staining, which obscures the actual color.

Figure XI-4-32: *Human (Homo sapiens) blood sm, 400X*

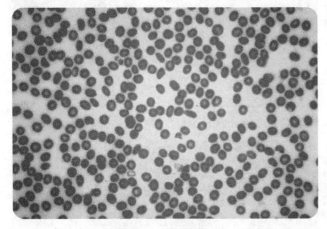

Figure XI-4-33: *Goldfish (Carassius auratus) blood sm, 400X*

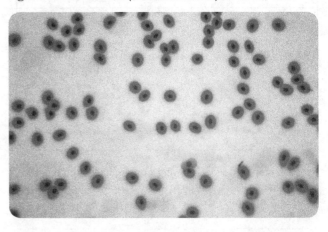

Figure XI-4-34: *Eel (Amphiuma congo) blood sm, 400X*

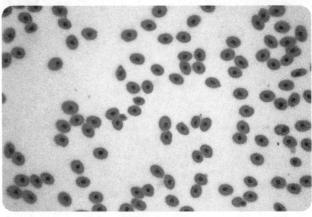

Figure XI-4-35: *Frog (Rana catesbeiana) blood sm, 400X*

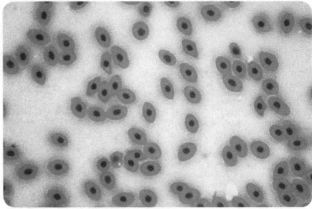

Cartilage is a semi-rigid special CT that is softer and more flexible than bone and harder and less flexible than muscle. Cartilage comprises cells called *chondrocytes* embedded in cavities called *lacunae* in an extracellular matrix of collagen fibers suspended in a gelatinous material called *chondrin*.

There are two subtypes of cartilage. *Hyaline cartilage* is the harder and more rigid, and is found as "padding" between the bearing surfaces where two bones would otherwise come into direct contact. *Elastic cartilage* is softer and more flexible, and is found in structures such as the larynx, external ear, and nose.

Examine prepared slides of cartilage in whole-mount as well as sections of hyaline and flexible cartilage. Figures XI-4-36 through XI-4-38 show examples.

Figure XI-4-36: *Boar (Sus scrofa) cartilage wm showing chondrocytes in lacunae, 100X*

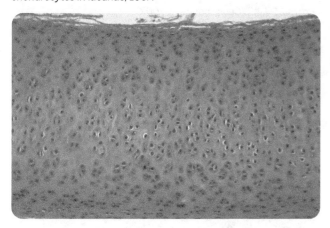

Figure XI-4-37: *Hyaline cartilage sec, 100X*

Figure XI-4-38: *Elastic cartilage sec, 100X*

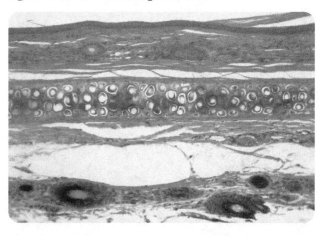

Bone is made up of bone cells called *osteocytes* embedded in lacunae in the extracellular matrix. The hardness and rigidity of bone relative to cartilage results from the composition of that matrix. In cartilage, the matrix is made up of collagen fibers suspended in chondrin, a gel-like material. In bone, the matrix is made up of collagen fibers suspended in an extremely hard and stiff substrate called *bone mineral*, which is made up of crystalline calcium salts that geologists call *hydroxyapatite*. Yep, a rock. (Well, technically, a mineral.)

Compact bone (also called *cortical bone*, *hard bone*, or *dense bone*) is extremely dense, strong and rigid, and forms the hard shell of bones. *Spongy bone* (also called *trabecular bone* or *cancellous bone*) is much less dense and relatively weak, and forms parts of the interior of bones, protected by compact bone.

Bone grows by forming *lamellae*, which are thin, concentric layers that resemble the growth rings of trees. Lamellae form around an interconnected network of thin longitudinal channels called *Halversian canals*, with the lamellae forming protective sheaths around the blood vessels and nerves that run through the Halversian canals.

Examine prepared slides of whole-mount and section slides of compact bone and spongy bone. Figures XI-4-39 through XI-4-43 show examples.

Figure XI-4-39: *Human (Homo sapiens) ground compact bone wm, 100X*

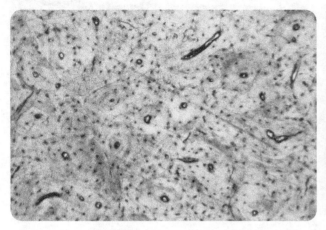

Figure XI-4-41: *Human (Homo sapiens) ground hard bone (decalcified tooth) sec, 100X*

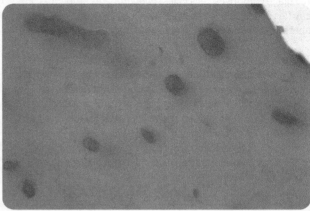

Figure XI-4-40: *Ground hard bone sec, 100X*

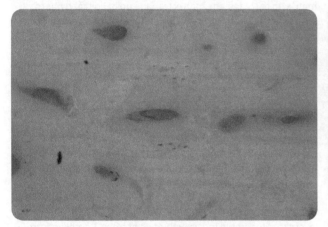

Figure XI-4-42: *Dense bone sec, 100X*

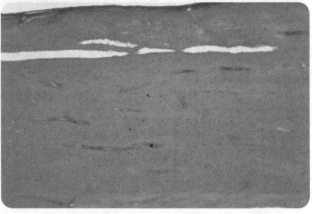

Figure XI-4-43: *Boar (Sus-scrofa) spongy bone sec, 100X*

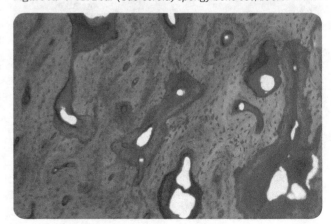

PROCEDURE XI-4-3: OBSERVING MUSCLE TISSUES

The defining characteristic of muscle tissue is that it is able to contract and relax. There are three types of muscle tissue:

- *Skeletal muscle tissue*, also called *voluntary muscle tissue*, is under conscious control, and powers all voluntary movements of the organism. Skeletal muscles range in size and power from the tiny muscles used to control movement of the eyes to the large, powerful muscles of the legs and buttocks. Skeletal muscles connect to bones via tendons, which anchor the muscle to the bone. Contractile cells in voluntary muscles are organized in regular parallel bundles, an arrangement called *striated muscle*. Smooth muscle tissue is made up of tubular muscle cells called *myofibers* or *myocytes*, which contain bundles of contractile filaments called *myofibrils*, which in turn are made up of chains of *sarcomeres*, the basic building block of striated muscle cells.

- *Cardiac muscle tissue*, found only in the heart, is very similar in structure and appearance to skeletal muscle tissue, with the primary difference that cardiac muscles are not under the voluntary control of the organism. Contractile cells in cardiac muscles are striated like those of skeletal muscles, although the arrangement of contractile cells in cardiac muscle is a branching network rather than the parallel arrangement in skeletal muscles.

- *Smooth muscle tissue*, also called *involuntary muscle tissue*, is not under conscious control. Unlike striated muscles, which are adapted for contracting intensely for short periods, smooth muscle is adapted to contract with much less power, but over very long periods. Smooth muscle is found within the walls of internal organs, where it performs such functions as dilation and contraction of blood vessels, peristalsis, and so on.

Examine prepared slides of whole-mount and section slides of skeletal, cardiac, and smooth muscle tissues. Figures XI-4-44 through XI-4-51 show examples.

Figure XI-4-44: *Dog (Canis lupus familiaris) teased skeletal (striated) muscle fibers wm, 100X*

Figure XI-4-45: *Skeletal (striated) muscle sec, 100X*

Figure XI-4-46: *Striated skeletal muscle ls, 100X*

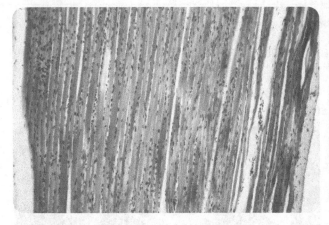

Figure XI-4-47: *Striated skeletal muscle cs, 100X*

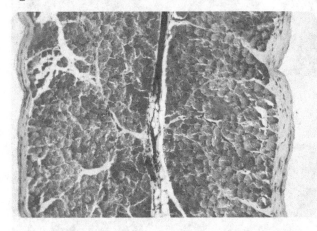

Figure XI-4-48: *Mammalian cardiac muscle sec, 100X*

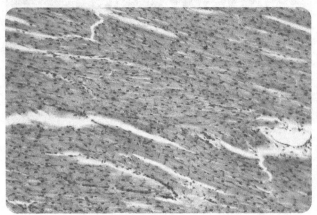

Figure XI-4-49: *Human cardiac muscle sec, 100X*

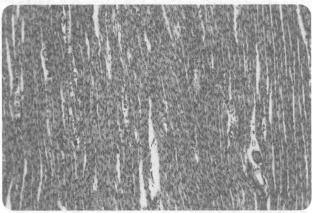

Figure XI-4-50: *Smooth muscle sec, 100X*

Figure XI-4-51: *Mouse (Mus musculus) tail cs, 40X*

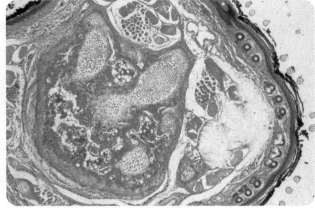

PROCEDURE XI-4-4: OBSERVING NERVOUS TISSUES

The nervous system—the brain, spinal cord, and nerves—is made up of nervous tissue. Nervous tissue is made up of large, complex cell groups called *neurons*, which transmit *nerve impulses*, and supporting cells—including *astrocytes*, *Schwann cells*, and other *neuroglia cells*—that support and work in conjunction with the neurons. Figure XI-4-52 shows a neuron and its associated structures.

Figure XI-4-52: *The structure of a neuron*

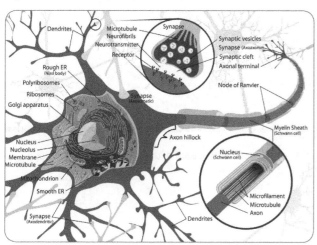

Figure XI-4-53: *Dog (Canis lupus familiaris) cerebellum sec, 100X*

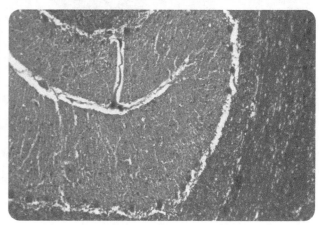

Figure XI-4-54: *Pika (Ochotona sp.) spinal cord cs, 40X*

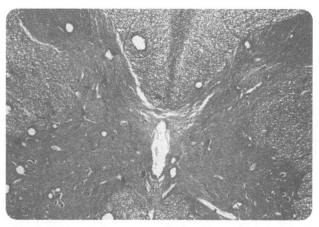

In addition to the cell body, neurons have two types of cytoplasmic extensions that transmit impulses. A *dendrite* is a short extension that normally carries inbound impulses to the cell body from nearby cells and axons. An *axon* is a long extension that normally carries impulses from the cell body to remote dendrites and cells.

Nerves, found in the peripheral nervous system, are groups of neuron cell fibers (peripheral axons without cell bodies) bound by loose connective tissue called the *endoneurium*, surrounded by a sheath of dense connective tissue called the *epineurium*. *Neural pathways*, which occur in the central nervous system, are made up of bundles of neurons (including cell bodies) surrounded by myelin sheathes, and connect distant parts of the central nervous system with each other.

Examine prepared slides of whole-mount and section slides of nervous tissues. Figures XI-4-53 through XI-4-55 show examples.

Figure XI-4-55: *Pika (Ochotona sp.) motor nerve ending wm, 100X*

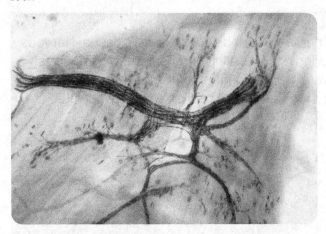

REVIEW QUESTIONS

Q1: How can you discriminate visually between stratified epithelium and pseudostratified epithelium? Why might a stratified epithelium appear pseudostratified in a sectioned slide?

_____

_____

_____

_____

Q2: In sections, which types of epithelial cells resemble fried eggs, squares, and thin rectangles, respectively?

_____

_____

_____

_____

_____

Q3: What difference characterizes mammalian erythrocytes from the erythrocytes of other vertebrates?

Q4: Of the four major components of human blood—plasma, erythrocytes, leukocytes, and platelets—which, if any, are not properly classified as cells? Why?

Q5: How can you distinguish mature erythrocytes produced by mammals from those produced by other vertebrates?

Q6: What purposes do leukocytes and platelets serve?

Q7: What difference did you note in the sections of hyaline and flexible cartilage?

Q8: What similarities and differences exist between the extracellular matrices of cartilage and bone?

_____

_____

_____

_____

Q9: Which types of muscle tissue are classified as involuntary? As striated?

_____

_____

_____

_____

Q10: How do striated muscle tissue and smooth muscle tissue differ in terms of power and endurance?

_____

_____

_____

_____

Q11: How are dendrites and axons similar and different?

_____

_____

_____

_____

Q12: How do nerves differ from neural pathways?

_____

_____

_____

_____

_____

Index

W

X

Z

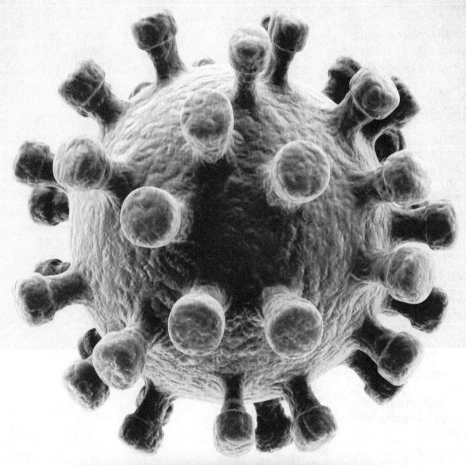